SPRINGER PROCEEDINGS IN PHYSICS 131

SPRINGER PROCEEDINGS IN PHYSICS

Please view available titles in *Springer Proceedings in Physics* on series homepage
http://www.springer.com/series/361/

Joachim Peinke • Martin Oberlack
Alessandro Talamelli
Editors

Progress in Turbulence III

Proceedings of the iTi Conference in Turbulence 2008

With 166 Figures

Springer

Editors
Prof. Dr. Joachim Peinke
Faculty of Physics
Turbulence, Wind Energy, and Stochastics (TWiSt)
Carl v. Ossietzky University
26111 Oldenburg
Germany
E-mail: peinke@uni-oldenburg.de

Prof. Dr. Martin Oberlack
Fachgebiet für Strömungsdynamik
Fachbereich Maschinenbau
Gebäude L5 / 01
Petersenstraße 13
64287 Darmstadt
Germany
E-mail: oberlack@fdy.tu-darmstadt.de

Prof. Alessandro Talamelli
Department of Mechanical
Nuclear Constructions, Aeronautics and of Metallurgy
University of Bologna
Via Zamboni, 33
40126 Bologna
Italy
E-mail: talamelli@unibo.it

ISSN 0930-8989
ISBN 978-3-642-26196-1 e-ISBN 978-3-642-02225-8
DOI 10.1007/978-3-642-02225-8
Springer Heidelberg Dordrecht London New York

Cover design: SPi Publisher Services

Printed on acid-free paper

Springer is part of Springer Science+Business Media (www.springer.com)

Preface

This third issue on "progress in turbulence" is based on the third ITI conference (ITI interdisciplinary turbulence initiative), which took place in Bertinoro, North Italy. Researchers from the engineering and physical sciences gathered to present latest results on the rather notorious difficult and essentially unsolved problem of turbulence. This challenge is driving us in doing basic as well as applied research. Clear progress can be seen from these contributions in different aspects. New sophisticated methods achieve more and more insights into the underlying complexity of turbulence. The increasing power of computational methods allows studying flows in more details. Increasing demands of high precision large turbulence experiments become aware. In further applications turbulence seem to play a central issue. As such a new field this time the impact of turbulence on the wind energy conversion process has been chosen.

Beside all progress our ability to numerically calculate high Reynolds number turbulent flows from Navier-Stokes equations at high precision, say the drag coefficient of an airfoil below one percent, is rather limited, not to speak of our lack of knowledge to compute this analytically from first principles. This is rather remarkable since the fundamental equations of fluid flow, the Navier-Stokes equations, have been known for more than 150 years. These difficulties go alongside with the mathematical difficulties to prove existence and uniqueness of the solution of the Navier-Stokes equations in three dimensions, which has been defined as one of the seven Millennium problems in mathematics by the Clay Institute in Cambridge, Massachusetts. Interesting enough, the corresponding problem in 2D has already been solved in the 1960th. Still, even if we would be able to prove the full 3D problem it remains unclear how to construct concrete solutions for a specific technical or scientific problem.

For this reason further research will be conducted in rather diverse fields of turbulence. This spans from pure mathematical analysis over turbulence physics to applied turbulence research. In the last decades this has led to a broad diversification of turbulence research where contact between different sub-communities sometimes has been lost. This has been the stimulation of the interdisciplinary turbulence initiative which started in 1999 as a cooperation between physicists and engineers working in turbulence and funded by the German science foundation DFG. This initiative now goes on with the sequence of ITI conferences, the fourth on is planned for September 2010.

The structure of the present book is as such that contributions have been bundled according to covering topics i.e. I Turbulence, II Experimental Methods, III Wind Energy, IV Modeling & Simulation & Mathematics, V Particle Laden Flows, VI Convection & Boundary Layer, VII Special Flows and VIII Vortex.

At this point we would like to thank all authors for their contributions to this proceedings and the referees giving critical comments to the contributions and therewith considerably raising the scientific quality. We would like to thank Thomas Ditzinger from Springer for his patience during the production of the book and BMBF (Ferderal Ministry of Education and Research in Germany) for the financial support for the Wind Energy session. Finally we gratefully acknowledge Julia Gottschall and Michael Hölling handling the author and referee communication and integrating the book to its final form.

J. Peinke
M. Oberlack
A. Talamelli
(Oldenburg, Darmstadt and Forli, 2009)

Contents

Turbulence

Experimental Methods

Wind Energy

Modelling and Simulation and Mathematics

Particle Laden Flows

Convection and Boundary Layer

Special Flows

Fully Developed Turbulence with Diminishing Mean Vortex Stretching and Reduced Intermittency

R.E.E. Seoud and J.C. Vassilicos

Abstract. Fluid turbulence determines drag characteristics of land/air/seaborne vehicles, mixing and reaction rates in chemical reactors, industrial mixers, burners, and complex non-equilibrium phenomena such as flame reignition and extinction. It is central to weather and environmental predictions, cloud precipitation and albido, atmospheric and oceanic transport and ocean-atmosphere interactions. Central to three-dimensional turbulent fluid flows is the vortex-stretching mechanism which generates the multitude of eddy sizes that make simulations prohibitive. Studies aiming at understanding the intrinsic dynamics of fully developed turbulence have concentrated on homogeneous isotropic turbulence in the wind tunnel [1, 2, 3], where the turbulence dynamics result purely from the vortex-stretching mechanism in isolation. Here we report the discovery of a new type of homogeneous isotropic three-dimensional fluid turbulence where the average vortex stretching diminishes and the level of small-scale intermittency remains constant as the turbulence intensifies. Hence, the deepest of all turbulence properties, vortex stretching and intermittency, can be tampered with.

Vortex stretching is the non-linear mechanism intrinsic to all fluid mechanics. It generates eddying motions with vorticity in all three-dimensional turbulent fluid flows (there are only few exceptions where turbulent flows are not three-dimensional). When the turbulence is intensified and the viscous forces are reduced relative to the inertial forces which cause vortex stretching and thereby vortical turbulence, the total amount of vorticity in the flow increases. This is the well-known vortex stretching mechanism, central to all

R.E.E. Seoud · J.C. Vassilicos
Turbulence, Mixing and Flow Control Group, Department of Aeronautics

J.C. Vassilicos
Institute for Mathematical Sciences, Imperial College London,
London, SW7 2BY, UK

turbulence theories, phenomenology and modelling [1, 4, 5]. No laboratory and field measurements of fluid turbulence nor any numerical simulations have ever produced evidence to the contrary. One of the major problems in modern mathematics and mathematical fluid mechanics concerns the regularity of solutions of the Navier-Stokes equations which govern fluid flow; this problem is directly related to this vortex stretching mechanism without which regularity would have been trivially established long ago.

Recently, mathematical analysis and numerical simulations of the Navier-Stokes equations forced in a multiscale, say fractal, way [6, 7] reached the conclusion that fractal-forced fluid turbulence can have fundamentally different properties than all other fluid turbulence. Laboratory experiments [8] where turbulence was generated in a wind tunnel by fractal grids support this conclusion and, in fact, show that the resulting decaying three-dimensional turbulence is homogeneous and isotropic to a satisfactory approximation. The fundamentally different property singled out by these wind tunnel experiments is this: fractal-generated homogeneous and isotropic turbulence requires only one length-scale for its scaling, whereas conventional small-scale turbulence requires two, an integral auto-correlation length-scale and a viscous microscale [1, 4, 5]. As a result, the time t-dependent energy spectrum $E(k,t)$ of the fractal-generated decaying turbulence takes the form $E(k,t) = u'^2(t)l(t)f(kl)$ throughout the range of excited wavenumbers k: u' is the rms turbulence velocity, $l(t)$ is this one length-scale and $f(kl)$ is a dimensionless function of the dimensionless wavenumber kl. Consequently, and as indeed observed [8], the integral auto-correlation length-scale L and the Taylor microscale λ are both proportional to $l(t)$ in the fractal-generated decaying turbulence of this wind tunnel study [8]. Direct numerical simulations of fractal-forced Navier-Stokes turbulence [7] also reach the conclusion that $E(k,t) = u'^2(t)\lambda(t)f(k\lambda)$. The possibility that a type of isotropic homogeneous decaying turbulence might exist with a scaling based on a single length-scale for the entire energy spectrum was suggested by George [9]. What was not predicted by George [9] is that such turbulence can be generated by fractal grids.

In homogeneous isotropic turbulence, the average of the third power of $\frac{\partial u}{\partial x}$, i.e. $\langle(\frac{\partial u}{\partial x})^3\rangle$ where u is the longitudinal turbulent fluctuating velocity in the direction x of the mean flow, is a direct measure of the generation rate of enstrophy [1, 4, 5] (enstrophy is the average of half the square of the vorticity in the turbulence). Divided by $\langle(\frac{\partial u}{\partial x})^2\rangle^{3/2}$, it gives the derivative skewness, $S \equiv \langle(\frac{\partial u}{\partial x})^3\rangle/\langle(\frac{\partial u}{\partial x})^2\rangle^{3/2}$ which is a normalised dimensionless measure of the average rate of enstrophy generation by vortex stretching in homogeneous isotropic turbulence. Direct numerical simulations [10, 11] and laboratory experiments [1, 12, 13] of homogeneous isotropic turbulence give values of S between -0.3 and -0.55. The negative sign of S is a reflection of the enhancement of vorticity by the vortex stretching mechanism, and indeed all numerically calculated and all experimentally measured values of S are negative [1, 4, 5]. As the Reynolds number (a measure of the ratio between inertial and viscous forces in the turbulence) is increased beyond the reach of these numerical simulations and laboratory experiments, the negative value of S grows in magnitude above 0.55, as for example in high Reynolds number atmospheric data [14]. Compilations of measurements from a variety of high-Reynolds number turbulent flows [15] indicate that S seems to be an increasing power law function of the derivative flatness $F \equiv \langle (\frac{\partial u}{\partial x})^4 \rangle / \langle (\frac{\partial u}{\partial x})^2 \rangle^2$, usually interpreted as being a measure of small-scale intermittency, and that $F \sim Re_\lambda^q$ where q lies between 0.3 and 0.4 and $Re_\lambda \equiv u'\lambda/\nu$ is the Taylor microscale-based Reynolds number where ν is the kinematic viscosity of the fluid (see figure 1).

The scale-by-scale energy balance of homogeneous isotropic decaying turbulence is $\frac{\partial}{\partial t}E(k,t) = T(k,t) - 2\nu k^2 E(k,t)$ where $\int_0^\infty k^2 T(k)\,\mathrm{d}k$ is the rate of change of the average enstrophy $\int_0^\infty k^2 E(k)\,\mathrm{d}k$ as a result of nonlinear, vortex stretching, interscale interactions [4]. Being a dimensionless measure of the average rate of enstrophy generation by vortex stretching, S is related to these quantities as follows [4]: $S = -(135/98)^{1/2} \frac{\int_0^\infty k^2 T(k)\,\mathrm{d}k}{(\int_0^\infty k^2 E(k)\,\mathrm{d}k)^{3/2}}$. If, as recently discovered both numerically and experimentally, the energy spectrum of fractal-generated turbulence scales with a single length-scale $l(t)$ which can be taken as either $L(t)$ or $\lambda(t)$ by virtue of the fact that they are all proportional to each other, then the form $E(k,t) = u'^2 \lambda f(k\lambda)$ can be inserted into the scale-by-scale energy balance so as to determine the scaling with Reynolds number of $T(k,t)$, and thereby of S, for fractal-generated homogeneous isotropic and decaying turbulence. Making use of Taylor's well-known relation [1, 5] $\frac{\mathrm{d}}{\mathrm{d}t}u'^2 \sim -\nu u'^2/\lambda^2$, the result is

$$-S = A Re_\lambda^{-1} + \frac{B}{u'}\frac{\mathrm{d}}{\mathrm{d}t}\lambda, \tag{1}$$

where A and B are dimensionless constants. Unlike all other known homogeneous and isotropic turbulent flows, S, and therefore the average rate of enstrophy generation by vortex stretching, decreases with increasing magnitude of turbulence fluctuations u' and increasing Reynolds number.

To test this remarkable prediction, we carried out laboratory experiments with planar fractal grids fitted across the entrance of a wind tunnel's test section as in earlier works [1, 2, 3] which used grids made of regular rectangular arrays of bars, to generate turbulence. We experimented with the three planar fractal square grids of Seoud & Vassilicos [8] in the same open-circuit wind tunnel (4.2 m length test section, maximum speed, when empty, of 33 m/s and background turbulence intensity of 0.4%) and the same AALab AN-1005 constant-temperature anemometer and single wire system (detailed information concerning our experimental apparatus can be found in their paper). These fractal grids are all space-filling (fractal dimension [17] $D_f = 2$), have the same blockadge ratio $\sigma = 25\%$, and the same number of fractal iterations, $N = 4$. A complete planar description of these grids requires a minimum of five parameters [17], including the tunnel's cross sectional size $T = 0.46$ m. With $D_f = 2, N = 4, \sigma = 25\%$ and $T = 0.46$ m, four of the five parameters are set to

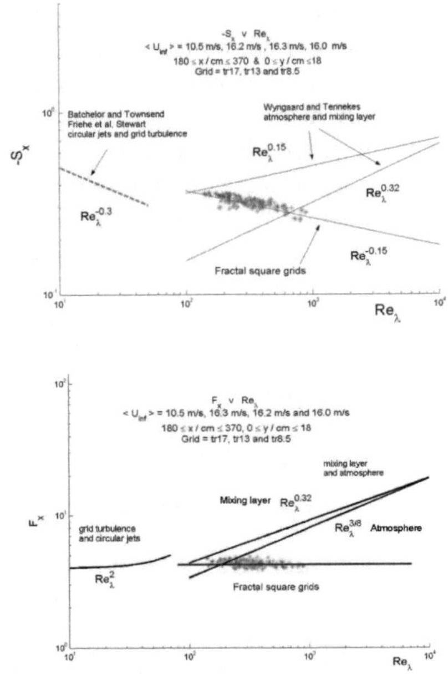

Fig. 1 Log-log plots of (a) $-S$ and (b) F versus Re_λ. The straight lines without data represent typical trends observed in various turbulent flows [15]. The references in these plots and further details can be found in van Atta & Antonia [15]. Both plots contain 168 data points obtained from our $6 \times 7 \times 4 = 168$ data sets. We estimated $\langle(\frac{\partial u}{\partial x})^n\rangle$ for $n = 2, 3, 4$ from the structure functions $\langle(u(x+r) - u(x))^n\rangle$ where $x = Ut$ and time-increments equal r/U (by virtue of Taylor's frozen turbulence hypothesis) and where the averages were taken over time records long enough for statistical convergence. Specifically, we approximate $\langle(\frac{\partial u}{\partial x})^n\rangle/\langle(\frac{\partial u}{\partial x})^2\rangle^{n/2}$ by $\langle(u(x+r)-u(x))^n\rangle/\langle(u(x+r)-u(x))^2\rangle^{n/2}$ with values of r small enough for these ratios to be independent of r, yet large enough for probe resolution requirements determined by examination of $E_{11}(k_1)$, the power spectrum of the longitudinal velocity fluctuations in the streamwise direction (we made sure that r was larger than the inverse of the highest value of k_1 for which $k_1^4 E_{11}(k_1)$ does not increase [14] with k_1). Measured frequency spectra are interpreted as wavenumber spectra using Taylor's frozen turbulence hypothesis where the frequency f determines the wavenumber $k_1 = 2\pi f/U$. Our hot wire's time-resolution is such that we fully resolve up to well above wavenumbers $k_1\eta = 1$ at $U_\infty \approx 10.5\,\mathrm{m/s}$ (η is the Kolmogorov microscale $(\nu^3/\epsilon)^{1/4}$ where $\epsilon = 15\nu\langle(\frac{\partial u}{\partial x})^2\rangle$), and up to $k_1\eta \approx 0.9$ at $U_\infty \approx 16\,\mathrm{m/s}$. We checked that the main contribution to the integral $\int k_1^2 E_{11}(k_1)\,\mathrm{d}k_1$ always comes from the integration range $k_1\eta \leq 0.4$ and we estimated that the unresolved scales result in this integral being underestimated by about 4% in our worst, highest speed, case. This integral equals $\langle(\frac{\partial u}{\partial x})^2\rangle$ and we verified that our estimations of $\langle(\frac{\partial u}{\partial x})^2\rangle$ from $\langle(u(x+r) - u(x))^2\rangle/r^2$ (for r larger than the inverse of the highest resolved value of k_1 yet small enough for $\langle(u(x+r) - u(x))^2\rangle/r^2$ to be approximately independent of r) are in reasonable agreement with $\int k_1^2 E_{11}(k_1)\,\mathrm{d}k_1$. In fact, time-resolution requirements in this present type of turbulence [8] where the energy spectrum $E_{11}(k_1) = u'^2\lambda f(k_1\lambda)$ need to be specified only in terms of $k_1\lambda$, not $k_1\eta$, and the energy is overwhelmingly captured by all wavenumbers smaller than $k_1\lambda = 25$ all resolved here. The hot wire's spatial resolution is less important for accurate measurements of S than its time resolution [16]. Still, our wire's length lies between $\lambda/5$ and $\lambda/10$, and λ does not vary much with Re_λ in this present type of turbulence [8]. Our estimations of Re_λ were made using $\lambda^2 = u'^2/\langle(\frac{\partial u}{\partial x})^2\rangle$.

equal values for all three grids. These grids were designed to vary by only one parameter, the thickness ratio t_r of the largest square's cross-stream thickness (from 14.2 to 19.2 mm, left to right in figure 2) to the smallest squares' cross-stream thickness (from 1.7 to 1.1 mm left to right in figure 2).

The turbulence intensity generated by these grids builds up as the turbulence is convected downstream till a distance x_{peak} from the grid is reached where the turbulence intensity peaks and then decays [17, 8]. In the decay region (streamwise distance x from the grid larger than x_{peak}) and in the central region of the tunnel (at the very least comparable in lateral extents to the largest square on the grids, see figure 2), the turbulence is, to a satisfactory approximation, homogeneous and isotropic by a number of measures [8, 17], and the Taylor microscale λ and the longitudinal and lateral integral length-scales L_u and L_v remain approximately constant [17, 8] with varying x and varying U_∞, the mean flow velocity upstream of the grid. Varying t_r

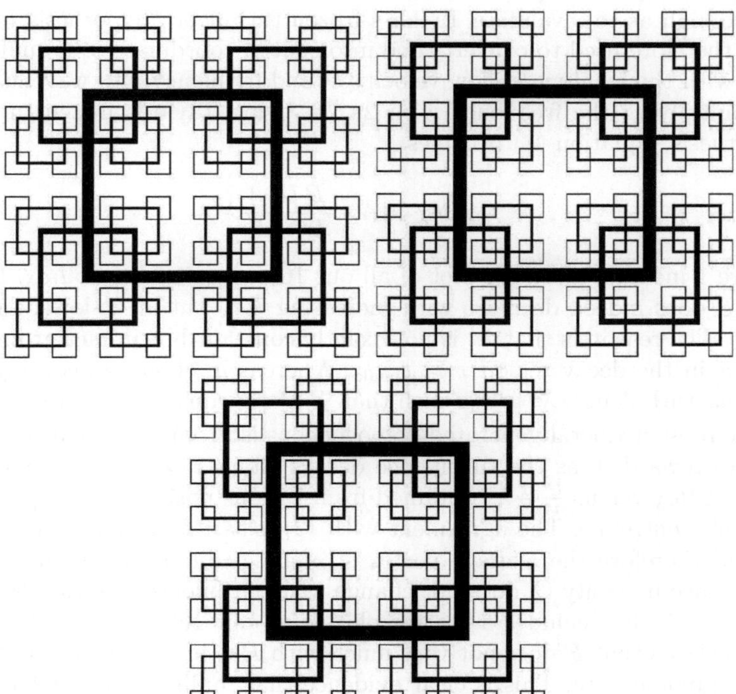

Fig. 2 Scaled diagrams of the three space-filling square grids each of overall size 0.46 m so as to fit in the tunnel's cross-section $T = 0.46$ m: from left to right, $t_r = 8.5, 13.0, 17.0$; $x_{\text{peak}} \approx 1.8$ m, 1.45 m, 1.2 m. Note the surprisingly large variations in x_{peak} even though the differences between the three grids are small to the eye. For a complete description of these grids see Hurst & Vassilicos [17], e.g. the streamwise thickness of all the squares is 5 mm.

as in figure 2 also leads to negligible changes to these length-scales and to the effective mesh size [17] M_{eff}.

We took single wire anemometry measurements of the far-stream homogeneous isotropic and decaying turbulence generated by each one of the three grids shown in figure 2 at six values of x/x_{peak} between above 1 and about 3 (i.e. x/M_{eff} between above 50 and about 110) and at seven values of the lateral coordinate y (defined horizontally with respect to the tunnel and figure 2) between 0 and 18 cm for each value of x (in each grid shown in figure 2 the size of the largest square is just below 24 cm). The wind tunnel speed was set at U_∞ close to 16 m/s for all these measurements, and an additional identical campaign of measurements was also taken with $U_\infty = 10.5$ m/s for the one grid $t_r = 17$. Hence, $6 \times 7 \times 4 = 168$ data sets were collected in total, each for the longitudinal velocity as a function of time. Our turbulence intensities are much higher than those obtained in similar and even larger wind tunnels where the turbulence was generated by regular grids [1, 2, 3]; as a result, our values of Re_λ reach up to nearly 900. However our turbulence intensities are not so high as to invalidate Taylor's frozen turbulence hypothesis which we have therefore used to convert time into spatial coordinate x by multiplying time with the local mean flow velocity U and frequency into wavenumber k_1 by multiplying the frequency with $2\pi/U$. Using Taylor's frozen turbulence hypothesis, equation (1) becomes

$$-SRe_\lambda = A + \frac{B}{2}\frac{U}{\nu}\frac{\mathrm{d}}{\mathrm{d}x}\lambda^2. \tag{2}$$

Figure 1 includes a scatter plot of all our 168 values of S and Re_λ. It shows that S does indeed decrease with increasing Re_λ but at a slower rate than Re_λ^{-1}. The reason is that λ^2 is not exactly constant but in fact grows slowly with x in the decay region $x > x_{\text{peak}}$. Whereas in classical decaying homogeneous turbulence, λ^2 grows such that $\frac{\mathrm{d}}{\mathrm{d}x}\lambda^2$ remains constant during decay, in the present fractal-generated homogeneous isotropic turbulence, $\frac{\mathrm{d}}{\mathrm{d}x}\lambda^2$ decreases towards 0 as the turbulence decays along $x > x_{\text{peak}}$. In figure 3 we plot $-SRe_\lambda$ versus $\frac{U}{\nu}\frac{\mathrm{d}}{\mathrm{d}x}\lambda^2$ obtained from various positions $x > x_{\text{peak}}$ on the tunnel's centreline. The agreement with (2) is satisfactory and implies that S, and therefore the average vortex stretching, diminishes with increasing turbulence intensity (λ does not change with turbulence intensity [8, 17, 1]).

Figure 1 also includes a scatter plot of all our 168 values of F and Re_λ which shows that F does not vary much with Re_λ in the range $100 < Re_\lambda < 1000$ explored here. This is clear evidence that, unlike all other known turbulent flows where this measure of small-scale intermittency increases with Reynolds number, it does not in the decaying homogeneous isotropic turbulence generated by fractal square grids. This may be a consequence of the fact that the ratio of outer to inner length-scales in this turbulence is itself

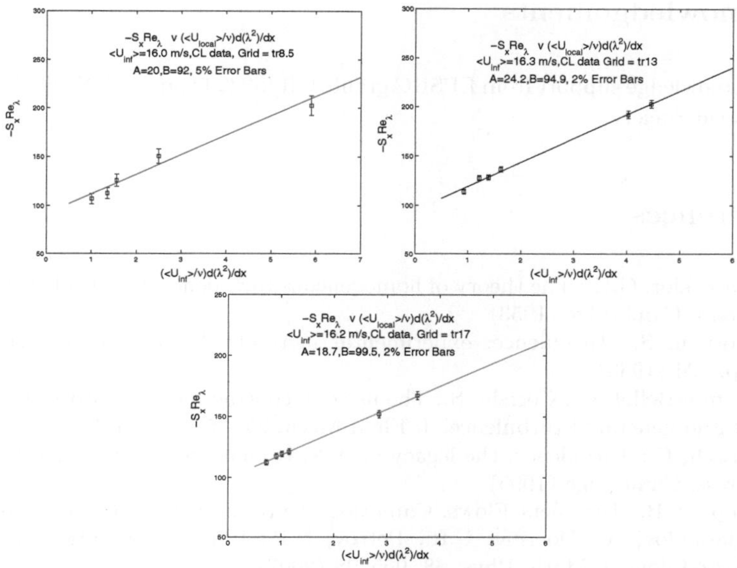

Fig. 3 $-SRe_\lambda$ versus $\frac{U}{\nu}\frac{\mathrm{d}}{\mathrm{d}x}\lambda^2$ for each one of the three fractal square grids, $t_r = 8.5, 13.0, 17.0$ from left to right. The data points correspond to different centreline positions x all larger than x_{peak} ($x = 1.8\,\text{m}$ for $t_r = 13.0, 17.0$ and $x = 2.1\,\text{m}, 2.8\,\text{m}, 3\,\text{m}, 3.2\,\text{m}, 3.7\,\text{m}$ for all t_r). $U_\infty \approx 16\,\text{m/s}$. S was calculated as explained in figure caption 1 and λ^2 was calculated from $\lambda^{-2} = \int k_1^2 \frac{E_{11}(k_1)}{u'^2}\,\mathrm{d}k_1$. The values of λ^2 at different centreline positions x were best fitted by $\lambda^2 \sim (x - x_{\text{peak}})^s$ with $s = 0.03$ and this form was then used to calculate $\frac{\mathrm{d}}{\mathrm{d}x}\lambda^2$ at different positions x. The straight lines represent equation (2) with best fit values of A between 19.2 and 24.1 and of B between 92.5 and 99.8. The coefficient of determination R^2 is above 0.99 for the fits of the data corresponding to $t_r = 13, 17$ and 0.97 for $t_r = 8.5$.

independent of Reynolds number [8]. We verified that the probability density functions of $u(x + r) - u(x)$ adopt non-Gaussian shapes as r reaches values smaller than λ but that these shapes do not change significantly with Re_λ.

In summary, it is possible to tamper with the deepest of all properties of homogeneous isotropic turbulence: vortex stretching and small-scale intermittency. This discovery provides a valuable unprecedented handle for understanding fluid turbulence and poses an immediate challenge to all existing turbulence models, theories and phenomenologies [1, 4, 5] as none predicts our observations that it is possible to create a fluid turbulence where vortex stretching decreases and the level of small-scale intermittency remains the same as the turbulence is intensified. This possibility to cap intermittency and dampen vortex stretching opens new horizons for many applications where mixing, combustion and turbulent flow control are important.

Acknowledgements

We acknowledge support from EPSRC grants GR/S23292 and GR/S82947 and from the Royal Society.

References

1. Batchelor, G.K.: The theory of homogeneous turbulence. Cambridge University Press, Cambridge (1953)
2. Corrsin, S.: Turbulence: experimental methods. In: Handbook der Physik, pp. 524 (1963)
3. Comte-Bellot, G., Corrsin, S.: The use of a contraction to improve the isotropy of grid-generated turbulence. J. Fluid Mech. 25, 657–682 (1966)
4. Frisch, U.: Turbulence: the legacy of A.N. Kolmogorov. Cambridge University Press, Cambridge (1995)
5. Pope, S.B.: Turbulent Flows. Cambridge University Press, Cambridge (2000)
6. Cheskidov, A., Doering, C.R., Petrov, N.P.: Energy dissipation in Fractal-Forced flow. J. Math. Phys. 48, 065208 (2007)
7. Mazzi, B., Vassilicos, J.C.: Fractal-generated turbulence. J. Fluid Mech. 502, 65–87 (2004)
8. Seoud, R.E., Vassilicos, J.C.: Dissipation and decay of fractal-generated turbulence. Phys. Fluids 19, 105108 (2007)
9. George, W.K.: The decay of homogeneous turbulence. Phys. Fluids A 4(7), 1492–1509 (1992)
10. Vincent, A., Meneguzzi, M.: The spatial structure and statistical properties of homogeneous turbulence. J. Fluid Mech. 225, 1–25 (1991)
11. de Bruyn Kops, S.M., Riley, J.J.: Direct numerical simulation of laboratory experiments in isotropic turbulence. Phys. Fluids 10(9), 2125–2127 (1998)
12. Tavoularis, S., Bennett, J.C., Corrsin, S.: Velocity-derivative skewness in small Reynolds number, nearly isotropic turbulence. J. Fluid Mech. 88, 63 (1978)
13. Monin, A.S., Yaglom, A.M.: Statistical Fluid Mechanics, vol. 2. MIT Press, Cambridge (1975)
14. Wyngaard, J.C., Tennekes, H.: Measurements of the small-scale structure of turbulence at moderate Reynolds numbers. Phys. Fluids 13, 1962–1969 (1970)
15. van Atta, C.W., Antonia, R.A.: Reynolds number dependence of skewness and flatness factors of turbulent velocity derivatives. Phys. Fluids 23, 252–257 (1980)
16. Burattini, P., Lavoie, P., Antonia, R.A.: Velocity derivative skewness in isotropic turbulence and it measurement with hot wires. Exp. Fluids (to appear)
17. Hurst, D., Vassilicos, J.C.: Scalings and decay of fractal-generated turbulence. Phys. Fluids 19, 035103 (2007)

Spectral and Physical Forcing of Turbulence

Zafer Zeren and Benoît Bédat

Abstract. Spectral and physical forcing algorithms to generate stationary isotropic turbulence are compared in terms of characteristics of turbulence. Homogeneous isotropic turbulence is first validated with the stochastic forcing algorithm using a different shell of wavenumbers. Linear forcing algorithm is then applied to the generated stochastic field to keep the characteristics of turbulence stationary in time. Linearly forcing is very fast because there is no need to transform from a spectral to physical space. However, levels of fluctuations are very high that very long duration for simulations needed to achieve robust statistics.

1 Introduction

In the fields of turbulence, mixing and two-phase flows, physical models can be tested and verified using stationary turbulence with a certain level of confidence [2]. Stationary turbulence is obtained by adding a divergence-free forcing term to the momentum equations for homogeneous and isotropic turbulence.

Basically, there are deterministic (linear) and stochastic (spectral) approaches to calculate the forcing term [3],[5]. Stochastic forcing assumes a periodic spectral domain to transform the coefficients to physical space. However, linear forcing detaches from these costly operations and restriction [7] by simply assuming an initial isotropic field.

Zafer Zeren
Institut de mecanique des fluides de Toulouse, Allée du Professeur Camille Soula 31400, Toulouse, France
e-mail: zeren@imft.fr

Benoît Bédat
Institut de mecanique des fluides de Toulouse, Allée du Professeur Camille Soula 31400, Toulouse, France
e-mail: bedat@imft.fr

2 Stochastic and Linear Forcing

In stochastic forcing, forcing term f_i is calculated using Uhlenbeck-Ornstein stochastic processes governed by the stochastic Langevin equation given as:

$$b_i(t+\Delta t) = b_i(t)(1-\Delta t/T_F) + e\left(2{\sigma_F}^2\Delta t/T_F\right) \tag{1}$$

where T_F is the forcing time and σ_F is the standard deviation of the stochastic processes [3]. Random number e, of Gaussian distribution, is of zero mean and unity variance. Important parameter is the range of the wavenumbers at which equation (1) is calculated and added to the momentum equation. This range directly forms a shell in spectral space and controls the length of the large scales. The coefficients b_i normally do not satisfy the divergence-free condition so they are projected by a proper projection operator [3]. After projection, inverse Fast Fourier Transform takes place to calculate the force f_i in physical space. In the present work, two wavenumber shells are tested $]0,\sqrt{8}k_0]$ and $[2k_0, 6k_0]$ where k_0 is the fundamental wavenumber $k_0 = 2\pi/L_{box}$ [4].

In linear forcing, on the other hand, f_i is calculated as $f_i = Bu_i$ where u_i is the fluctuation velocity. B is calculated as $B = \varepsilon/3u_{rms}^2$ where $u_{rms}^2 = \langle u_i u_i\rangle/3$. The idea behind is that writing the transport equation for fluctuation velocity, the production term is proportional to the fluctuation velocity [5]. B is then simply the indicator of the balance between the production and dissipation. Prescription of this value and keeping it constant in time is equivalent to imposition of an eddy turnover time [7].

3 Results and Discussion

With the stochastic forcing, the computation starts with a zero velocity field. Due to the forcing, turbulent kinetic energy increases and reaches to a plateau after 5 eddy turnover time (see figure 1, for the second shell) where viscous dissipation by the small eddies equals the production of the turbulent kinetic energy. It is clear that two different shells generate close levels of energies, however, the large scales differ in length. This is reflected in the turbulent Reynolds number of the turbulence as seen in table 1. The linear forcing requires a non-null initial velocity field. Two fields have been tested; one obtained with the stochastic forcing and one generated with a Passo Pouquet spectrum. As seen in figure 1 linear forcing amplifies the initial field according to the value of B, the energy level adjusts itself to higher level after almost 20 eddy turnover times with stronger oscillations.

As expected, the integral length scale (see table 2) are larger for the shell 1 than the shell 2. The shell 2 is more suitable for performing statistics and shows better isotropy at large scales as shown by the ratio L_f/L_g. If the longitudinal integral scales for the linear forcing are roughly equal to the one obtained with the shell 2, the ratio L_f/L_g tends to unity where the deviations from the Karman-Howarth's relation are observed.

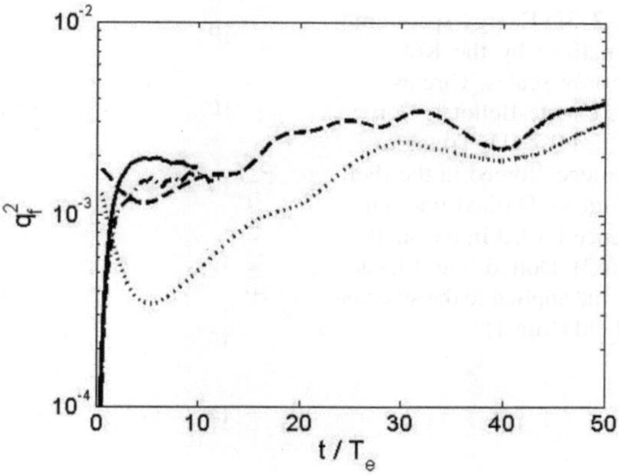

Fig. 1 Average turbulent kinetic energy Solid line: Stochastic forcing (Sim 2 of table 1), Dash-dot line: Stochastic forcing (Sim 1 of table 1), Long-dashed line: Linear forcing applied to stochastic field (Run 1 of table 2 Short-dashed line: Linear forcing applied to Passo-pouquet spectrum (Run 2 of table 2

Table 1 Simulations with stochastic forcing, T_E: Eulerian time scale, T_e: eddy turnover time, T_L: Lagrangian time scale, Reynolds number defined as: $Re_L = u'L_f/\nu$

Simulation	$[k_{min}, k_{max}]$	T_F	σ_F	T_E	T_e	T_E/T_L	L_f/L_{box}	L_f/L_g	Re_L	$k_{max}\eta$	η/L_{box}
Sim 1	$]0, 2\sqrt{2}k_0]$	20	8.10E-5	23	41	0.77	0.2	2.22	83	2.4	0.0083
Sim 2	$[2k_0, 6k_0]$	20	8.10E-5	19	19	0.8	0.1	2.01	50	1.5	0.0055

Figure 2 displays the kinetic energy spectra with comparison to the experimental data of Comte-Bellot and Corrsin [1]. In large wavenumbers, all spectra follow the experimental measurements perfectly. However, for small wavenumbers, different behaviors are observable. The shape of the spectrum for Sim 2 is typical of the one observed experimentally. The energy contained in the largest scale is lower than Sim 1 (shell 1) which puts a part of the energy to the largest scale. At this point, it is necessary to mention that the smoothing scheme of Overholt and Pope [6] is used for the stochastic forcing simulations to obtain smoother spectrum at large wavenumbers.

Spectrum of the Run 2 is averaged over the almost stationary state (Dotted line). Until its own largest wavenumber, it follows the experimental measurements and forms the inertial subrange longer than the stochastic forcing simulations (see the figure).

Table 2 Simulations with linear forcing

Simulation	B	N_x^3	Init. Cond.	L_f/L_{box}	L_f/L_g	$k_{max}\eta$	Re_L
Run 1	0.018	128	Stochastic (Sim 2)	0.1192	1.1742	1.37	190
Run 2	0.014	128	Passo-Pouquet	0.1143	1.0518	1.5	121

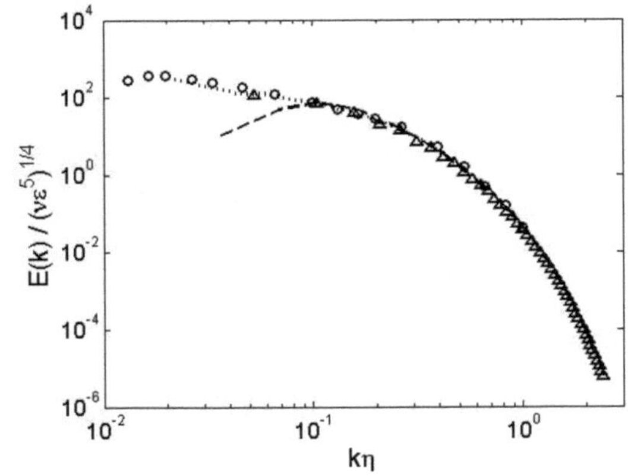

Fig. 2 3D Energy spectrum normalized by the Kolmogorov scales, Circles: Data Comte-Bellot & Corrsin $Re_\lambda = 60.7$ [1], Triangles: turbulence forced in the shell 1 (Sim 1), Dashed line: turbulence forced in the shell 2 (Sim 2), Dotted line: Linear scheme applied to the stochastic field (Run 1)

4 Conclusion

Two different methods of forcing have been compared in terms of their efficiency in generating isotropic turbulence. The results have shown that stochastic method generates smoother levels of oscillations with improved isotropy. Linear forcing, on the other hand, is an efficient method with no calculations in spectral space. However, large fluctuations requires longer simulations for robust statistics.

References

1. Comte-Bellot, G., Corrsin, S.: Simple Eulerian time correlation of full narrow band velocity signals in isotropic turbulence. J. Fluid Mech. 375, 235–263 (1971)
2. Couzinet, A., et al.: Numerical study and lagrangian modelling of turbulent heat transport. Flow Turbulence Combust 80, 37–46 (2008)
3. Eswaran, V., Pope, S.: An examination of forcing in direct numerical simulations of turbulence. Computers & Fluids 16, 257–278 (1988)
4. Fevrier, P.: Etude numerique des effets de concentration preferentielle et de correlation spatiale entre vitesses de particules solides en turbulence homogene isotrope stationnaire. PhD thesis, Institut National Polytechnique de Toulouse (2000)
5. Lundgren, T.S.: Linearly forced isotropic turbulence. Annual Research Briefs (Center for Turbulence Research), pp. 461–473 (2003)
6. Overholt, M., Pope, S.B.: A deterministic forcing scheme for direct numerical simulations of turbulence. Computers & Fluids 27(1), 11–28 (1988)
7. Rosales, C., Meneveau, C.: Linear forcing in numerical simulations of isotropic turbulence: physical space implementations and convergence properties. Physics of Fluids 17, 095106 (2005)

Fractal-Generated Turbulent Scaling Laws from a New Scaling Group of the Multi-Point Correlation Equation

Martin Oberlack and George Khujadze

Abstract. Investigating the multi-point correlation (MPC) equations for the velocity and pressure fluctuations in the limit of homogeneous turbulence a new scaling symmetry has been discovered. Interesting enought this property is not shared with the Euler or Navier-Stokes equations from which the MPC equations have orginally emerged. This was first observed for parallel wall-bounded shear flows (see [2]) though there this property only holds true for the two-point equation. Hence, in a strict sense there it is broken for higher order correlation equations. Presently using this extended set of symmetry groups a much wider class of invariant solutions or turbulent scaling laws is derived for homogeneous and homogeneous-isotropic turbulence which is in stark contrast to the classical power law decay arising from Birkhoff's or Loitsiansky's integrals. In particular, we show that the experimentally observed specific scaling properties of fractal-generated turbulence (see [1, 4]) fall into this new class of solutions. Due to this specific grid a breaking of the classical scaling symmetries due to a wide range of scales acting on the flow is accomplished. This in particular leads to a constant integral and Taylor length scale downstream of the fractal grid and the exponential decay of the turbulent kinetic energy along the same axis. These particular properties can only be conceived from MPC equations using the new scaling symmetry. The latter new scaling law may have been the first clear indication towards the existence of the extended statistical scaling group. Though the latter is not obvious from the instantaneous Euler or Navier-Stokes equations it is directly implied.

1 Multi-Point Equation in the Limit of Homogeneous Turbulence

We investigate the symmetry and invariance structure of the infinite set of multi-point correlation (MPC) equations for the velocity and pressure fluctuations $\mathbf{u}(\mathbf{x},t)$

Martin Oberlack · George Khujadze
Chair of Fluid Dynamics, Technische Universität Darmstadt, 64289 Darmstadt, Germany
e-mail: oberlack@fdy.tu-darmstadt.de, khujadze@fdy.tu-darmstadt.de

and $p(\mathbf{x},t)$ respectively in the limit of homogeneous turbulence (for the full equation of inhomogenous turbulence see [3])

$$\frac{\partial R_{i_{\{n+1\}}}}{\partial t} + \sum_{l=1}^{n} \left(- \left. \frac{\partial P_{i_{\{n\}}[0]}}{\partial r_{m_{(l)}}} \right|_{[m_{(l)} \mapsto i]} + \frac{\partial P_{i_{\{n\}}[l]}}{\partial r_{i_{(l)}}} \right) \tag{1}$$

$$- \nu \sum_{l=1}^{n} \sum_{m=1}^{n} \left(\frac{\partial^2 R_{i_{\{n+1\}}}}{\partial r_{k_{(m)}} \partial r_{k_{(l)}}} + \frac{\partial^2 R_{i_{\{n+1\}}}}{\partial r_{k_{(l)}} \partial r_{k_{(l)}}} \right)$$

$$+ \sum_{l=1}^{n} \left(- \frac{\partial R_{i_{\{n+2\}}[i_{(n+1)} \mapsto k_{(l)}]}[\mathbf{x}_{(n+1)} \mapsto \mathbf{x}]}{\partial r_{k_{(l)}}} + \frac{\partial R_{i_{\{n+2\}}[i_{(n+1)} \mapsto k_{(l)}]}[\mathbf{x}_{(n+1)} \mapsto \mathbf{r}_{(l)}]}{\partial r_{k_{(l)}}} \right) = 0,$$

where n varies from 1 to ∞. The system (1) is extended by its corresponding continuity equations of the form

$$\sum_{j=1}^{n} \frac{\partial R_{i_{\{n+1\}}[i_{(0)} \mapsto k_{(j)}]}}{\partial r_{k_{(j)}}} = 0 \ , \quad \frac{\partial R_{i_{\{n+1\}}[i_{(l)} \mapsto k_{(l)}]}}{\partial r_{k_{(l)}}} = 0 \quad \text{for} \quad l = 1, \ldots, n \ . \tag{2}$$

In the latter equations the MPC tensor is defined as

$$R_{i_{\{n+1\}}} = R_{i_{(0)} i_{(1)} \ldots i_{(n)}} = \overline{u_{i_{(0)}}(\mathbf{x}_{(0)}) \cdot \ldots \cdot u_{i_{(n)}}(\mathbf{x}_{(n)})} \ , \tag{3}$$

and the four variations of it appearing in (1) and (2) are given by

$$R_{i_{\{n+1\}}[i_{(l)} \mapsto k_{(l)}]} = \overline{u_{i_{(0)}}(\mathbf{x}_{(0)}) \cdot \ldots \cdot u_{i_{(l-1)}}(\mathbf{x}_{(l-1)}) u_{k_{(l)}}(\mathbf{x}_{(l)}) u_{i_{(l+1)}}(\mathbf{x}_{(l+1)}) \cdot \ldots \cdot u_{i_{(n)}}(\mathbf{x}_{(n)})} \ , \tag{4}$$

$$R_{i_{\{n+2\}}[i_{(n+1)} \mapsto k_{(l)}]}[\mathbf{x}_{(n+1)} \mapsto \mathbf{x}_{(l)}] = \overline{u_{i_{(0)}}(\mathbf{x}_{(0)}) \cdot \ldots \cdot u_{i_{(n)}}(\mathbf{x}_{(n)}) u_{k_{(l)}}(\mathbf{x}_{(l)})} \ , \tag{5}$$

$$R_{i_{\{n\}}[i_{(l)} \mapsto \emptyset]} = \overline{u_{i_{(0)}}(\mathbf{x}_{(0)}) \cdot \ldots \cdot u_{i_{(l-1)}}(\mathbf{x}_{(l-1)}) u_{i_{(l+1)}}(\mathbf{x}_{(l+1)}) \cdot \ldots \cdot u_{i_{(n)}}(\mathbf{x}_{(n)})} \ , \tag{6}$$

$$P_{i_{\{n\}}[l]} = \overline{u_{i_{(0)}}(\mathbf{x}_{(0)}) \cdot \ldots \cdot u_{i_{(l-1)}}(\mathbf{x}_{(l-1)}) p(\mathbf{x}_{(l)}) u_{i_{(l+1)}}(\mathbf{x}_{(l+1)}) \cdot \ldots \cdot u_{i_{(n)}}(\mathbf{x}_{(n)})} \ , \tag{7}$$

t is time and the correlation distance is defined according to

$$\mathbf{r}_{(l)} = \mathbf{x}_{(l)} - \mathbf{x}_{(0)} \quad \text{with} \quad l = 1, \ldots, n \ . \tag{8}$$

2 Symmetries of the Multi-Point Equation

In the limit of $|\mathbf{r}| \gg \eta_K$, i.e. for length scales beyond the viscosity dominated Kolmogorov scale [3], we find a new scaling symmetry $\mathbf{G_{s3}}$ of the system (1). The system also admits the classical scaling groups $\mathbf{G_{s1}}$ and $\mathbf{G_{s2}}$ representing the independent scaling of space and time

$$\mathbf{G_{s1}}: \ \tilde{t} = t,\ \tilde{r}_{i_{(l)}} = r_{i_{(l)}} e^{a_1},\ \tilde{R}_{ij} = R_{ij} e^{2a_1},\ \tilde{R}_{ijk} = R_{ijk} e^{3a_1}, \cdots, \tag{9}$$

$$\mathbf{G_{s2}}: \ \tilde{t} = e^{a_2} t,\ \tilde{r}_{i_{(l)}} = r_{i_{(l)}},\ \tilde{R}_{ij} = R_{ij} e^{-2a_2},\ \tilde{R}_{ijk} = R_{ijk} e^{-3a_2}, \cdots, \tag{10}$$

$$\mathbf{G_{s3}}: \ \tilde{t} = t,\ \tilde{r}_{i_{(l)}} = r_{i_{(l)}},\ \tilde{R}_{ij} = R_{ij} e^{a_3},\ \tilde{R}_{ijk} = R_{ijk} e^{a_3}, \cdots. \tag{11}$$

The latter scaling symmetries are Lie groups which define transformations that leave the differential equation under analysis invariant if written in the new variables and independent of the group parameters.

It is important to note that $\mathbf{G_{s3}}$ is clearly distinct i.e. linearly independent from the classical scaling groups in fluid mechanics. Interesting enough this property is not shared with the Euler or Navier-Stokes equations from which the MPC equations have originally emerged. Hence it is a purely statistical property of the equations (1) and subsequently referred to as statistical scaling group (SSG).

This was first observed for parallel wall-bounded shear flows (see [2]) though there this property only holds true for the two-point equation. Hence, in a strict sense there it is broken for higher order correlation equations. In fact, the new scaling group $\mathbf{G_{s3}}$ is due to the linearity of (1) which arise out of the assumption of homogeneity. At the same time linearity of (1) implies that there is the general superposition group admitted by (1) a property shared by all linear differential equations.

Beside the above symmetries the system (1) admits the classical symmetry translation in time

$$\mathbf{G_t}: \ \tilde{t} = t + a_4,\ \tilde{r}_{i_{(l)}} = r_{i_{(l)}},\ \tilde{R}_{ij} = R_{ij},\ \tilde{R}_{ijk} = R_{ijk}, \cdots. \tag{12}$$

and a translation in correlation space which is also a group not admitted by Euler or Navier-Stokes equations (and not to be mistaken for the classical translation in space)

$$\mathbf{G_t}: \ \tilde{t} = t,\ \tilde{r}_{i_{(l)}} = r_{i_{(l)}} + a_{i_{(l)}},\ \tilde{R}_{ij} = R_{ij},\ \tilde{R}_{ijk} = R_{ijk}, \cdots. \tag{13}$$

Note that the latter is broken due to the Schwarz inequality in correlation space, e.g. for the two-point tensor for homogeneous turbulence $R_{\alpha\beta}(\mathbf{r})^2 \leq R_{\alpha\alpha}(0) R_{\beta\beta}(0)$.

3 Invariant Solutions and Turbulent Decay Scaling Laws

Classical theories on decaying turbulence such as Birkhoff's and Loitsyansky's integral entirely rely on the groups $\mathbf{G_{s1}}$ and $\mathbf{G_{s2}}$. There these two groups give rise to a one-parameter family of similarity solution where e.g. the turbulent kinetic energy decays and the integral length scale increases according to a power law. The exponent is usually settled by one of the above proposed conserved integrals.

Presently using the above extended set of symmetry groups (9)-(12) a much wider class of invariant solutions or turbulent scaling laws is derived for homogeneous turbulence. For this we need to define an invariant solution, usually called similarity solution, employing the following three concepts.

(i) Any Lie symmetry group, and the groups (9)-(12) with the group parameters a_1-a_4 are among those, have an equivalent infinitesimal representation defined by

$$\tilde{x} = x + \xi(x,y)a + O(a^2) \quad \text{and} \quad \tilde{y} = y + \eta(x,y)a + O(a^2) \ . \tag{14}$$

with x and y the vector of independent and dependent variables respectively and $\tilde{x}$ and $\tilde{y}$ refer to the corresponding variables in the transformed space, a is the group parameter and the infinitesimals of the Lie symmetry are defined as $\xi(x,y) = \left.\frac{\partial \tilde{x}}{\partial \varepsilon}\right|_{\varepsilon=0}$ and $\eta(x,y) = \left.\frac{\partial \tilde{y}}{\partial \varepsilon}\right|_{\varepsilon=0}$. Lie's first theorem states that the infinitesimals are sufficient for the recovery of the full symmetry transformation (see e.g. [5]).

(ii) The condition of invariance, i.e. implementing any symmetry transformation into its associated equation leaves the equation unchanged written in the new variables, has an infinitesimal correspondent. This is defined by

$$\left.\left[X\mathscr{F}(x,y,y^{(1)},\ldots,y^{(m)})\right]\right|_{\mathscr{F}=0} = 0 \tag{15}$$

with $\mathscr{F} = 0$ the set of equations under investigation, here (1), and X is given by

$$X = \xi_i \frac{\partial}{\partial x_i} + \eta_j \frac{\partial}{\partial y_j} \ . \tag{16}$$

(iii) Suppose the Lie symmetries of an equation are given, as is the case here, the invariant solution $\theta(x)$ is defined according to the condition

$$X\left(y - \theta(x)\right) = 0 \quad \text{on} \quad y = \theta(x) \ . \tag{17}$$

The latter condition constitutes a hyperbolic partial differential equation which may be solved using method of characteristic. The corresponding ordinary differential equation

$$\frac{\mathrm{d}x_1}{\xi_1} = \frac{\mathrm{d}x_2}{\xi_2} = \cdots = \frac{\mathrm{d}y_1}{\eta_1} = \frac{\mathrm{d}y_2}{\eta_2} = \cdots \tag{18}$$

is referred to as invariant surface condition of Lie group theory which leads to self-similar or, more general, invariant solutions.

For the present case the infinite set of equations (1), which in the turbulence community called scaling laws, we obtain the invariant surface condition

$$\frac{\mathrm{d}t}{a_2 t + a_4} = \frac{\mathrm{d}r_{(i)}}{a_1 r_{(i)}} = \frac{\mathrm{d}R_{(ij)}}{[2(a_1 - a_2) + a_3]R_{(ij)}} = \cdots \tag{19}$$

with the group parameters a_1-a_4 descending from the groups (9)-(12) and here written in infinitesimal form and the indices in brackets denote no summation but instead each component is to be taken separately.

We note that any solution for an arbitrary set of group parameters of the latter system allows for an invariant solution of (1). Two different cases may be distinguished. Firstly we consider the case without the new SSG (11). Further assuming $a_1 \neq a_2$ and $a_4 \neq 0$ we find the following invariants of the system (19)

$$\hat{\mathbf{r}}_{(1)} = \frac{\mathbf{r}_{(1)}}{(t+t_0)^n} \quad , \quad R_{ij}(\mathbf{r}_{(1)},t) = (t+t_0)^{-m}\hat{\mathscr{R}}_{ij}\left(\hat{\mathbf{r}}_{(1)}\right) \quad , \quad \ldots \tag{20}$$

with $n = a_1/a_2$, $t_0 = a_4/a_2$ and $m = 2(1-a_1/a_2)$. The variables $\hat{\mathbf{r}}_{(1)}$ and $\hat{\mathscr{R}}_{ij}$ are the constants of integration of the invariant surface condition (19) or in other word the invariants of the system. They are to be taken as new independent and dependent variables of the system (1) leading to a similarity reduction depicting the classical power law behavior. Therein $m = 6/5, n = 2/5$ and $m = 10/7, n = 2/7$ respectively correspond to Birkhoff's and Loitsiansky's integrals.

For the present purpose of primarily understanding the scaling behaviour of fractal generated turbulence we need to consider both the breaking of the two classical scaling groups due to external symmetry breaking quantities to be detailed from a physical point of view below. Hence, we set $a_1 = a_2 = 0$. Further, for the present case a non-zero a_3 related to the new scaling group is needed in order to allow for the construction of an invariant solution at all. Hence, employing the latter two informations into equation (19) we observe two important conclusions.

First, due to $a_2 = 0$ and combining the first and the last term in (19) we have an exponential scaling of the two- and MPC with time. Second, because $a_1 = 0$ we have no scaling of space and hence any $\mathbf{r}$ itself is an invariant. Following the methodology above this leads to a similarity solution for the infinite set of MPC tensors where the first term in the row, i.e. the two-point tensor, has the following form

$$\hat{\mathbf{r}}_{(1)} = \mathbf{r}_{(1)} \quad , \quad R_{ij}(\mathbf{r}_{(1)},t) = e^{-t/t_0}\mathscr{R}_{ij}\left(\hat{\mathbf{r}}_{(1)}\right) \quad , \quad \ldots \tag{21}$$

where $\mathscr{R}_{ij}$ is the similarity variable of the reduced set of MPC equations independent of time and $t_0 = -a_4/a_3$.

In order to compare (20) and (21) to experimentally observable one-point quantities we introduce the Reynolds stress tensor $\overline{u_i u_j}$ and the integral length scale L_i as functionals of R_{ij}

$$\overline{u_i u_j} = R_{ij}(\mathbf{r}=0,t) \quad , \quad L_i = \frac{1}{2K}\int R_{kk}(\mathbf{r})\,\mathrm{d}r_i \ . \tag{22}$$

Employing these definitions and implementing (20) and (21) into the latter we respectively obtain the rather different turbulent scaling laws

$$\overline{u_i u_j} \sim (t+t_0)^{-m} \quad , \quad L_i \sim (t+t_0)^n \tag{23}$$

and

$$\overline{u_i u_j} \sim e^{-t/t_0} \quad , \quad L_i \sim const.. \tag{24}$$

In Hurst and Vassilicos, (2007) it was first reported that fractal-generated turbulence in a wind tunnel experiment may lead to an exponential decay law for the turbulent kinetic energy according to (24) and it was more fully consolidated in Seoud and Vassilicos, (2007). For certain cases they also find a constant integral length scale (22) (and also Taylor length scale) downstream of the fractal grid. A variety of different fractal grids were employed for the experiment such as cross-grids, square-grids and I-grids. Therein t_r is the thickness ratio defined as the scaling factor between the largest to the smallest bar thicknesses.

The data for the turbulent kinetic energy and the integral length scale showing both the behaviour according to (24) are given in figure 1 and 2. Three key results may be taken from 1 and 2. First, we observe that only the higher thickness ratios allow for the establishment of an exponential decay law i.e. beginning with $t_r = 8.5$ and higher the new scaling laws is clearly visible. Second, the development of the exponential decay downstream of the grid becomes faster for increasing thickness ratios and is the largest for $t_r = 17$. Third, the constant integral length scale downstream of the grid appears to be less sensitive to the thickness ratio.

The physical interpretation of the latter results and in particular the fact that for large t_r the new scaling laws are established faster are due to the fact that a broad bandwidth of external scales have been imposed on the flow which are symmetry breaking. In the present case we have in fact only one scaling group of space. In the wind tunnel experiment however the scaling of time, here denoted by a_2, may be re-interpreted as another scaling group of space due to the relation $\tau = a/U$ where a is any fractal grid length scale and U is the constant mean velocity in the wind tunnel. As a result we observe a symmetry breaking of a_1 and a_2 i.e. $a_1 = a_2 = 0$ which directly leads to the given multi-point scaling law (21) or the related one-point scaling law (24).

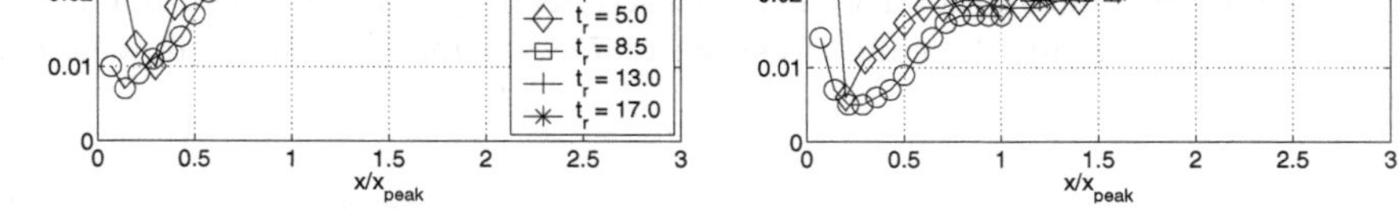

Fig. 1 Longitudinal L_u and lateral L_v integral scales as functions of x/x_{peak} approaching a constant downstream of the grid. Space-filling fractal square grids at $U = 10m/s$ in the $T = 0.46m$ wind tunnel have been used. t_r defines the scaling factor between the largest to smallest bar thicknesses. All results are taken from [1].

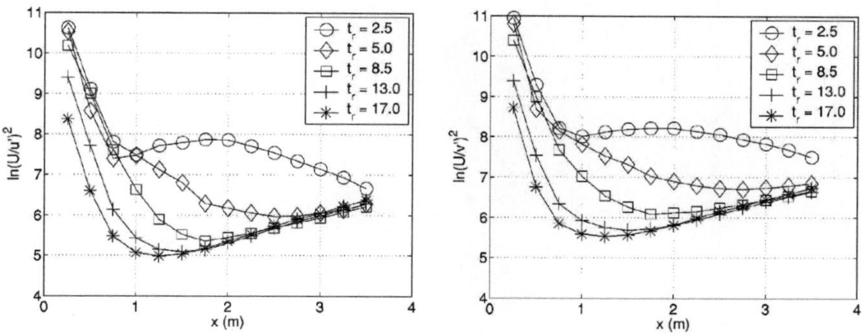

Fig. 2 $\ln[(U/u)^2]$ and $\ln[(U/v)^2]$ as functions of x (in meters) for all five space-filling fractal square grids revealing the straight line in agreement with equation (22) on which the turbulence decay curves generated by all these grids eventually asymptote to. t_r defines the scaling factor between the largest to smallest bar thicknesses. All results are taken from [1].

4 Conclusions

Two significant conclusions may be drawn from the present results. First, the above elaborated extended statistical scaling properties which go beyond the Euler and Navier-Stokes have been observed, though somewhat indirect, in the fractal grid turbulence experiments for the first time.

Second, none of the existing turbulence models in particular RANS models i.e. two equation and Reynolds stress transport model, admit three scaling groups. Hence they are all incapable to mimic the above presented behaviour. In the infinite Reynolds number limit essentially all RANS models admit two scaling groups some even less.

The authors are deeply indebted to Ch. Vassilicos for very helpful discussion on fractal generated turbulence and leaving the figures 1 and 2 to their disposition.

References

1. Hurst, D., Vassilicos, J.C.: Scalings and decay of fractal-generated turbulence. Phys. Fluids 19, 035103 (2007)
2. Khujadze, G., Oberlack, M.: DNS and scaling laws from new symmetry groups of ZPG turbulent boundary layer flow. TCFD 18, 391–411 (2004)
3. Oberlack, M.: Symmetry, Invariance and Selfsimilarity in Turbulence, Habilitation Thesis, RWTH Aachen, Germany (2000)
4. Seoud, R.E., Vassilicos, J.C.: Dissipation and decay of fractal-generated turbulence. Phys. Fluids 19, 105108 (2007)
5. Bluman, G.W., Kumei, S.: Symmetries and Differential Equations. Applied Mathematical Science. Springer, Heidelberg (1989)

Investigation of the Conditional Scalar Dissipation Rate Across a Shear Layer Using Gradient Trajectories

Juan Pedro Mellado, Lipo Wang, and Norbert Peters

Abstract. The dependence on the lateral distance y to the center-plane and on time t of the average of the scalar dissipation rate $\chi = 2D|\nabla Z|^2$ conditioned on the scalar Z has been investigated in a temporally-evolving shear layer using a direct numerical simulation. As the inviscid scaling of the unconditional mean dissipation rate is approached, the conditional mean $\bar{\chi}_Z$ follows also an approximate self-similar behavior within the time interval considered in the simulations and $\bar{\chi}_Z$ shows two strong peaks at a distance of about one vorticity thickness from the center-plane, with dissipation values about 3 times the one in the central region. It is showed that this spatial variation is introduced by the nonturbulent/turbulent transition regions, which are identified by means of gradient trajectories of the scalar field.

1 Introduction

The scalar dissipation rate $\chi = 2D|\nabla Z|^2$ and its relation with the scalar Z, D being the corresponding molecular diffusivity, has received considerable attention in turbulence research [1, 2]. In particular for combustion, this relation between χ and Z plays a crucial role and enters in almost all of the existing nonpremixed turbulent combustion models [3, 4], commonly in the form of the first conditional moment $\bar{\chi}_Z$. However, the behavior of $\bar{\chi}_Z$ in free turbulent flows is not well understood yet and data available in the literature is not conclusive. Free turbulent flows add the phenomenon of external intermittency and the nonturbulent/turbulent transition regions [5, 6], and therefore a possible spatial variation of the conditional statistics. Experimental data from nonreacting flows [7, 8, 9] indicate a strong spatial variation, whereas laboratory measurements in reacting flows [10, 11] show a weaker

Juan Pedro Mellado · Lipo Wang · Norbert Peters
Institut für Technische Verbrennung, RWTH Aachen, Templergrabenstr. 64, 52056 Aachen, Germany
e-mail: `jpmellado@itv.rwth-aachen.de`, `l.wang@itv.rwth-aachen.de`, `n.peters@itv.rwth-aachen.de`

dependence. Analysis based on direct numerical simulations (DNS) is scarce; it has been suggested [12, 13] that the conditional mean scalar dissipation rate has a weak dependence on the lateral distance and the curve $\bar{\chi}_Z(Z)$ shows a double-hump profile [cf. Fig. 2 (a) below] when the total volume is employed to calculate the average. This paper addresses this issue and studies in detail the variation of $\bar{\chi}_Z$ using a passive scalar from a DNS of a temporally-evolving shear layer.

2 Results

The numerical algorithm follows that of [13], though the resolution is doubled up to $\eta/\Delta x \simeq 1.2$, η being the Kolmogorov scale and Δx the uniform grid spacing, in order to correctly detect the gradient trajectories [14]. The grid size is $1024 \times 256 \times 1024$, the Reynolds number at the end of the simulation is about 1500 based on the vorticity thickness δ_ω and the Schmidt number is one. The configuration is represented in Fig. 1, where the scalar gradient is shown. Statistics depend on the distance to the center plane y and on time t, the averages being taken over the horizontal planes xOz.

The mean scalar dissipation rate conditioned on the scalar, $\bar{\chi}_Z(Z,y,t)$, is shown in Fig. 2 (b) for the final time of the simulation. This figure shows contour plots of the conditional mean normalized by the center-plane value of the unconditional mean dissipation $\bar{\chi}_c(t)$, the abscissas being the conditioning scalar Z and the ordinates the distance y to the center plane normalized by the instantaneous vorticity thickness. There is a clear *strong lateral variation* of the conditional mean, over a factor of 3 and mainly beyond $\delta_\omega(t)/2$, and the local profile $\bar{\chi}_Z(Z)$ develops a conspicuous peak between $Z = 0.2$ and 0.4 as we move towards the outer stream $Z = 0$, and a symmetrical one as the second stream $Z = 1$ is approached. When the conditional mean is averaged across y by using the whole volume in the post-processing of the data, these two peaks are the origin of the two-humps observed in Fig. 2 (a). However, the spatial variation is then missed and, more importantly, the maximum of the conditional scalar dissipation is strongly under-predicted, as observed by comparing the normalized values in Fig. 2 (a), below 1.5, with the real ones observed in Fig. 2 (b), greater than 3.5. With respect to the time variation, it was observed that the quantity

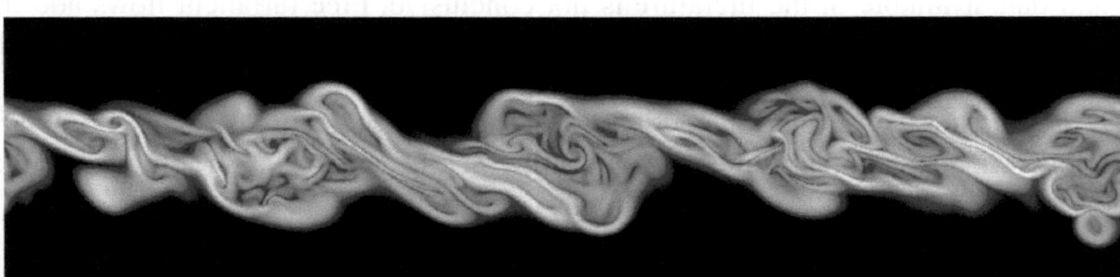

Fig. 1 Logarithm of the scalar gradient field on a xOy plane at the final time of the simulation

$\bar{\chi}_Z(Z, y/\delta_\omega(t), t)/\bar{\chi}_c(t)$ tends towards an approximate constant behavior as depicted in Fig. 2 (b), although the interval of Reynolds number covered by the simulation is too small to draw any conclusion with respect to the details of the peaks.

The previous result shows that the statistical dependence between the scalar dissipation rate χ and the scalar Z itself varies across the shear layer, and the possible link with external intermittency of this variation (which increases as we move away from the center-plane) has been further investigated by means of *gradient trajectories*. A *partition of the flow* that allows the identification of the turbulent/nonturbulent transition zone is proposed based on the type of the gradient trajectory that passes through each point. Gradient trajectory analysis has been already successfully applied to study homogeneous turbulence [14]. This partition is sketched in Fig. 3. If the gradient trajectory starting from any given point joints one minimum with one maximum, like case A in that figure, then that point is said to belong to the turbulent zone. On the other hand, if the trajectory connects one extremum with one of the two outer homogenous streams, then the point is said to belong to the upper or the lower turbulence interface, cases B and C in Fig. 3, respectively. This differentiation among zones improves that based on the intermittency function because it is nonlocal and does not relay on an ad-hoc threshold.

Figure 4 shows isocontours of the normalized conditional mean scalar dissipation rate like in Fig. 2 (b) but restricted to the upper turbulence interface defined as explained in the previous paragraph (the conditional mean inside the lower interface is

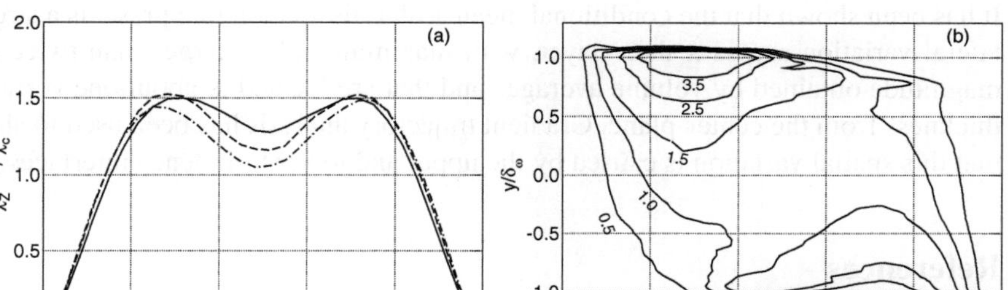

Fig. 2 Conditional mean scalar dissipation rate $\bar{\chi}_Z$ normalized by the unconditional value at the center-plane $\bar{\chi}_c$ at the final time of the simulation: (a) computed over the whole volume; (b) Contours of the function $\bar{\chi}_Z(Z, y/\delta_\omega)$, showing the strong spatial variation.

Fig. 3 Flow partition based on gradient trajectories (see text for details). Solid line separates turbulent zone from turbulence interface. Dashed line indicates threshold in the gradient to define the conventional intermittency function.

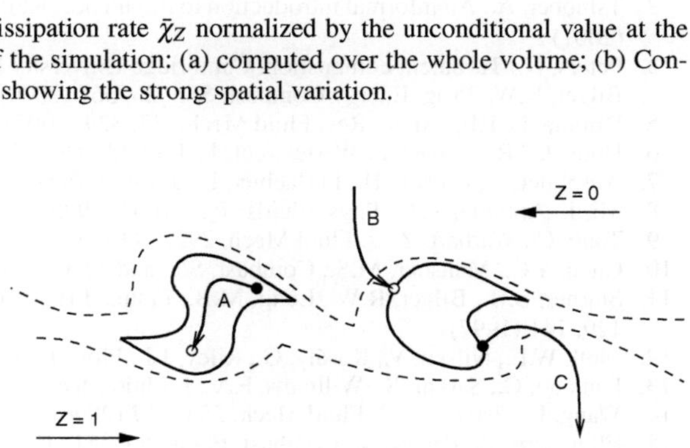

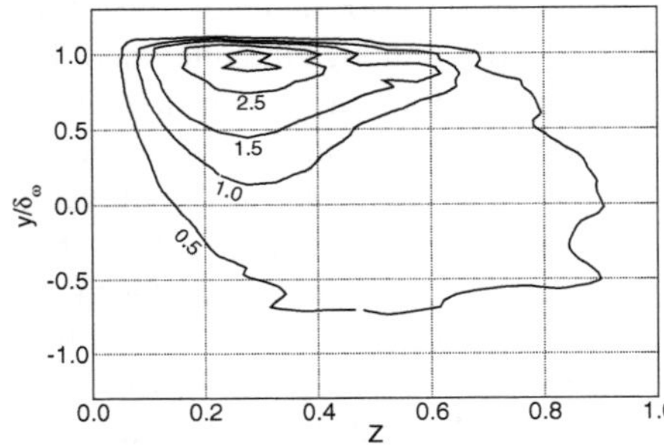

Fig. 4 Contours of the function $\bar{\chi}_Z(Z, y/\delta_\omega)$ as in Fig. 2 (b) but restricted to the upper turbulence interface defined by the gradient trajectories joining the upper external laminar flow with the turbulent zone in the center of the layer. The same figure of the lower layer is symmetric with respect to $y = 0$ and $Z = 0.5$.

the appropriately symmetrical figure of Fig. 4). Comparison between Fig. 2 (b) and Fig. 4 indicates that the turbulence interface represents most of the mixing region in terms of conditional scalar dissipation, and the statistics inside it contribute significantly to the total one, confirming the hypothesis that these nonturbulent/turbulent transition zones determine some statistics to a large extent [15]. In particular, the result confirms that the origin of the aforementioned peaks in Fig. 2 (b) is in the turbulent interface between the turbulent zones and the outer laminar regions.

3 Conclusions

It has been shown that the conditional mean scalar dissipation rate presents a strong lateral variation across a shear layer, with maximum values larger than twice the magnitude obtained by volume averages and that are located at about one vorticity thickness from the center-plane. Gradient trajectory analysis has been used to show that this spatial variation is caused by the upper and lower turbulence interfaces.

References

1. Warhaft, Z.: Annu. Rev. Fluid Mech. 32, 203 (2000)
2. Tsinober, A.: An informal introduction to turbulence. Kluwer Academic Publishers, Dordrecht (2001)
3. Peters, N.: Turbulent combustion. Cambridge University Press, Cambridge (2000)
4. Bilger, R.W.: Prog. Energy Combust. Sci. 26, 367 (2000)
5. Dimotakis, P.E.: Annu. Rev. Fluid Mech. 37, 329 (2005)
6. Hunt, J.C.R., Eames, I., Westerweel, J.: J. Fluid Mech. 554, 449 (2006)
7. Anselmet, F., Djeridi, H., Furlachier, L.: J. Fluid Mech. 280, 173 (1994)
8. Mi, J., Antonia, R.A.: Phys. Fluids 7(7), 1665 (1995)
9. Tong, C., Warhaft, Z.: J. Fluid Mech. 292, 1 (1995)
10. Chen, Y.C., Mansour, M.S.: Combust. Sci. and Tech. 126, 291 (1997)
11. Starner, S.H., Bilger, R.W., Long, M.B., Frank, J.H., Marran, D.F.: Combust. Sci. and Tech. 129, 141 (1997)
12. Mell, W.E., Nilsen, V., Kosály, G., Riley, J.J.: Phys. Fluids 6(3), 1331 (1994)
13. Pantano, C., Sarkar, S., Williams, F.A.: J. Fluid Mech. 481, 291 (2003)
14. Wang, L., Peters, N.: J. Fluid Mech. 554, 457 (2006)
15. Effelsberg, E., Peters, N.: Combust. Flame 50, 351 (1983)

‘Rational’ Turbulence Models?

Robert Rubinstein and Stephen L. Woodruff

Abstract. A procedure to construct turbulence models is outlined beginning with the simplest case, a model for weak time-dependent perturbations of homogeneous isotropic turbulence, and ending with some models for inhomogeneous turbulence. The approach combines features of Yoshizawa’s two-scale direct interaction approximation and the Hilbert expansion of kinetic theory.

1 Models for Homogeneous Isotropic Turbulence

A turbulence model is a finite dimensional system that mimics a broad class of turbulent flows. Since turbulent motion couples infinitely many scales of motion, the possibility of modeling poses an interesting and important theoretical question. Assume that in the Lin equation

$$\dot{E}(\kappa,t) = P(\kappa,t) - \frac{\partial}{\partial \kappa}\mathscr{F}(\kappa,t) - 2\nu\kappa^2 E(\kappa,t) \qquad (1)$$

the flux $\mathscr{F}$ is a functional of the energy spectrum E. Under statistically steady forcing $P(\kappa)$, a time-independent spectrum $E_0(\kappa) = E_0(\kappa; a_1, a_2, \cdots)$ satisfies the equations of motion. Modeling focuses on the possibility that the parameters a_i are functions of time. But $E(\kappa,t) = E_0(\kappa, a_1(t), a_2(t), \cdots)$ does not satisfy Eq. (1), because the left side, $(\partial E_0/\partial a_1)\dot{a}_1 + \cdots$ is nonzero. If these terms are small, say because $a_i(t) = a_i(\tau)$, with a slow time variable τ, then they can be balanced by a perturbation E_1. Such a perturbation requires a small, slowly varying perturbation $P_1(\kappa,\tau)$. E_1 satisfies the linear integral equation

Robert Rubinstein
Newport News, VA, USA
e-mail: arubin@hra.org

Stephen L. Woodruff
CAPS, Florida State University, Tallahassee, FL, USA
e-mail: slwoodruff@gmail.com

$$\frac{\partial E_0}{\partial a_1}\dot{a}_1 + \cdots = P_1(\kappa,\tau) - \frac{\partial}{\partial\kappa}\mathscr{L}[E_1] - 2\nu\kappa^2 E_1(\kappa,t) \tag{2}$$

where $\mathscr{L} = (\delta\mathscr{F}/\delta E)_{E_0}$ is the Fréchet derivative of $\mathscr{F}$ evaluated at E_0. This linear integral equation can only be solved for E_1 if the *compatibility conditions*

$$\int d\kappa\, \Psi_j(\kappa)\frac{\partial E}{\partial a_1}\dot{a}_1 + \cdots = \int d\kappa\, \Psi_j(\kappa)P_1(\kappa) - 2\nu\int d\kappa\, \kappa^2\Psi_j(\kappa)E_1(\kappa) \tag{3}$$

are satisfied, where the Ψ_j are solutions of $\mathscr{L}^\dagger\,[\partial\Psi/\partial\kappa] = 0$ and $\dagger$ indicates the adjoint. The structure of this equation implies that one condition is the energy balance obtained from $\Psi \equiv 1$.

These conditions provide a finite number of equations for the parameters a_i. In the case of most interest, there are two parameters a_i, which could be taken as a length scale L and the dissipation rate ε. A determinate model is obtained only if there are exactly two compatibility conditions. For classical models, including the Kovaznay, Ellison, and Heisenberg models, the linearized transfer equation $\partial/\partial\kappa(\delta\mathscr{F}/\delta E)\,[E_1] = 0$ always has only one solution, in which the constant flux solution is simply shifted in the inertial range. To obtain a two-equation model, the adjoint equation must have more null solutions than the direct equation. This is possible because the linearized transfer operator is (1) not self-adjoint and (2) infinite dimensional. Whereas the Heisenberg model does give a two-equation model [1], the Kovaznay and the Ellison models give only one compatibility condition. Adding time-scale evolution to the model does not change the conclusion. A more interesting case would be the linearized transfer for EDQNM.

In contrast to kinetic theory, turbulence has many more relevant reference states than the steady state. For example, in self-similar decay $E(\kappa,t) = k(t)L(t)\psi(x)$ with $x = L\kappa$ and one possibility is $k(t) = A^{2/5}t^{-6/5}$, $L(t) = A^{1/5}t^{2/5}$ where A is a constant. Then $-\psi + \frac{1}{3}(\psi + x\psi') = \partial/\partial x\mathscr{F}[\psi]$. The similarity solution is a consequence of scaling invariance. If this invariance is weakly broken by a small, slowly varying production term $P(k,t) = (k^{3/2}/L)\pi(x,\tau)$ then $k(t) = A(\tau)^{2/5}t^{-6/5}$, $L(t) = A(\tau)^{1/5}t^{2/5}$. Let ψ_0 correspond to the self-similar energy spectrum with slowly varying $A(\tau)$, and introduce the perturbation ψ_1. Then

$$\frac{1}{A}\tau\frac{dA}{d\tau}[\frac{3}{5}\psi_0 + \frac{1}{5}x\frac{d\psi_0}{dx}] - \pi = \mathscr{L}[\psi_1] \tag{4}$$

where

$$\mathscr{L}[\phi] = \frac{\partial}{\partial x}\left(\frac{\delta\mathscr{F}}{\delta\psi}\right)_{\psi_0}[\phi] + \phi - \frac{1}{3}\frac{\partial}{\partial x}(x\phi) \tag{5}$$

The equation governing the evolution of A due to the production perturbation π is found as above as the compatibility equation to solve Eq. (4) for ψ_1. Under the perturbation hypotheses made here, a one-equation model is appropriate, not a two-equation model. At this point, the question of which models lead to exactly one compatibility condition is unanswered.

2 Models for Inhomogeneous Turbulence

In the model of [2], the basic descriptor of the isotropic part of the turbulence is the local energy spectrum $E(\mathbf{x}, \kappa)$ satisfying

$$\dot{E} + U_n \partial E/\partial x_n = 2\nu_t[E]S^2 - (\partial/\partial\kappa)\mathscr{F}[E] - \mathbf{V}[E]$$
$$+a\nabla_{\mathbf{x}} \cdot (\nu_t[E]\nabla_{\mathbf{x}} E) - c(\partial/\partial\kappa)(\kappa\nu_t[E])S^2 \tag{6}$$

where $\nu_t[E]$ is a scale-dependent eddy viscosity, S^2 is the square of the mean strain rate, $\mathbf{V}$ denotes viscous dissipation and diffusion, and a and c are model constants. Assume that a homogeneous steady state $E(\kappa) = kLE(L\kappa)$ is perturbed by weak spatially and temporally varying shear, so that the spectrum becomes *locally* homogeneous: $E_0(\mathbf{x}, \kappa, t) = k(\mathbf{x},t)L(\mathbf{x},t)\psi(L(\mathbf{x},t)\kappa)$. Because E_0 does not satisfy Eq. (6), add a correction term E_1 satisfying

$$\dot{E}_0 + U_n \partial E_0/\partial x_n - 2\nu_t[E_0]S^2 - a\nabla_{\mathbf{x}} \cdot (\nu_t[E_0]\nabla_{\mathbf{x}} E_0) + c(\partial/\partial\kappa)\kappa\nu_t[E_0]S^2$$
$$= -\partial/\partial\kappa(\delta\mathscr{F}/\delta E)_{E_0}[E_1] - \mathbf{V}[E_1] \tag{7}$$

As in [1], the compatibility conditions to solve this linear integral equation for E_1 are obtained by multiplying the left side by the null solutions of the adjoint of the right side and integrating over κ. The structure of the model of [2] is such that the compatibility conditions are generated simply from the linearized transfer of the homogeneous flux model $\mathscr{F}[E]$ (as in the homogeneous problem, the viscous contributions to (7) can be ignored). The model equations will have the standard form containing advection, production, destruction, and diffusion terms. Although one equation will be an energy balance, the second will be nonstandard [1]. Note also the contribution from the last term on the left side of (7).

Because the model of [2] makes very sweeping assumptions and consequently seems over-simplified, we reconsider the problem of inhomogeneous turbulence from a more general viewpoint. First consider inhomogeneous turbulence in free space without boundaries. The flow region admits arbitrary translations. Given any two-point field $U(\mathbf{x}, \mathbf{x}')$, the transformation $\mathbf{l} \to U(\mathbf{x}+\mathbf{l}, \mathbf{x}'+\mathbf{l})$ defines a representation of the translation group of space. The invariant quantities (occurrences of the identity representation) are the homogeneous fields. Consider the decomposition of this representation into its irreducible components given by the usual prescription: multiplication by characters and invariant integration over the group. In this very simple case, this defines the quantities

$$U(\mathbf{x}, \mathbf{x}'; \mathbf{K}) = \int d\mathbf{l}\, U(\mathbf{x}+\mathbf{l}, \mathbf{x}'+\mathbf{l}) \exp(i\mathbf{K}\cdot\mathbf{l}) \tag{8}$$

as Fourier transforms.

Two useful properties are $U(\mathbf{x}, \mathbf{x}') = \int d\mathbf{K}\, U(\mathbf{x}, \mathbf{x}'; \mathbf{K})$ and $U(\mathbf{x}+\mathbf{a}, \mathbf{x}'+\mathbf{a}; \mathbf{K}) = U(\mathbf{x}, \mathbf{x}'; \mathbf{K}) \exp(-i\mathbf{K}\cdot\mathbf{a})$ so that

$$U(\mathbf{x},\mathbf{x}') = \int d\mathbf{K}\, U^\dagger(\mathbf{x}-\mathbf{x}';\mathbf{K})\exp(i\mathbf{K}\cdot\mathbf{x}') \tag{9}$$

where $U^\dagger(\mathbf{x}-\mathbf{x}';\mathbf{K}) = U(\mathbf{x}-\mathbf{x}',\mathbf{0};\mathbf{K})$. Note also that $\mathbf{K}=0$ is the homogeneous part of the random field. Now let U_{ij} denote the two-point velocity correlation. In terms of the corresponding fields $U^\dagger_{ij}(\mathbf{x}-\mathbf{x}';\mathbf{K})$, the correlation equation has the structure $\dot{U}^\dagger_{ij}(\mathbf{r};\mathbf{K}) = -[\frac{1}{2}\partial/\partial r_p + iK_p]T^\dagger_{pij}(\mathbf{r};\mathbf{K}) + \cdots$ where $\mathbf{r}=\mathbf{x}-\mathbf{x}'$ and $T^\dagger_{pij}(\mathbf{r};\mathbf{K})$ is formed from the three-point function by analogy to $U^\dagger$. We will obtain a definite theory by introducing closure models for $T^\dagger$. Inhomogeneous turbulence evolution is expressed in terms of the coupled evolution of the quasi-homogeneous quantities $U^\dagger_{ij}(\mathbf{r};\mathbf{K})$ for different $\mathbf{K}$. Expressing closure theories in terms of $U^\dagger(\mathbf{r};\mathbf{K})$ connects to the previous section, since the reference homogeneous field corresponds to $\mathbf{K}=0$, and the inhomogeneous corrections to small, nonzero $\mathbf{K}$.

This approach is suitable as an approximation for a localized region of turbulent flow sufficiently far from walls or other boundaries. To treat wall effects more explicitly, consider turbulence in a half-space. The flow region admits translations in the $\mathbf{x}=(x,y)$ plane and scale transformations $z\to\lambda z$ with $\lambda>0$. Again, we observe that the transformation $U(\mathbf{x},\mathbf{x}';z,z') \to U(\mathbf{x}+\mathbf{l},\mathbf{x}'+\mathbf{l};\lambda z,\lambda z')$ defines a representation of the group of translations in the (x,y) plane and scalings in z (multiplicative group of positive reals). The invariants are now fields which are homogeneous in the (x,y) plane and scale invariant (homogeneous of degree zero) in z,z'. The decomposition into irreducible representations is again given by the projection

$$U(\mathbf{x},\mathbf{x}';z,z':\mathbf{k};\lambda) = \int d\mathbf{a}\int \frac{d\alpha}{\alpha} U(\mathbf{x}+\mathbf{a},\mathbf{x}'+\mathbf{a};\alpha z,\alpha z')\exp i(\mathbf{a}\cdot\mathbf{k})\alpha^{i\lambda} \tag{10}$$

which is a Fourier transformation in $\mathbf{a}$ and a Mellin transformation in α. The wall boundary condition is not an issue; if it were, a half-range Fourier expansion would be more natural. The explanation is that flows like the boundary layer are not completely self-similar if we consider them all the way to the wall. This formulation allows us to perturb about a self-similar, nonhomogeneous reference state [3].

References

1. Woodruff, S.L., Rubinstein, R.: Multiple-scale perturbation analysis of slowly evolving turbulence. J. Fluid Mech. 565, 95–103 (2006)
2. Harlow, B.D., et al.: Spectral transport model for turbulence. Theor. Comput. Fluid Mech. 8, 1–35 (1996)
3. Oberlack, M., Guenther, S.: Shear-free turbulent diffusion - classical and new scaling laws. Fluid. Dyn. Res. 33, 453 (2003)

An Approximation of the Invariant Measure for the Stochastic Navier-Stokes

Björn Birnir

1 Introduction

Kolmogorov's statistical theory of turbulence is based on the existence of the invariant measure of the Navier-Stokes flow. Recently the existence of the invariant measure was established in the three-dimensional case [2]. It was established for uni-directional flow in [1] and for rivers in [3]. Below we will discuss how one can try to go about approximating the invariant measure in three dimensions.

2 Fluid Flow and Ito Diffusion

A reasonable model for the motion of an Eulerian fluid particle is given by the stochastic ordinary differential equation (SODE)

$$dX_t = -u(X_t,t)dt + \sqrt{2\nu}dB_t$$

Here $u(x,t)$ is the fluid velocity and we expect the fluid particle to move upstream with velocity u and also to move randomly. This random motion is modeled by the second term, where B_t is Brownian motion and dB_t models the white noise affecting the motion of the particle. There is noise in any fluid flow and we expect fluid particles to diffuse under influence of the noise. The above equation is the equation of Ito's diffusion X_t with the generator

$$A = \nu\Delta - u(x,t)\cdot\nabla$$

The backward Kolmogorov equation, corresponding to X_t and A, is the dissipative Burger's equation, or the Navier-Stokes equation without pressure

$$\frac{\partial u}{\partial t} + u\cdot\nabla u = \nu\Delta u$$

Björn Birnir
Dept. of Math., Center for Complex and Nonlinear Science and CNLS, UC Santa Barbara

$$u(x,0) = f(x)$$

This initial value problem has the *implicit* solution

$$u(x,t) = E[f(X_t)]$$

where

$$X_t = X_0 - \int_0^t u(X_s,s)ds + \sqrt{2\nu}B_t$$

Notice that this is not the *explicit* solution of Burger's equation that is obtained by the Cole-Hopf transformation.

Anologously the Navier-Stokes equation for fully-developed turbulent flow, with periodic boundary conditions, can be written in the form

$$\frac{\partial u}{\partial t} + u \cdot \nabla u = \nu \Delta u - \nabla p + \sum_{k \neq 0} h_k^{1/2} db_t^k e_k(x) \tag{1}$$

$$u(x,0) = f(x)$$

In laminar flow the driving term

$$f(x,t) = \sum_{k \neq 0} h_k^{1/2} db_t^k e_k(x)$$

is absent, but in fully developed turbulence the small ambient white noise is magnified into large turbulent noise which is modeled by f, see [1, 2]. The coefficients $h_k^{1/2} \in \mathbb{R}^3$ decay as $k \to \infty$, the $e_k(x)$ are Fourier components, that each come with their independent Brownian motion b_t^k, and we have imposed periodic boundary conditions in $x \in \mathbb{T}^3$. Thus the large turbulent noise is modeled by (independent) white noise in time in all directions, in function space, but the decay of the coefficients $h_k^{1/2}$ makes this noise colored in space. The color is characteristic for turbulent noise in three dimensions.

3 The Approximation

We can proceed further if we now project onto the space of divergence-free vectors eliminating the pressure gradient. Let P denote the projection operator, then we will model the difference between the projection of the inertial terms and the intertial terms themselves as

$$P[u \cdot \nabla u] - u \cdot \nabla u \approx \sum_{k \neq 0} g_k^{1/2} db_t^k e_k(x) \cdot \nabla u$$

This expression is of course not exact, but the modeling is motivated by numerical simulations where an analogous difference the "eddy viscosity", is shown to depend on the gradient ∇u.

Now the initial value problem (1) can be written in the form

$$\frac{\partial u}{\partial t} + w \cdot \nabla u = \nu \Delta u + \sum_{k \neq 0} h_k^{1/2} db_t^k e_k(x) \tag{2}$$

$$u(x,0) = f(x)$$

where

$$w(x,t) = u + \sum_{k \neq 0} g_k^{1/2} db_t^k e_k(x)$$

and we use the same notation for the divergence free $k \cdot h_k^{1/2} = 0$ vectors as for the original $h_k^{1/2}$. Then introducing the Ito diffusion

$$dX_t = -w(X_t,t)dt + \sqrt{2\nu} dB_t$$

we can write the solution of (2) of the form

$$u(x,t) = E[f(X_t)] + \sum_{k \neq 0} h_k^{1/2} \int_0^t E[e_k(X_{t-s})] db_s^k$$

Now by Girsanov's theorem, see [4], we can rewrite u in the form

$$u(x,t) = E[f(B_t)M_t] + \sum_{k \neq 0} h_k^{1/2} \int_0^t E[e_k(B_{t-s})M_{t-s}] db_s^k \tag{3}$$

where

$$M_t = exp\{-\int_0^t w(B_s,s) \cdot dB_s - \frac{1}{2}\int_0^t |w(B_s,s)|^2 ds\}$$

This implies that (2) has the invariant measure

$$d\mu = \lim_{t \to \infty} M_t d\, [\mathcal{N}(0,\sqrt{2\nu}) * \mathcal{N}(0,Q_\infty)] \tag{4}$$

where the variance Q_∞ is

$$Q_\infty^{-1} = \sum_{k \neq 0} \frac{h_k}{2\nu\lambda_k}$$

the coefficients being $h_k = |h_k^{1/2}|^2$. The statistical theory of (2) is determined by the invariant measure (4).

We can also write the approximate invariant measure in terms of densities

$$d\mu = \lim_{t \to \infty} e^{\{-\int_0^t w(x,s)\cdot dx - \frac{1}{2}\int_0^t |w(x,s)|^2 ds\}} \frac{e^{-\frac{|x|^2}{2\nu}}}{\sqrt{2\nu}} dx \prod_{k \neq 0} \frac{e^{-\frac{h_k \hat{u}_k^2}{2\nu\lambda_k}}}{\sqrt{2\nu\lambda_k / h_k}} d\hat{u}_k$$

where $\hat{u}_k$ are the Fourier coefficients of u.

In numerical simulation and fluid experiments the approximate velocity w will have similar statistical properties as the real velocity u. Thus w can be approximated by simulated or measured values of the fluid velocity u itself.

Acknowledgement. The author gratefully acknowledges helpful discussion with José A. Carrillo and Filippo Santambrogio. This research was conducted by partial support from the UC Santa Barbara Academic Senate Research Grant and with travel support from IPAM.

References

1. Birnir, B.: Turbulence of a unidirectional flow. In: Proceedings of the Conference on Probability, Geometry and Integrable Systems, MSRI, December 2005, vol. 55. MSRI Publications, Cambridge Univ. Press (2007), `http://repositories.cdlib.org/cnls/`
2. Birnir, B.: The Existence and Uniqueness of Turbulent Solution of the Stochastic Navier-Stokes Equation (2008) (submitted), `http://repositories.cdlib.org/cnls/`
3. Birnir, B.: Turbulent Rivers. Quarterly of Applied Mathematics 66, 565–594 (2008)
4. Oksendal, B.: Stochastic Differential Equations. Springer, New York (1998)

Spatial Multi-Point Correlations in Inhomogeneous Turbulence

R. Stresing*, M. Tutkun, and J. Peinke

Abstract. We examine the spatial Markov properties of all three velocity components in inhomogeneous turbulence. We use measurement data of the axisymmetric far wake behind a disk at $Re = 2 \cdot 10^4$, measured simultaneously with cross hot wire probes at twelve different distances from the flow axis. We show that the velocity components and Reynolds stresses can be approximated by Markov processes for large enough separations perpendicular to the flow direction. Our results indicate that the n-point correlations of the velocity components and the Reynolds-stresses in inhomogeneous turbulence might be approximated by a stochastic process governed by a Fokker-Planck equation, which could be the basis of a stochastic closure of the Reynolds averaged momentum equations.

1 Introduction – Closure Problem and Stochastic Processes

The central problem of turbulence is to determine the n-point probability density functions (pdfs) of the velocity field. One mathematical formulation of this problem is the infinite Friedmann-Keller system of differential equations for all possible moments of these pdfs [1]. Any finite subsystem of this system is always unclosed, as, for example, are the Reynolds equations:

$$\frac{\partial \rho \bar{u}_i}{\partial t} + \frac{\partial}{\partial x_j}\left(\rho \bar{u}_i \bar{u}_j + \rho \overline{u'_i u'_j}\right) = -\frac{\partial \bar{p}}{\partial x_i} + \rho \nu \nabla^2 \bar{u}_i, \tag{1}$$

R. Stresing · J. Peinke
University of Oldenburg, D-26111 Oldenburg, Germany
e-mail: robert.stresing@uni-oldenburg.de

M. Tutkun
Norwegian Defense Research Establishment, NO-2027 Kjeller, Norway

* Corresponding author.

where the velocity components have been split into their mean and fluctuating parts, $u_i = \bar{u}_i + u'_i$. Turbulence models based on the Reynolds equations focus on expressions for the Reynolds stresses $\tau_{ij} = -\rho\overline{u'_i u'_j}$. We propose a solution based on stochastic process equations for the Reynolds stresses.

If the pdfs of the Reynolds stresses, $p(u'_i u'_j)$, or the joint pdf of the fluctuations, $p(u'_1, u'_2, u'_3)$, are known, the Reynolds stresses can be calculated:

$$\overline{u'_i u'_j} = \int_{-\infty}^{\infty} u'_i u'_j \, p(u'_i u'_j) \, \mathrm{d}\, u'_i u'_j = \int_{-\infty}^{\infty}\int_{-\infty}^{\infty}\int_{-\infty}^{\infty} u'_i u'_j \, p(u'_1, u'_2, u'_3) \, \mathrm{d}u'_1 \mathrm{d}u'_2 \mathrm{d}u'_3. \tag{2}$$

We want to know the n-point joint pdfs of the components of the velocity fluctuations or the Reynolds stress tensor, $p(\phi_n, r_n; \ldots; \phi_1, r_1)$, where ϕ stands for u_i or τ_{ij}, and the points r_k are assumed to be equally spaced and lie on a straight line. If the stochastic process for the spatial evolution of the quantity ϕ has Markov properties, that is, if

$$p(\phi_n, r_n | \phi_{n-1}, r_{n-1}; \ldots; \phi_1, r_1) = p(\phi_n, r_n | \phi_{n-1}, r_{n-1}), \tag{3}$$

the n-point joint pdf of can be expressed by a product of conditional pdfs:

$$p(\phi_n, r_n; \phi_{n-1}, r_{n-1}; \ldots; \phi_1, r_1) = p(\phi_n, r_n; \phi_{n-1}, r_{n-1}) \ldots p(\phi_2, r_2; \phi_1, r_1) p(\phi_1, r_1). \tag{4}$$

The stochastic process for the conditional pdfs can be described by a Kramers-Moyal expansion. If the fourth-order Kramers-Moyal coefficient $D^{(4)}$ is zero, the expansion truncates after the second term[1] and becomes a Fokker-Planck equation:[2,3]

$$\frac{\partial}{\partial r} p(\phi, r | \phi_0, r_0) = \left[-\frac{\partial}{\partial \phi} D^{(1)}(\phi, r) + \frac{\partial^2}{\partial \phi^2} D^{(2)}(\phi, r) \right] p(\phi, r | \phi_0, r_0), \tag{5}$$

which is equivalent to the corresponding Langevin equation (Itō formalism) for the quantity $\phi(r)$ itself:

$$\frac{\partial}{\partial r} \phi(r) = D^{(1)}(\phi, r) + \sqrt{D^{(2)}(\phi, r)}\, \Gamma(r), \tag{6}$$

where $\Gamma(r)$ represents Gaussian white noise, and the drift and diffusion functions $D^{(1)}$ and $D^{(2)}$ are defined as:

$$D^{(k)}(\phi, r) = \lim_{\Delta r \to 0} \frac{1}{k! \Delta r} \int_{-\infty}^{+\infty} (\tilde{\phi} - \phi)^k \, p(\tilde{\phi}, r - \Delta r | \phi, r) \, \mathrm{d}\tilde{\phi}. \tag{7}$$

[1] Pawula's theorem states that if $D^{(4)} = 0$, then $D^{(k)} = 0$ for all $k > 2$.

[2] The present work is based on previous investigations of the n-*scale* joint pdfs of the velocity *increments* on different scales [3, 4, 5, 6]. It has been shown for several different flow types over a wide range of Reynolds numbers, that the scale-to-scale evolution of the pdfs of the increments can be described by a Fokker-Planck equation.

[3] For simplicity, we write ϕ and ϕ_0 instead of ϕ_n and ϕ_{n-1} in eq. (5) to eq. (7).

Note that averaging over eq. (6) for $\phi = \tau_{ij}$ and $r = x_j$ gives an expression for the term $\frac{\partial}{\partial x_j}\overline{u_i' u_j'}$ in eq. (1)[4].

2 Experimental Results – Markov Properties

As an example for an inhomogeneous turbulent flow we analyze cross hot-wire measurements in the axisymmetric wake of a disc with diameter $D = 20$ mm, taken simultaneously at twelve distances r_i from the flow axis at a downstream distance $x/D = 50$, at Reynolds number $Re = 20400$ [2]. We denote the component of the velocity fluctuations in the direction of the main flow as u, the transverse compontent as v, and the azimuthal (tangential) component as w. We only look at the three probe positions closest to the axis of the disk: $r_1 = 0.67\delta_*$, $r_2 = \delta_*$, and $r_3 = 1.33\delta_*$, where r is zero on the axis, and $\delta_* = 42$ mm is the transverse length scale.

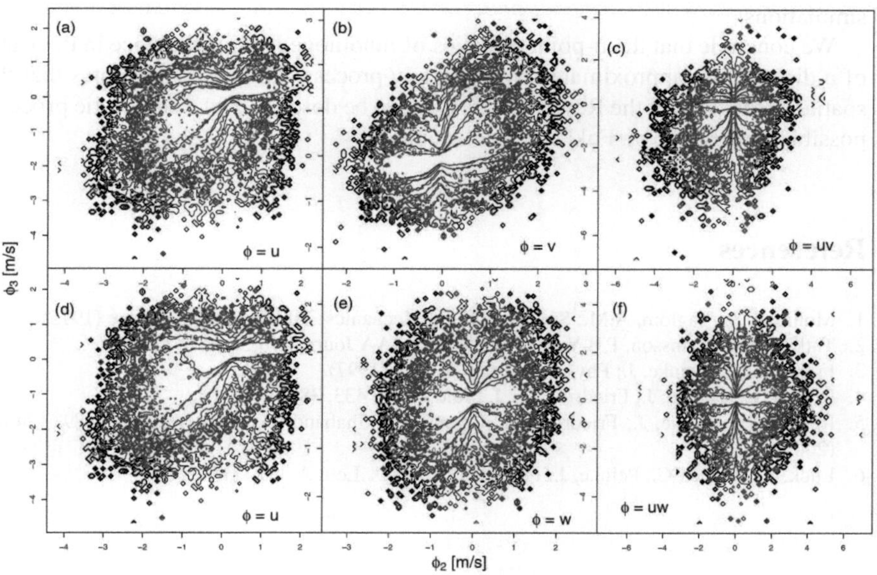

Fig. 1 Contour plots of $p(\phi_3, r_3|\phi_2, r_2)$ (black) and $p(\phi_3, r_3|\phi_2, r_2; \phi_1, r_1)$ (red), with (a,d) $\phi_i = u_i$, (b) $\phi_i = v_i$, (c) $\phi_i = u_i v_i$, (e) $\phi_i = w_i$, and (f) $\phi_i = u_i w_i$. In all plots, $\phi_1 = 0 \pm \sigma_\phi/8$, except in (d), where $\phi_1 = -1 \pm \sigma_\phi/6$.

We examine the Markov properties of the measurement data by a graphical inspection of both sides of

$$p(\phi_3, r_3|\phi_2, r_2; \phi_1, r_1) = p(\phi_3, r_3|\phi_2, r_2), \tag{8}$$

[4] Incompressibility is assumed and ρ is omitted.

which is a simplification of eq. (3), sufficient for finite data sets. Fig. 1 shows the right and left hand sides of eq. (8), where ϕ_i stands for the variables u_i, v_i, w_i, $u_i v_i$, and $u_i w_i$ at the position r_i.

In all cases, the general shapes of the conditional pdfs of eq. (8) agree very well. There are some differences especially around $\phi_2 = 0$ (the "notch" in the pdfs), which can be explained by the presence of more quiescent, or quasi-laminar phases of the flow, which do not have Markov properties. The distance between the probes is $\Delta r = 14$ mm $= \delta_*/3$, and the Integral length in the direction of u is $L_u = 23$ mm at r_1 (inner position), and $L_u = 43$ mm at r_3 (outer position).

Having shown that the data has approximate Markov properties for sufficiently large spatial separations, we can state that the stochastic process for $\frac{\partial}{\partial r} p(\phi_i, r_i | \phi_{i-1}, r_{i-1})$ can be approximated by a Kramers-Moyal expansion. If, furthermore, the fourth-order Kramers-Moyal coefficient is zero, the process follows a Fokker-Planck equation. As we cannot calculate the (spatial) Kramers-Moyal coefficients directly from our data, this problem is left for future studies on the basis of multi-point measurements with smaller separations Δr, or on the basis of numerical simulations.

We conclude that the n-point statistics of inhomogeneous turbulence in the wake of a disk can be approximated by a Markov process. This result indicates that the spatial derivatives of the Reynolds stresses can be described by a stochastic process, possibly governed by a Fokker-Planck equation.

References

1. Monin, A.S., Yaglom, A.M.: Statistical Fluid Mechanics. MIT Press, Cambridge (1975)
2. Tutkun, M., Johansson, P.B.V., George, W.K.: AIAA Journal 46(5), 1118 (2008)
3. Friedrich, R., Peinke, J.: Phys. Rev. Lett. 78, 863 (1997)
4. Renner, C., Peinke, J., Friedrich, R.: J. Fluid Mech. 433, 383 (2001)
5. Renner, C., Peinke, J., Friedrich, R., Chanal, O., Chabaud, B.: Phys. Rev. Lett. 89, 124502 (2002)
6. Lück, S., Renner, C., Peinke, J., Friedrich, R.: Phys. Lett. A 359, 335 (2006)

Statistical Properties of Velocity Increments in Two-Dimensional Turbulence

Michel Voßkuhle, Oliver Kamps, Michael Wilczek, and Rudolf Friedrich

Abstract. The multiple-point probability density $f(v_1,r_1;v_2,r_2;\ldots v_N,r_N)$ of velocity increments v_i at different length scales r_i is investigated in a direct numerical simulation of two-dimensional turbulence. It has been shown for experimental data of three-dimensional turbulence, that this probability density can be represented by conditional probability densities in form of a Markov chain [1]. We have extended this analysis to the case of two-dimensional forced turbulence in the inverse cascade regime.

1 Introduction

Turbulence is commonly believed to exhibit universal statistical properties. This applies to the three- as well as to the two-dimensional case. In three dimensions the universal state is characterized by the direct energy cascade, which cascades energy from large scales down to smaller ones. Whereas in two-dimensional turbulence two cascades can be found: the enstrophy cascade and the inverse energy cascade. The latter one transports, in contrast to the three-dimensional case, energy injected at small scales towards larger scales. The main quantity of interest when describing those universal states of turbulence are the longitudinal velocity inrcrements $v(r,t)$ at scale r, defined by

$$v(r,t) = \frac{r}{r} \cdot [u(x+r,t) - u(x,t)]\,,$$

where $u(x,t)$ is the velocity field at time t and location x. Due to homogeneity and stationarity of the turbulence statistics the statistical properties of the longitudinal velocity increment do not depend on the reference point x and time t.

M. Voßkuhle
Institute for Theoretical Physics, Westfälische Wilhelms-Universität Münster,
Wilhelm-Klemm-Str. 9, 48149 Münster, Germany
e-mail: m.vosskuhle@uni-muenster.de

A statistical description of the inverse cascade is formally conveyed in the N-point probability density function (pdf) $f(v_1,r_1;v_2,r_2;\ldots v_N,r_N)$, where all r_i are within the inertial range. To simplify matters we choose in the following $r_{i+1} < r_i$. From this pdf the conditional probability density function $p(v_1,r_1|v_2,r_2;\ldots v_N,r_N)$ may be obtained via

$$p(v_1,r_1|v_2,r_2;v_3,r_3;\ldots v_N,r_N) = \frac{f(v_1,r_1;v_2,r_2;v_3,r_3;\ldots v_N,r_N)}{f(v_2,r_2;v_3,r_3;\ldots v_N,r_N)}.$$

The N-point pdf simplifies to a product of two-point conditionals pdfs, if the governing stochastic process is a Markov process, i.e. if the conditional pdfs fulfill the Markov property

$$p(v_1,r_1|v_2,r_2;\ldots v_N,r_N) = p(v_1,r_1|v_2,r_2),$$

for all $N \geq 3$ and every set of length scales $r_1,\ldots r_N$. Often one finds that this property only holds for distances $r_i - r_{i+1}$, $i = 1,\ldots N-1$ that are larger than a certain length, which is frequently referred to as Markov–Einstein length l_{ME} (see e.g. [3]). As a consequence the multiple-point pdf $f(v_1,r_1;\ldots v_N,r_N)$ can be evaluated on scales larger than l_{ME} via a Markov chain

$$f(v_1,r_1;\ldots v_N,r_N) = p(v_1,r_1|v_2,r_2)\cdots p(v_{N-1},r_{N-1}|v_N,r_N)f(v_N,r_N).$$

2 Numerical Treatment and Statistical Analysis

We have numerically solved the forced two-dimensional Navier–Stokes equation on a doubly periodic square domain of 2π side length and 1024^2 grid points. To obtain a statistically stationary flow we apply a small-scale forcing with a spatiotemporal correlation function $\langle f(x+r,t), f(x,t')\rangle \sim \delta(t-t')\exp(-r^2/2l_c^2)$, where l_c is the correlation length. The solution yields an inverse cascade with a well-defined inertial range and an energy spectrum $E(k) \approx k^{-5/3}$ (for a detailed description of the numerics see [2]). In the following L shall denote the integral scale and λ the Taylor scale of this flow.

We have evaluated the statistics of the longitudinal velocity increments from the numerical solution and checked the Markov property. A rigorous proof for the validity of this property would afford an investigation of $p(v_1,r_1|v_1,r_2;\ldots v_N,r_N)$ for all $N \geq 3$ and every set of length scales $r_1,\ldots r_N$. This is clearly not possible. The obtainable data allowed for an investigation of the conditional probability densities $p(v_1,r_1|v_2,r_2)$ and $p(v_1,r_1|v_2,r_2;v_3,r_3)$ for different length scales r_1, r_2 and r_3. One example for these pdfs is displayed in Fig 1.

Comparison of the two conditional pdfs shows that they coincide for all v_3 given the differences $r_1 - r_2$ and $r_2 - r_3$ are large enough. To quantify this result we have calculated the mean correlation of cuts through the conditional pdfs

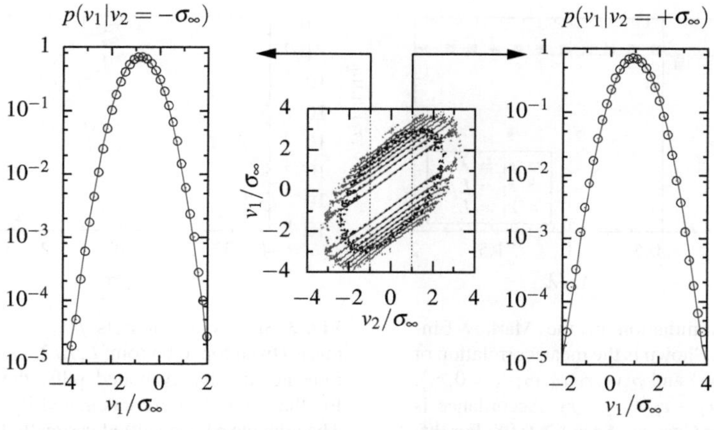

Fig. 1 The Graph in the middle shows the contour plots of the conditional pdfs $p(v_1,r_1|v_2,r_2)$ (solid lines) and $p(v_1,r_1|v_2,r_2;v_3 = 0,r_3)$ (dashed lines) for $r_1 = \lambda + L$, $r_2 = \lambda + L/2$, and $r_3 = \lambda$. At the sides, cuts through those pdfs at $v_2 = \pm\sigma_\infty$ respectiveley are shown. The two pdfs clearly coincide.

$$\mathrm{Corr}(r_1,\Delta r,v_3) = \left\langle \frac{\int p(v_1,r_1|v_2,r_2)p(v_1,r_1|v_2,r_2;v_3,r_3)\mathrm{d}v_1}{\sqrt{\int p(v_1,r_1|v_2,r_2)^2\mathrm{d}v_1}\sqrt{\int p(v_1,r_1|v_2,r_2;v_3,r_3)^2\mathrm{d}v_1}} \right\rangle_{v_2},$$

which proved to be a good measure for their accordance. Fig 2 shows that the coincidence occurs for $\Delta r = r_1 - r_2 = r_2 - r_3 > 0.4\lambda = l_{ME}$.

As a test for self-consistency of our results we have computed the one-point pdfs from a Markov Chain within the inertial range, i.e. we have calculated $f(v_i,r_i)$ via

$$\begin{aligned} f(v_i,r_i) &= \int \mathrm{d}v_{i+1}\cdots\mathrm{d}v_N\, f(v_i,r_i;v_{i+1},r_{i+1};\ldots v_{N-1},r_{N-1};v_N,\lambda) \\ &= \int \mathrm{d}v_{i+1}\cdots\mathrm{d}v_N\, p(v_i,r_i|v_{i+1},r_{i+1})\cdots p(v_{N-1},r_{N-1}|v_N,\lambda)f(v_N,\lambda), \quad (1) \end{aligned}$$

for different r_i with the step size $r_{i+1} - r_i = l_{ME}$. Fig 3 compares the directly evaluated pdfs and the calculated ones. Their coincidence for all r_i within the inertial range affirms the validity of the Markov property.

3 Conclusion

Our results give evidence that the evolution of the pdfs of velocity increments within the inertial range can be described as a Markov process. This extends former statistical analyses of three-dimensional turbulence to the case of two-dimensional turbulence in the inverse cascade regime. For both cases a finite Markov–Einstein length

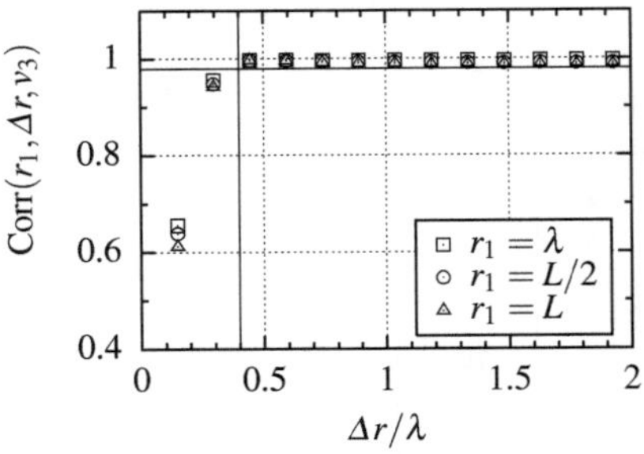

Fig. 2 Determination of the Markov–Einstein length. Shown is the mean correlation of $p(v_1, r_1|v_2, r_2)$ and $p(v_1, r_1|v_2, r_2; v_3 = 0, r_3)$, with $\Delta r = r_1 - r_2 = r_2 - r_3$. Accordance is supposed for $\mathrm{Corr}(r_1, \Delta r, v_3) \geq 0.98$. For different v_3 similar results are obtained.

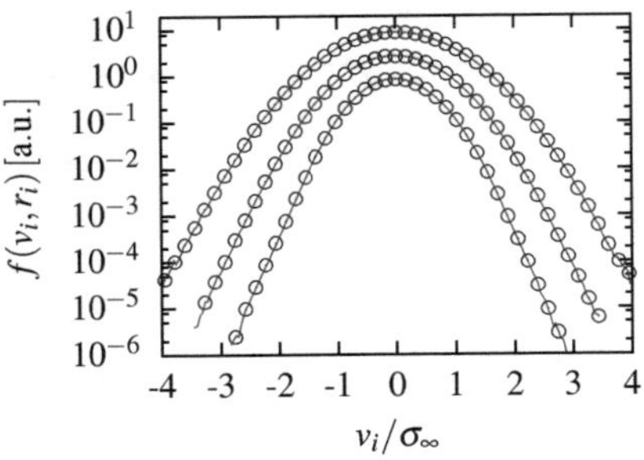

Fig. 3 Shown are the pdfs $f(v_i, r_i)$ for various r_i (from top to bottom: L, $L/2$, 2λ). Lines indicate directly evaluated pdfs, the values for the symbols were calculated by Eq (1). The pdfs have been shifted vertically for clarity of presentation.

within the range of the Taylor scale is found. These results suggest that the Markov property and the finite Markov–Einstein length might be universal properties of turbulent flows.

The existence of a Markov property demonstrates that the longitudinal velocity increment statistics at scale r_i can be determined from those at scale r_{i+1}. This process can be iterated and holds for the whole inertial range. It can be seen as a manifestation of the turbulent cascade.

As an application our results provide an approach to the modeling of increment statistics in turbulent flows in terms of Langevin and Fokker–Planck equations.

Acknowledgement. M. Voßkuhle acknowledges the financial support of the *Center for Nonlinear Science* (CeNoS), Münster.

References

1. Friedrich, R., Peinke, J.: Description of a Turbulent Cascade by a Fokker-Planck Equation. Phys. Rev. Lett. (1997), doi:10.1103/PhysRevLett.78.863
2. Kamps, O., Friedrich, R.: Lagrangian statistics in forced two-dimensional turbulence. Phys. Rev. E (2008), doi:10.1103/PhysRevE.78.036321
3. Lück, S., Peinke, J., Friedrich, R.: The Markov-Einstein coherence length—a new meaning for the Taylor length in turbulence. Phys. Lett. A (2006), doi:10.1016/j.physleta.2006.06.053

Enstrophy Transfers Study in Two-Dimensional Turbulence

Patrick Fischer and Charles-Henri Bruneau

Abstract. Two-dimensional turbulence admits two different ranges of scales: a direct enstrophy cascade from the injection scale to the small scales and an inverse energy cascade at large scales. It has already been shown in previous papers that vortical structures are responsible for the transfers of energy upscale while filamentary structures are responsible for the forward transfer of the enstrophy. Here we introduce an original wavelet-based new mathematical tool, the interaction function, for studying the space localization of the enstrophy fluxes. It is based on two-dimensional orthogonal wavelet decompositions of the two terms involved in the transport term in the Navier-Stokes equations.

1 The Interaction Function

The enstrophy flux $\Pi_Z(k)$ derives from the nonlinear transport term in the vorticity equation written in Fourier space

$$\Pi_Z(k) = \int_k^{+\infty} T_Z(k')dk' \tag{1}$$

where the enstrophy transfer function $T_Z(k)$ is obtained by angular integration of $N_Z(\omega) = \widehat{\omega^*(\mathbf{k})}.(\widehat{\mathbf{v}\cdot\nabla)\omega(\mathbf{k}})$. We can replace the Fourier transform by a wavelet

Patrick Fischer
Université Bordeaux 1, Institut de Mathématiques de Bordeaux, INRIA Team MC^2, CNRS UMR 5251, 33405 Talence Cedex, France
e-mail: Patrick.Fischer@math.u-bordeaux1.fr

Charles-Henri Bruneau
Université Bordeaux 1, Institut de Mathématiques de Bordeaux, INRIA Team MC^2, CNRS UMR 5251, 33405 Talence Cedex, France
e-mail: Charles-Henri.Bruneau@math.u-bordeaux1.fr

transform leading to a different representation of the enstrophy transfers in the flow. This original representation is called the enstrophy interaction function. The term $N_Z(\omega)$ introduced above is in fact a scalar product, in Fourier space, between the vorticity field ω and its transported field $(\mathbf{v}\cdot\nabla)w$. If the transported field is spectrally close to the initial vorticity field then $N_Z(\omega)$ will be large but if it is very different or even orthogonal then $N_Z(\omega)$ will be insignificant. So, the term $N_Z(\omega)$ measures the correlation, in Fourier space, between the transported and the initial vorticity fields. This correlation is then used to compute the transfer function T_Z and the enstrophy flux Π_Z. By using a Fourier transform, we obtain a description of the enstrophy transfers through the scales, but all the information about the space localization of these transfers is completely lost. However, it is well known, from the classical theory of two-dimensional turbulence and from numerical experiments, that the direct enstrophy cascade takes place from the injection scale to the smallest scales. The enstrophy cascade is thus essentially a small scales phenomenon and may be localized in space [4]. The method we propose in this paper is based on a two-dimensional wavelet transform and leads to a space-scale description of the enstrophy tranfers (see [5] for many references about wavelets applied to turbulence). It consists in replacing the usual Fourier transform in the computation of $N_Z(\omega)$ by a wavelet transform. The interaction function for the enstrophy transfer is obtained through a three steps process, and may be summarized by the formula:

$$IF_Z = WT^{-1}\left[WT(\omega).WT((\mathbf{v}\cdot\nabla)\omega)\right], \tag{2}$$

where WT denotes the two-dimensional wavelet transform and WT^{-1} its inverse transform. The result of this last computation is by definition the interaction function and is denoted by IF_Z. It can be conveniently represented onto a contour plot of the vorticity field. The same process can be used for the energy transfers using the corresponding data.

2 Experimental Setup and Numerical Results

The experiments consist in the numerical simulation of a two-dimensional channel flow perturbed by an horizontal array of cylinders. Two vertical arrays of additional cylinders have been added in order to increase the number of vortices in the flow, and thus to enhance the turbulent behavior of the flow [3]. The length of the rectangular channel Ω is four times its width L and the Reynolds number based on the cylinders diameter is $Re = 50,000$. The injection scale k_{inj} is then given by the diameter of cylinders $L/8$ and consequently the injection scale is around $k_{inj} = 8$.

The penalization method is used to solve the flow around the obstacles. Consequently the Brinkman-Navier-Stokes equations are solved in the whole channel Ω including the solid obstacles Ω_s and the fluid domain Ω_f [2]. The numerical results obtained through such direct numerical simulation can be compared to those obtained by soap film experiments where the flow is perturbed by analogous arrays of small cylinders [3].

The computation of the enstrophy flux has been performed on 100 snapshots of the flow. To compute the enstrophy flux, we select a square of size $L = 1$ located at the end of the channel as domain of analysis. The cutting process to select this domain creates many discontinuities in the velocity and vorticity fields at the boundaries, and thus introduces high frequency Fourier coefficients. This phenomenon is well known from people using the classical FFT algorithm and has been described in [1]. We avoid this problem by using a windowed Fourier transform that removes the spurious coefficients created by the discontinuities.

In order to study the interactions occuring into the enstrophy cascade, the interaction function has been computed for few snapshots with strong direct enstrophy fluxes. Results corresponding to snapshot 85 are reported here. The vorticity field corresponding to this snapshot is given in Figure 1(a). Various structures can be observed in this vorticity field. According to previous studies [2], we already know that the vorticity filaments are responsible for the inverse enstrophy cascade, but we don't have yet any information about the space localization of the interactions leading to this cascade. The enstrophy interaction function, represented in Figure 2(a), allows us to get this kind of information. The contrast and the colormap have been chosen such that only the most important coefficients are noticeable. Tuning the contrast would make appear more colored regions corresponding to weak enstrophy fluxes. In this snapshot, a region with strong values around $(90,285)$ can be detected. This zone corresponds to interactions between two vortices of opposite signs and where most of the enstrophy cascade occur. This can be verified by computing the total enstrophy flux and the partial one corresponding to this region. The selection of the desired region is made with a gaussian mask applied to the velocity and vorticity fields. The total and partial enstrophy fluxes are given in Figure 2(b) and the selected area in the vorticity field in Figure 2(a). As expected, the partial enstrophy flux almost fits the total enstrophy flux proving that the main interactions creating the enstrophy cascade effectively take place in this region. The other parts of the domain represent overall a very small amount of enstrophy transfers. This computation has been repeated many times, and the same results were observed.

3 Conclusion

We propose in this paper an original mathematical tool for studying two dimensional turbulent flows. This object, called the interaction function, describes the local interactions leading to the enstrophy flux. The major result is that the enstrophy flux is not an homogeneous phenomenon spread over the whole flow but a local phenomenon corresponding to local interactions. We observed that in most of the cases more than 80% of the enstrophy flux take place in less than 15% of the flow surface.

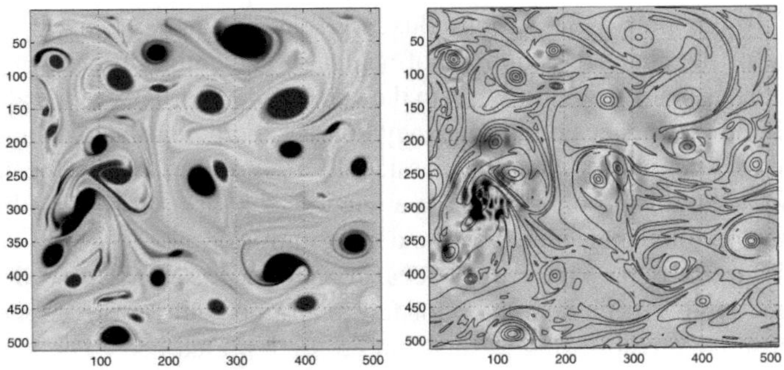

(a) *Vorticity field* (b) *Enstrophy interaction function*

Fig. 1 Vorticity field and enstrophy interaction function for snapshot 85

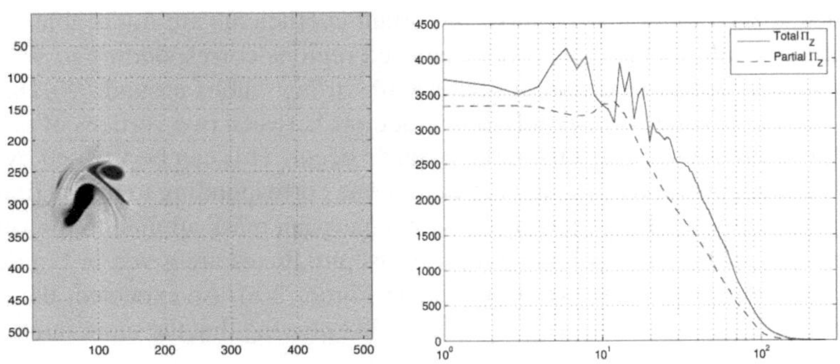

(a) *Partial vorticity field* (b) *Enstrophy fluxes*

Fig. 2 Partial vorticity field and enstrophy fluxes for snapshot 85

References

1. Bruneau, C.H., Fischer, P.: Influence of the filtering tools on the analysis of two-dimensional turbulent flows. Computers and Fluids (2008), doi:10.1016/j.comp?uid.2008.01.035
2. Bruneau, C.H., Fischer, P., Kellay, H.: The structures responsible for the inverse energy and the forward enstrophy cascades in two-dimensional turbulence. Europhys. Lett. 78 (2007), doi:10.1209/02955075/78/34,002
3. Bruneau, C.H., Kellay, H.: Coexistence of two inertial ranges in two-dimensional turbulence. Phys. Rev. E 71, 046, 305 (2005)
4. Chen, S., Ecke, R.E., Eyink, G.R., Wang, X., Xiao, Z.: Physical mechanism of the two- dimensional enstrophy cascade. Phys. Rev. Lett. 91, 214, 501 (2003)
5. Farge, M.: Wavelets and turbulence (1988-2008), `http://wavelets.ens.fr/`

Two Point Velocity Difference Scaling along Scalar Gradient Trajectories in Turbulence

Lipo Wang

Abstract. In the context of dissipation element analysis of scalar fields in turbulence [1], the elongation of elements by the velocity difference at the minimum and maximum points was found to increase linearly with the length of an element. To provide a theoretical basis for this finding by analyzing two-point properties along the gradient trajectories, an equation for the mean product of the scalar gradient at two points along the same trajectory is derived. In the inertial range a balance similar to that from which Kolmogorov's $4/5$ law can be derived. While that law leads to a $1/3$ scaling for the velocity difference, by conditioning on gradient trajectories we obtain a linear relation between the velocity difference and the two-point's arclength on the same trajectory. Results from DNS show satisfactory agreement with the theoretical prediction.

1 Introduction

The nonlocal behavior of turbulent fields, which can be investigated through two-point correlation functions or structure functions, still remains one of the central topics in turbulence. The most prominent results derived from the Navier-Stokes equations are the Kolmogorov equation [2] and the Yaglom equation [3], based on the assumption of isotropy.

For turbulent scalar mixing problems, local alignment of the scalar gradient with the eigenvectors of strain rate tensor has been widely investigated [5] [6]. From a laminar diffusion equation, Batchelor [5] found the fact that locally scalar gradients would be inclined to compressive strain. From DNS results, the general understanding is that compressive strain is also responsible for large scalar dissipation [6].

Lipo Wang
Institut für Technische Verbrennung, RWTH-Aachen, Templergraben 64,52056, Aachen, Germany
e-mail: wang@itv.rwth-aachen.de

In an attempt to analyze the geometrical properties of scalar fields, the concept of dissipation elements has been proposed [1] [4]. In a given turbulent scalar field, starting from every grid point, trajectories in directions of ascending and descending gradients may be traced until reaching a local maximum and minimum point, respectively. A dissipation element is defined as the spatial region containing all the grid points from which the same pair of maximum and minimum points can be reached. While it has been found that the conditional scalar difference follows the $1/3$ Kolmogorov scaling [4], numerical data show that the velocity difference at the extremal points surprisingly seems to follow a linear scaling [1]. In order to explain this finding, it is meaningful to study the two-point correlation functions affixed to gradient trajectories. This means that the two point velocity difference to be obtained is conditioned by the direction of the scalar gradient. The results to be presented in this paper step further to provide a view of the non-local mean strain rate between two separated points along gradient trajectories.

2 Theory and Results

In a turbulent passive scalar field ϕ, the gradient direction is $\mathbf{n} = \nabla\phi/|\nabla\phi|$. The governing equation of ϕ can be written in the scalar form as

$$\frac{\partial \phi}{\partial t} + u_n \phi_{,n} = D(\partial^2\phi/\partial n^2 - \phi_{,n}\kappa), \tag{1}$$

where $u_n = \mathbf{u}\cdot\mathbf{n}$ is the velocity component of $\mathbf{u}$ along $\mathbf{n}$, $\phi_{,n} = \partial\phi/\partial n$ is the scalar gradient and $\kappa = -\nabla\cdot\mathbf{n}$ is the local curvature, respectively.
Multiplying the operator $\mathbf{n}\cdot\nabla$ on both sides of Eq. (1), one obtains

$$\mathbf{n}\cdot\frac{\partial(\nabla\phi)}{\partial t} + \mathbf{n}\cdot\nabla(u_n\phi_{,n}) = \frac{\partial\phi_{,n}}{\partial t} + \frac{\partial u_n\phi_{,n}}{\partial n} = D\frac{\partial(\partial^2\phi/\partial n^2 - \phi_{,n}\kappa)}{\partial n}. \tag{2}$$

Along the same trajectory considering the same equation at another point $\mathbf{x}'$ at n' and noticing that for homogenous turbulence the two-point correlation is only a function of l, we may have the following result

$$\frac{\partial[\overline{(u'_{n'} - u_n)\phi'_{,n'}\phi_{,n}}]}{\partial l} = D\frac{\partial}{\partial l}[2\frac{\partial}{\partial l}\overline{(\phi_{,n}\phi'_{,n'})} - \overline{\phi_{,n}\phi'_{,n'}(\kappa' - \kappa)}] - \frac{\partial\overline{\phi_{,n}\phi'_{,n'}}}{\partial t}. \tag{3}$$

For convenience, we introduce $C_{ij}(r,t) \equiv \overline{\phi_{,i}\phi'_{,j}}$ and $C_{nn'}(l,t) \equiv \overline{\phi_{,n}\phi'_{,n'}}$. Fig.1 (a) and (b) show the normalized function $f_{ij}(r,t) \equiv C_{ij}(r,t)/C_{ij}(0,t)$ (taking $f_{11}(r,t)$ for illustration) in the Cartesian coordinate and $g(l,t) \equiv C_{nn'}(l,t)/C_{nn'}(0,t)$ in the gradient trajectory coordinate, respectively, from two homogeneous shear turbulence DNS cases: case 1 with a Reynolds number $Re_\lambda = 116.5$, and case 2 with $Re_\lambda = 125.0$.

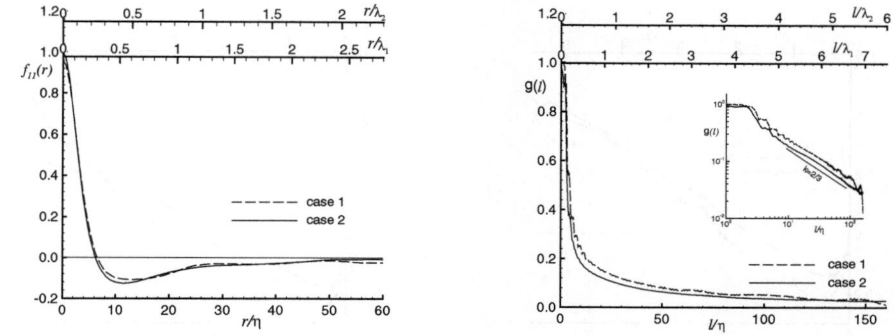

(a) a

(b) b

Fig. 1 The correlation function: (a) $f_{11}(r)$ in the Cartesian coordinate; (b) $g(l)$ along gradient trajectories.

It can be seen f_{11} is positive only at about the Kolmogorov scale η and then transits to negative, whereas $\overline{\phi_{,n}\phi_{,n'}}$ is always positive. This is the essential difference between the derivative product in these two coordinate systems.

The convective term and viscous term are the dominant term in inertial range and viscous range, respectively. In the inertial range $g \sim l^{-2/3}$. Assuming a steady g and generally let

$$g(l) \sim l^{-p}, \tag{4}$$

we can obtain

$$\frac{\overline{(u'_{n'} - u_n)\phi'_{,n'}\phi_{,n}}}{C_{nn'}(l,t)} \equiv f_1(l) \sim -\frac{1}{C_{nn'}(0,t)}\frac{dC_{nn'}(0,t)}{dt} l \sim \frac{\varepsilon}{k} l. \tag{5}$$

From Fig. 1(a) and (b), it shows that the correlation length between scalar gradients is of the order of η, much shorter than that of scalar itself, which can be understood from the fact that, under the frequent compressing and stretching perturbation from velocity field, the profile of scalar gradient may have 'ripples', not as smooth as that of scalar. By the same token, $u'_{n'} - u_n$ and $\phi'_{,n'}\phi_{,n}$ should also be correlated at scale $\sim \eta$. Thus in the inertial range ($\gg \eta$) it is reasonable to write

$$\overline{(u'_{n'} - u_n)\phi'_{,n'}\phi_{,n}} \sim \overline{(u'_{n'} - u_n)} \cdot \overline{\phi'_{,n'}\phi_{,n}}. \tag{6}$$

Therefore

$$\overline{(u'_{n'} - u_n)} \equiv f_2(l) \sim f_1(l) \sim l\frac{\varepsilon}{k}. \tag{7}$$

At small scales in the viscous range, the first term $\partial C_{nn'}(l,t)/\partial l$ is large negative. Along positive gradient directions on average curvature increases from negative to positive. Therefore $\overline{(u'_{n'} - u'_{n'})\phi'_{,n'}\phi_{,n}}$, needs to be negative in the small viscous range.

Numerical results of f_1 and f_2 from two DNS cases are shown in Fig. 2 (a) and (b). The linear tendency of f_1 and f_2 with respect to l appears clearly. Furthermore,

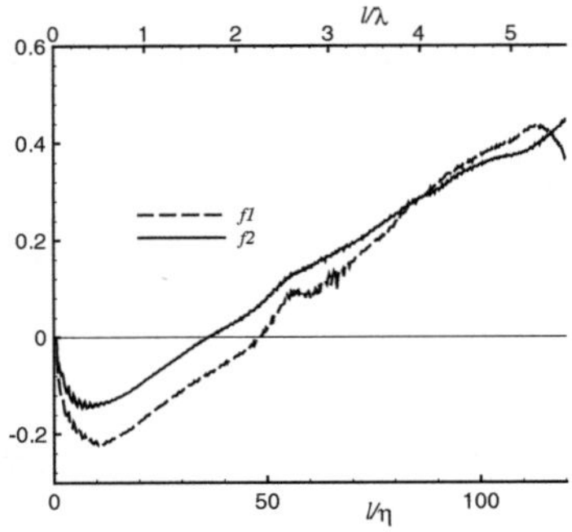

(a) a

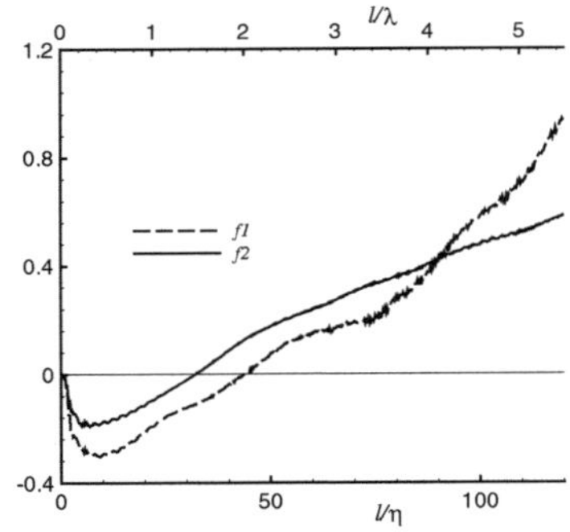

(b) b

Fig. 2 Structure functions along trajectories for: (a) case 1; (b) case 2.

the slope f_2 in Fig. 2 (a) and (b) are approximately proportional to the ratio of ε/k for the two DNS cases, which may work as a strong manifestation of the theoretical prediction of Eq. (7). In addition, in the viscous range $\overline{(u'_{n'} - u'_{n'})}$ is proved to be negative. The negative velocity or compressive strain for small scales is the direct effect of diffusivity to control the orientation preference of scalar gradient directions.

3 Conclusions

The mean strain rate along the orientation of dissipation elements, shown in Fig.8 in [1], is calculated from the velocity difference at two extremal points of an element. On average the strain rate is a positive constant, except for very short elements it becomes negative. The results presented in this paper provide a theoretical explanation of this finding. At large scales, on average scalar gradient trajectories will be stretched, instead of being compressed as at small scales.

Acknowledgement. The author acknowledges the motivation by Prof. N.Peters (RWTH-Aachen, Germany) and the helpful discussion with Dr. J.P.Mellado (RWTH-Aachen). This work is founded by Deutsche Forschungsgemeinschaft. HLRS (Stuttgart) is appreciated for accessing NEC SX-8 supercomputer.

References

1. Wang, L., Peters, N.: J. Fluid Mech. 608, 113, (2008)
2. Kolmogorov, A.N.: Dokl. Akad. Nauk SSSR 32, 19 (1941) (in Russian)
3. Yaglom, A.M.: Dokl. Akad. Nauk SSSR 69, 743 (1955) (in Russian)
4. Wang, L., Peters, N.: J. Fluid Mech. 554, 457 (2006)
5. Batchelor, G.K.: J. Fluid Mech. 5, 113 (1959)
6. Ashurst, W.T., et al.: Phys. Fluids 30(8), 2343 (1987)

Stochastic Analysis of Turbulence n-Scale Correlations in Regular and Fractal-Generated Turbulence

R. Stresing, J. Peinke, R.E. Seoud, and J.C. Vassilicos

Abstract. We present a stochastic analysis of turbulence data, which provides access to the joint probability of finding velocity increments at several scales. The underlying stochastic process in form of a Fokker-Planck equation can be reconstructed from given data. Intermittency effects are included. The stochastic process is Markovian for scale separations larger than the *Einstein-Markov coherence length* l_{EM}, which is closely related to the Taylor microscale λ. We extend our analysis to turbulence generated by a fractal square grid. We find that in contrast to other types of turbulence, like free-jet turbulence, the n-scale statistics of the velocity increments and the leading coefficients of the Fokker-Planck equation do not depend strongly on the Reynolds number.

1 Introduction

Standard statistical analysis of small-scale turbulence is based on two-point correlations and their dependence on the scale r. A central quantity is the velocity increment $u(r)$,

$$u(r) = v(x + (1-q)r, t) - v(x - qr, t) \tag{1}$$

where v denotes here the velocity component in r direction, and $q \in [0,1]$ is a parameter which controls the position of nested increments, e.g. left-bounded ($q = 0$) or centered ($q = 0.5$). A characterization of the scale dependent disorder of turbulence by means of two-point statistics like structure functions can be extended to multi-scale (point) statistics. It has been shown for several different turbulent flows

R. Stresing · J. Peinke
Institute of Physics, University of Oldenburg, Germany
e-mail: `robert.stresing@uni-oldenburg.de`

R.E. Seoud · J.C. Vassilicos
Department of Aeronautics,

J.C. Vassilicos
Institute for Mathematical Sciences, Imperial College London, UK

that these multi-scale statistics can be treated within the framework of Markov processes, and that the corresponding stochastic differential equation is the Fokker-Planck equation [1, 2, 3].

If the stochastic process for the evolution of the velocity increments from scale to scale[1] has Markov properties, i.e. if

$$p(u_1,r_1|u_2,r_2;\ldots;u_n,r_n) = p(u_1,r_1|u_2,r_2), \tag{2}$$

the n-scale joint probability density function (pdf) of the velocity increments can be expressed by a product of conditional pdfs:

$$p(u_1,r_1;u_2,r_2;\ldots;u_n,r_n) = p(u_1,r_1|u_2,r_2)\ldots p(u_{n-1},r_{n-1}|u_n,r_n)p(u_n,r_n). \tag{3}$$

The stochastic process for the conditional pdfs can be described by a Kramers-Moyal expansion. If the fourth-order Kramers-Moyal coefficient $D^{(4)}$ is zero, the expansion truncates after the second term and becomes a Fokker-Planck equation:

$$-\frac{\partial}{\partial r}p(u,r|u_0,r_0) = \left[-\frac{\partial}{\partial u}D^{(1)}(u,r) + \frac{\partial^2}{\partial u^2}D^{(2)}(u,r)\right] p(u,r|u_0,r_0), \tag{4}$$

where drift and diffusion functions $D^{(1)}$ and $D^{(2)}$ can be estimated as Kramers-Moyal coefficients pointwise by:

$$D^{(k)}(u,r) = \lim_{\Delta r\to 0} \frac{r}{k!\Delta r}\int_{-\infty}^{+\infty} (\tilde{u}-u)^k\, p(\tilde{u},r-\Delta r|u,r)\,\mathrm{d}\tilde{u}. \tag{5}$$

The experimental pdfs and conditional pdfs of the velocity increments can be reproduced by integration of the Fokker-Planck equation, including intermittency effects.

2 Results for Fractal-Generated Turbulence

We analyze hot wire measurement data from turbulence generated by a fractal square grid[2], measured for two different flow velocities at five different down-stream positions in the decay region, where the turbulence is small-scale homogeneous and isotropic [5, 6, 7]. We confirm the result of [6], that the Taylor microscale λ is almost independent of the flow velocity and the downstream position, so that we obtain an unusually wide range of Taylor-based Reynolds numbers Re_λ from 153 to 740.

We find that the stochastic process for the velocity increments has Markov properties for scale separations Δr greater than the *Einstein-Markov coherence length*

---

[1] Without loss of generality we assume $r_1 < r_2 < \ldots < r_n$.

[2] The grid has a thickness ratio $t_r = 17$, that is, the thickest bars are 17 times thicker than the thinnest bars of the grid. For more details see [5, 6, 7].

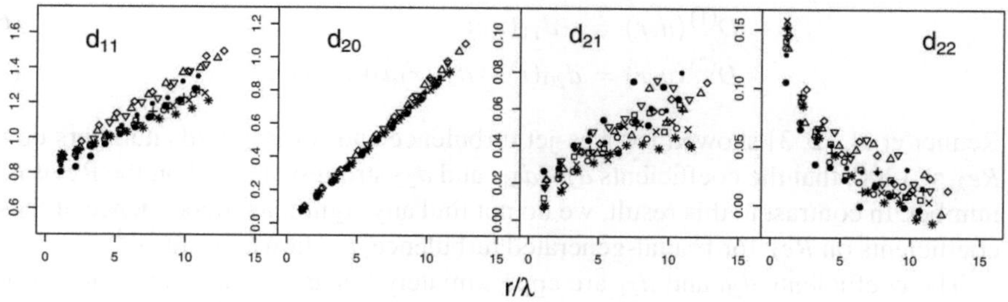

Fig. 1 Estimated coefficients d_{ij} from eq. (6) and (7) as functions of r/λ for the fractal square grid. The Reynolds numbers increase from $Re_\lambda = 153$ to 740 over the following sequence of symbols: $\bullet, \cdot, \bigtriangledown, \triangle, \diamond, \circ, \square, +, \times, *$.

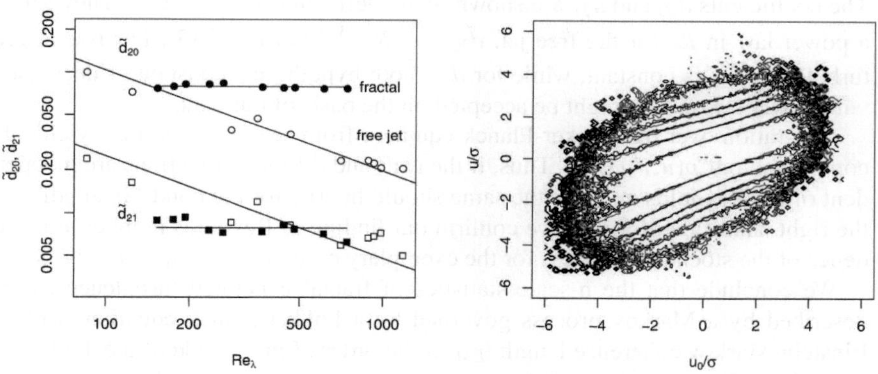

Fig. 2 Left: $\tilde{d}_{20}$ (circles) and $\tilde{d}_{21}$ (squares) from eq. (8) as functions of Re_λ for the fractal grid (solid symbols) and free jet (open symbols, taken from [3]). The open symbols are plotted together with the fit functions $\tilde{d}_{20} = 2.8Re^{3/8}$ and $\tilde{d}_{21} = 0.68Re^{3/8}$ from [3]. For the free jet data, we take $Re_\lambda = Re^{0.5233}$. Right: contour plots of $p(u,r|u_0,r_0)$ for $r_0 = 12l_{EM} \approx L$, $r = 6l_{EM}$, for the fractal grid for $Re_\lambda = 153, 366, 740$.

l_{EM}, and we find a constant ratio of $l_{EM}/\lambda = 0.59 \pm 0.02$, which is comparable to previous results for other turbulent flows.[3]

We determine the Kramers-Moyal coefficients performing a linear fit to the conditional moments defined in eq. (5) in the range $l_{EM} \leq \Delta r \leq 2l_{EM}$. The drift and diffusion functions can be approximated by a linear and a second-order function in u, respectively:

[3] For centered increments, we define l_{EM} as half the smallest distance $\Delta r = r_i - r_{i-1}$, for which eq. (2) still holds. It can be estimated by statistical tests like the Wilcoxon test (see [2]). Lück et al. [4] find $l_{EM}/\lambda \approx 0.8$, with some scattering between 0.6 and 1.0. Our rather low value of $l_{EM}/\lambda \approx 0.6$ can be explained by differences in the methods of calculation of the length scales: we calculate λ as in [6] from the power spectrum, and not from the increments as in [4], which in the present case leads to about 15% higher values of λ; and we determine l_{EM} from centered rather than left-bounded increments, which gives about 10% lower values of l_{EM}.

$$D^{(1)}(u,r) = -d_{11}(r)u \tag{6}$$

$$D^{(2)}(u,r) = d_{20}(r) - d_{21}(r)u + d_{22}(r)u^2 \tag{7}$$

Renner et al. [2, 3] showed for free-jet turbulence and for Reynolds numbers up to $Re_\lambda \approx 1200$, that the coefficients d_{20}, d_{21}, and d_{22} strongly depend on the Reynolds number. In contrast to this result, we do not find any significant dependence of these coefficients on Re_λ for fractal-generated turbulence, as shown in Fig. 1.[4]

The coefficients d_{20} and d_{21} are approximately linear functions in r for free-jet turbulence [2, 3], as well as for fractal-grid turbulence (Fig. 1), and thus can be approximated by:

$$d_{20}(r) = \tilde{d}_{20}\frac{r}{\lambda}, \quad d_{21}(r) = \tilde{d}_{21}\frac{r}{\lambda}, \tag{8}$$

The coefficients $\tilde{d}_{20}$ and $\tilde{d}_{21}$ are shown on the left hand side of Fig. 2. They follow a power law in Re for the free jet, $\tilde{d}_{20,21} \propto Re^{-3/8} \approx Re_\lambda^{-3/4}$ [3]. For fractal grid turbulence, $\tilde{d}_{20}$ is constant, while for $\tilde{d}_{21}$, both hypotheses (constant or decreasing with power law in Re) might be accepted on the basis of our data.

Integration over the Fokker-Planck equation from scale r_0 to scale r yields the conditional pdf $p(u,r|u_0,r_0)$. Thus, if the drift and diffusion functions are independent of the Reynolds number, the same should be true for the conditional pdfs. On the right hand side of Fig. 2, we confirm our finding of Reynolds number independence of the stochastic process for the exemplary case of $r_0 = 12 l_{EM}$, $r = 6 l_{EM}$.

We conclude that the n-scale statistics of fractal generated turbulence can be described by a Markov process governed by a Fokker-Planck equation, with an Einstein-Markov coherence length l_{EM} in the order of magnitude of the Taylor microscale λ. The coefficients of the stochastic process equations for fractal-generated turbulence, especially d_{20}, do not depend significantly on the Reynolds number, which differs substantially from the findings for free-jet turbulence in [2, 3]. Thus we propose to have found a new class of (nearly) Reynolds-number independent turbulence generated by boundary conditions of a fractal grid.

References

1. Friedrich, R., Peinke, J.: Phys. Rev. Lett. 78, 863 (1997)
2. Renner, C., Peinke, J., Friedrich, R.: J. Fluid Mech. 433, 383 (2001)
3. Renner, C., Peinke, J., Friedrich, R., Chanal, O., Chabaud, B.: Phys. Rev. Lett. 89, 124502 (2002)
4. Lück, S., Renner, C., Peinke, J., Friedrich, R.: Phys. Lett. A 359, 335 (2006)
5. Hurst, D., Vassilicos, J.C.: Phys. Fluids 19, 035103 (2007)
6. Seoud, R.E., Vassilicos, J.C.: Phys. Fluids 19, 105108 (2007)
7. Seoud, R.E., Vassilicos, J.C.: In: M. Oberlack, J. Peinke (eds.): Proc. iTi Conference on Turbulence III. Springer, Heidelberg (2009)

[4] We do have more scattering in the coefficients d_{ij}, which is probably due to the shorter data sets of $3 \cdot 10^6$ velocity values in contrast to $1.6 \cdot 10^7$ for the free jet data [3]. Note that our analysis is based on centered ($q = 0.5$) rather than left-bounded ($q = 0$) increments, as in [2, 3], which might lead to some systematic differences in the estimates of the coefficients d_{ij}. How far our results are robust against the change from centered to left-bounded increments, still has to be investigated.

Holographic PIV with Low Coherent Light – Recent Progress in 3D Flow Measurements

Gerd Gülker, Christian Steigerwald, and Klaus D. Hinsch

Abstract. Holographic PIV (HPIV) is a fully three-dimensional technique for measuring flows. A field of small tracer-particles is recorded in double-exposure holograms obtained with two reference waves. Particle displacements are extracted from CCD-based 3D-scans of real images with three-dimensional correlation methods. To cope with the problem of noise from out-of-focus particles the technique of light-in-flight holography has been introduced that utilizes properties of low coherent laser light. This drastically increases the signal-to-noise ratio in deep-field particle holography. Some experimental checks of these concepts are presented and first results from measurements in the wake of an airfoil are shown.

1 Principles of Holographic PIV

Holographic particle image velocimetry (HPIV) is a flow measurement technique, first introduced by Meng [1], which is capable of instantaneously measuring all three velocity components inside an extended measurement volume with high spatial resolution. Basically, double-exposure off-axis holograms of two consecutive flow states, seeded with appropriate particles, are recorded on a single recording medium (Fig. 1 left). To distinguish between these recordings two reference beams

Gerd Gülker
Institute of Physics, Carl von Ossietzky University, 26111 Oldenburg, Germany
e-mail: `gerd.guelker@uni-oldenburg.de`

Christian Steigerwald
Carl Zeiss SMT AG, Rudolf-Eber-Str. 2, 73447 Oberkochen, Germany
e-mail: `c.steigerwald@smt.zeiss.com`

Klaus D. Hinsch
Institute of Physics, Carl von Ossietzky University, 26111 Oldenburg, Germany
e-mail: `klaus.hinsch@uni-oldenburg.de`

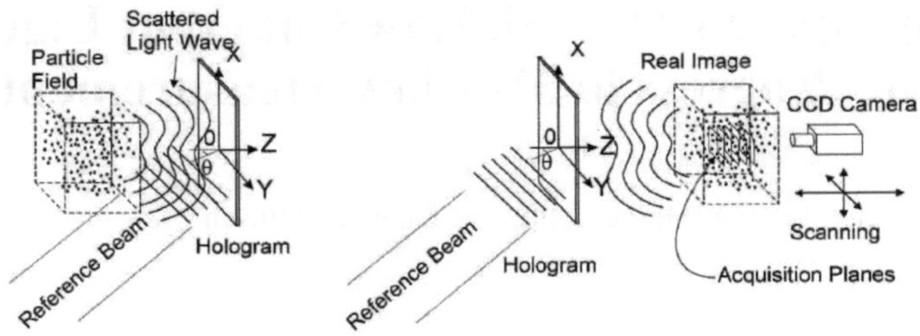

Fig. 1 Principles of off-axis HPIV: recording (left) on holographic film and digitization of the real image using a CCD camera (right); second reference beam omitted for clarity [1].

with different directions of incidence $\pm\Theta$ are used. In Fig. 1 the second reference beam is omitted for clarity.

After storage the two particle image fields are reconstructed and digitized separately for further evaluation. In our case real image reconstruction is being used and the digitization of the reconstructed images is done with a scanning CCD-target (Fig. 1 right). In analogy to 2D-PIV a 3D-vector field is calculated from the digitized acquisition planes applying a three dimensional intensity cross-correlation on appropriate sub-volumes. Well known fitting algorithms for sub-pixel accuracy have also been adapted to the three dimensional case.

2 Noise Problem and Light-in-Flight Configuration

The main problem of HPIV is that under extreme conditions, e.g. deep measurement volume, the in-focus particles of the digitized images are difficult to recognize. They are buried in speckle noise coming from the simultaneously reconstructed out-of-focus particles. This leads to a bad signal-to-noise-ratio which produces bad vectors during the evaluation procedure. To reduce the amount of speckle noise light-in-flight holography (LiF-HPIV) is used which significantly improves the signal quality. The basic idea of this concept is, contrarily to standard HPIV, to use laser light with low coherence in combination with an oblique incidence of the reference wave. As shown in Fig. 2, reference light incident from the left has to travel a longer path to the right side of the holographic plate than to the left. Object light scattered from particles is recorded only if its path length differs by no more than the coherence length l from that of the corresponding reference light. Thus, with proper alignment of the path lengths, object light from particles in a shell in the middle of the observed particle field can be made to be recorded only in a narrow horizontal band in the middle of the holographic medium. Particles from a front shell are recorded in a region on the left and from a rear shell in the right of the plate. Upon real-image reconstruction with an accordingly small aperture only a shell with a depth of d will

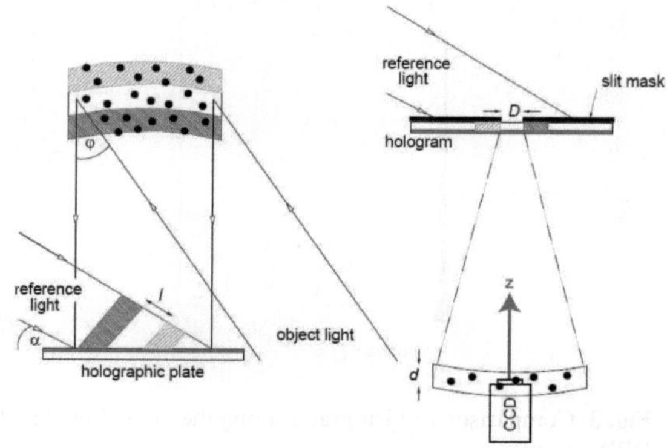

Fig. 2 Schematic of short coherence particle holography: recording (left) and real image reconstruction (right); l: coherence length, d: depth of reconstructed shell, D: aperture size.

show up (Fig. 2, right) [2]. Particles outside the shell are not reconstructed and can not contribute to the particle field. Thus the noise in LiF-HPIV is clearly reduced and the whole measurement volume can be explored by moving the aperture across the hologram.

3 Measurements

To demonstrate the capability of the modified technique a direct comparison between LiF-HPIV and traditional HPIV was performed. A laminar air flow with a mean velocity of 5.2 m/s in y-direction was produced in a small wind tunnel. For both holograms the recording conditions were kept as constant as possible. The same laser was used which can be operated in a short or in a long-coherence mode. Particle density was fixed to 110 particles/mm3. During evaluation the same representative part of the reconstructed particle field (15.4x12x29.1 mm3) is digitized and about 1,000 velocity vectors were determined within this volume. In Fig. 3 two characteristic examples of histograms of the measured flow velocity component are shown. While in case of LiF-HPIV a sharp peak is found located at the expected displacement the histogram of traditional HPIV shows a uniform distribution. This impressively shows the high performance of the light-in-flight technique in comparison to standard holographic recording. The maximum measurement volume is limited to about 50x50x40 mm3 and the accuracy is about 0,1%.

In a further application the method was used to measure the flow field around an airfoil with dimensions 15 mm high and 100 mm long. Again, a laminar incident flow in y-direction with a velocity of 5.2 m/s was produced. The Re-Number in this experiment was about 33,000. In Fig. 4 the measured flow field (about 40x40x30 mm3) in the wake of the airfoil is shown, normalized and with incident velocity subtracted (only each third vector shown for clarity). In addition a contour plane at z=2 mm is plotted to get a better insight of the flow situation. More than 80,000 velocity vectors were evaluated from only one hologram and the complete 3D velocity field could be determined simultaneously.

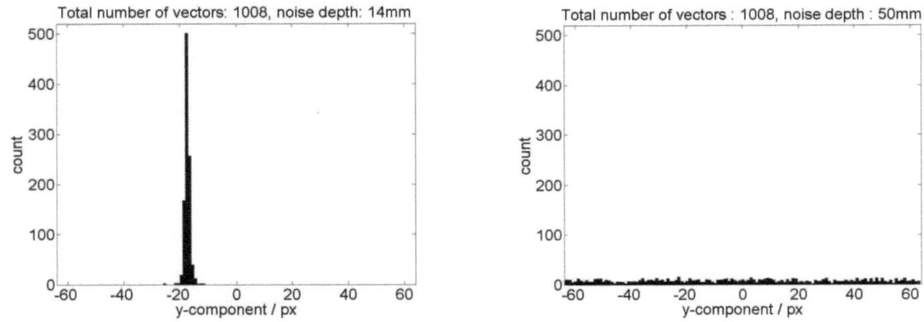

Fig. 3 Comparison of histograms along the main flow direction for LiF-HPIV (left) and traditional HPIV (right).

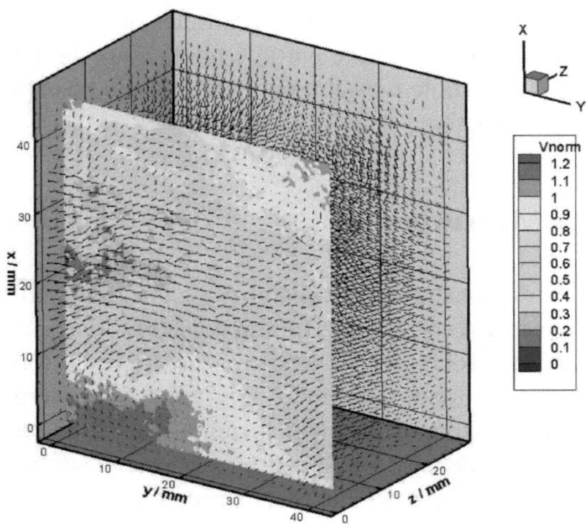

Fig. 4 Complete 3D velocity field from the wake of an airfoil; contour plane at $z = 2$mm.

4 Conclusions

Based on a simple flow configuration the outstanding performance of LiF-HPIV in comparison to HPIV under adverse measurement conditions was shown. It could be demonstrated that the modified method is very well suited to determine the complete 3D velocity field around an airfoil with high spatial resolution.

References

1. Pu, Y., Song, X., Meng, H.: Off-axis holographic particle image velocimetry for diagnosing particulate flows. Exp. Fluids (Suppl.), 117–128 (2000)
2. Hinsch, K.D., Herrmann, S.F.: Signal quality improvements by short-coherence holographic particle image velocimetry. Meas. Sci. Technol. 15, 622–630 (2004)

An Experimental Demonstration of Accelerated Tomo-PIV

N.A. Worth and T.B. Nickels

Abstract. Tomographic Particle Image Velocimetry (Tomo-PIV) is a promising new PIV technique. However, its high computational costs often make time-resolved measurements impractical. In this paper, a new preprocesseing technique is tested experimentally for the first time on a simple vortex ring. The new method produces very similar velocity and vorticity distributions to the standard iterated solution, at a fraction of the computational cost. Therefore, through this new technique, the processing of thousands of vector fields required for turbulent statistics can be made significantly more affordable; an important step in the development of Tomo-PIV.

1 Introduction

One of the characteristic features of turbulent flow is its three-dimensionality. Tomographic Particle Image Velocimetry (Tomo-PIV) is a promising new PIV technique capable of producing fully three-dimensional velocity fields[1]. However, the high computational costs of Tomo-PIV often make time-resolved measurements impractical[4]. These high costs stem from the use of an iterative algebraic approach, in which an initial volume intensity distribution is set, before iteratively updating the volume using a reconstruction algorithm (typically the Multiplicative Algebraic Reconstruction Technique (MART) algorithm is used). The most basic first guess is uniform intensity throughout the volume; a solution commonly employed in other investigations[1, 2]. However, this forces the algorithm to search

---

N.A. Worth
Cambridge University, UK
e-mail: naw30@cam.ac.uk

T.B. Nickels
Cambridge University, UK
e-mail: tbn22@cam.ac.uk

every voxel element in the discretised volume of interest during the initial iteration, which is extremely expensive.

Worth and Nickels[5] proposed a new preprocessing method to estimate the initial volume intensity distribution, thereby reducing the algorithm workload. The Multiplicative First Guess (MFG) calculates the current projection of each camera through the volume, which extends intensity from each pixel along its line of sight, creating a series of constant intensity streaks through the volume for each camera. Overlapping regions are multiplied to create an initial intensity field, thereby placing intensity only where all lines of sight converge. The method can be used to preprocess volumes before using an iterative algorithm, or as a stand alone technique to provide a rapid estimate of the volume intensity distribution (with a small accuracy penalty). A series of simulations on a simple angled vortex line flow field showed this technique can significantly reduce computational costs, and accelerate solution convergence[5].

In the current paper this new technique is tested experimentally for the first time on a simple laminar vortex ring flow field.

2 An Experimental Assessment of Accelerated Tomo-PIV

A computer controlled piston and cylinder arrangement (orifice size $D_0 = 20mm$) was fired at $50mm/s$ for a stroke length of $L_0 = 20mm$ to generate vortex rings ($Re = 1 \times 10^3$ and $L_0/D_0 = 1$) in a small water tank. The flow was seeded with $120\mu m$ Talismann particles to a density of 0.01 particles per pixel, which were illuminated in the volume of interest ($50 \times 50 \times 20mm$) by the clipped focused beam from a $250W$ lamp. The timescales of the flow were sufficiently slow, that particle blur was minimal, depite the use of relatively low shutter speeds ($125Hz$). The flow was captured using 3 Redlake Motionpro cameras (at $0° and \pm 45°$), with 60mm Nikon lenses at $f/\# = 8$, using a selected image area of 512×512 pixels. The cameras were calibrated by fitting a 3rd order polynomial Taylor series to a known 2 plane calibration object, giving a mean calibration error of 0.2 pixels.

The light intensity in the volume of interest was tomographically reconstructed for three cases: *1.* The MFG alone (MFG k=0); *2.* A uniform field initial guess and 5 MART iterations (UF k=5); *3.* The MFG and 5 MART iterations (MFG k=5). The volumes were discretised at 10 voxels/mm to maintain a pixel/voxel ratio of unity, with the single volume reconstruction times listed in Tab. 1. Employing a 3-pass recursive window-shifting method, with a final interrogation volume size of $42 \times 42 \times 42$ voxels and 50% overlap gave a spatial resolution of $2.1mm$. Spurious vectors were eliminated using a median criterion, and the velocity fields smoothed using a 2-pass box filter. The mean velocity error from calibration is 0.5mm/s. The uncertainty resulting from particle blur was evaluated as a function of flow velocity; approximated by the distance a particle travels during the exposure in each frame. The combined error is indicated at selected points on Fig. 1.

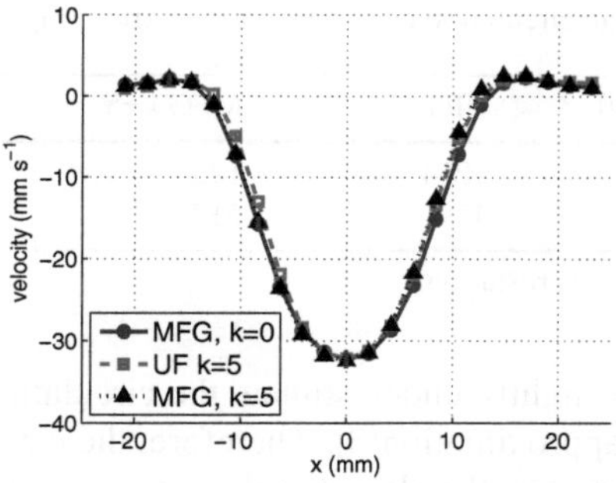

(a) Vertical velocity profile

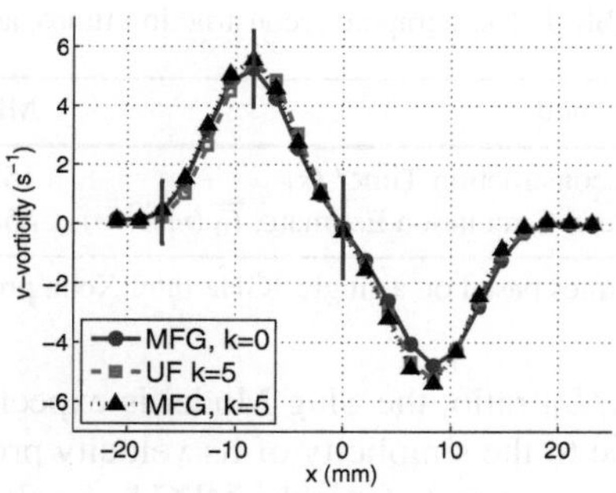

(b) v-vorticity profile

Fig. 1 Velocity and vorticity profile comparison.

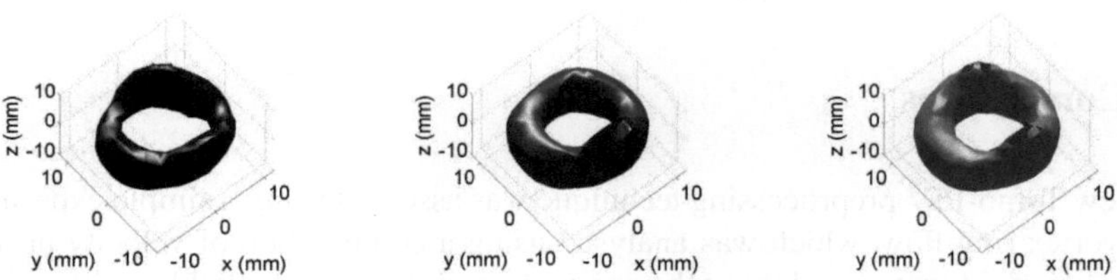

(a) MFG k=0 (b) UF k=5 (c) MFG k=5

Fig. 2 Vorticity magnitude isosurface comparison (65% maximum vorticity magnitude).

Fig. 1 shows a comparison of the *w*-velocity and *v*-vorticity component profiles along the x and z-planes of symmetry. The high velocity magnitude slug of fluid ejected from the cylinder, and the recirculating regions can be readily identified in Fig. 1(a), as can the vortex core regions in Fig. 1(b). The profiles are very similar for all cases, especially the MART iterated solutions. The MFG method alone also gives a relatively good estimate of the velocity and vorticity fields, which deviates from the iterated solutions by a maximum of 10%. This is most noticable in the recirculating region on Fig. 1(a) and in the vortex core region in Fig. 1(b), with a slightly lower magnitude peak vorticity predicted. The difference between the velocity profiles is of the order of the experimental error. High magnitude vorticity isosurfaces are plotted in Fig. 2, with the characteristic donut-shaped vortex core clearly identifiable in all cases. Despite the vortical structure being slightly thinner for the zero iteration case, reflecting the lower vorticity values predicted, the level of agreement between the iterated cases is good, especially given the sensitivity of the velocity gradients to variations in the velocity field.

The Slug Model[3] (Eq. 1) was used to predict a circulation of $\Gamma_{slug} = 500mm^2/s$, which was compared to line integral estimates performed at four locations around the ring vortex core (see Tab. 1). All schemes predict similar circulation values, which are close to the theoretical value and within 12% of each other. At the current

Table 1 Tomographic reconstruction times, and mean circulation estimates for the vortex ring

Method	MFG k=0	UF k=5	MFG k=5
Reconstruction Time[a] (s)	15	1054	123
Mean Circulation Estimate, $\overline{\Gamma_0}$ (mm^2/s)	458	479	517

[a] times based on a single 3GHz Intel Xeon processor workstation.

L_0/D_0 ratio, the Slug Model is expected to slightly underestimate the circulation, due to the simplicity of its velocity profile approximation[3]. Therefore, the circulation estimate from the MFG k=5 solution may be the closest to the actual value.

$$\Gamma_0 = \int_0^{T_0} \frac{1}{2} U_p^2(t)dt \quad (1)$$

3 Conclusions

A new Tomo-PIV preprocessing technique was assessed using a simple experimental vortex ring flow, which was analysed using a combination of velocity profiles, circulation estimates, and through examination of the vorticity field.

The main flow details are captured well using all processing variations. The deviations between velocity profiles are within the bounds of the experimental error, and even the more sensitive vorticity distributions are very closely matched. Comparing the Slug Model circulation prediction to line integral measurements confirms the similarity between schemes. The variation of the MFG method alone is consistent with the previous numerical study[5], and the advantage of this new method is demonstrated; reducing the computational workload by up to 70 times. Therefore, through this technique the processing of thousands of vector fields required for turbulent statistics can be made significantly more affordable; an important step in the development of Tomo-PIV.

References

1. Elsinga, G.E., et al.: Tomographic Particle Image Velocimetry. Exp. Fluids 41, 933–947 (2006)
2. Elsinga, G.E., et al.: Investigation of the three-dimensional coherent structures in a turbulent boundary layer with Tomographic-PIV. In: 45th AIAA Aerospace Sciences Meeting and Exhibit, Reno, Nevada (2007)
3. Lim, T.T., Nickels, T.B.: Vortex Rings. In: Green, S.I. (ed.) Fluid Vorticies. Kluwer Academic Publishers, Dordrecht (1995)
4. Schröder, A., et al.: Investigation of a Turbulent Spot Using Time-Resolved Tomographic PIV. In: 13th Int. Symp. on Applications of Laser Techniques to Fluid Mechanics, Lisbon, Portugal (2006)
5. Worth, N.A., Nickels, T.B.: Acceleration of Tomo-PIV by estimating the initial volume intensity distribution. Exp. Fluids 45, 847–856 (2008)

Using the 2D Laser-Cantilever-Anemometer for Two-Dimensional Measurements in Turbulent Flows

Michael Hölling and Joachim Peinke

Abstract. We present data measured with our new 2D Laser-Cantilever-Anemometer (2D LCA) in comparison to data acquired with a commercial x-wire anemometer. Measurements were taken in the wake of a cylinder with a diameter of $D = 1cm$ in a distance of $68D$. Stochastic analyses show good agreement between the measured data of both anemometers.

1 LCA - Basic Principle

The Laser-Cantilever-Anemometer (LCA) is based on the detection of the deflection of a tiny cantilever brought into the flow. The deflection is caused by the force F acting on the surface area A of the cantilever

$$F = c_d \cdot \frac{1}{2} \cdot \rho \cdot A \cdot v^2, \tag{1}$$

where c_d is the drag coefficient of the cantilever, ρ the density of air and v the flow velocity of the ambient medium. The deflection is measured by means of a laser pointer principle. A laser beam, generated by a laser diode, is focused on the tip of the cantilever. The reflected light hits a position sensitive detector (PSD) whose output signal is proportional to the position of the center of gravity of the incident light. This PSD element allows the detection of the movement of the incident light in one dimension (1D PSD). Therefore the measured signal is related to the deflection of the cantilever (see figure 1).

Michael Hölling
ForWind - Center for Wind Energy Research, University of Oldenburg, Oldenburg, Germany
e-mail: michael.hoelling@uni-oldenburg.de

Joachim Peinke
ForWind - Center for Wind Energy Research, University of Oldenburg, Oldenburg, Germany
e-mail: peinke@uni-oldenburg.de

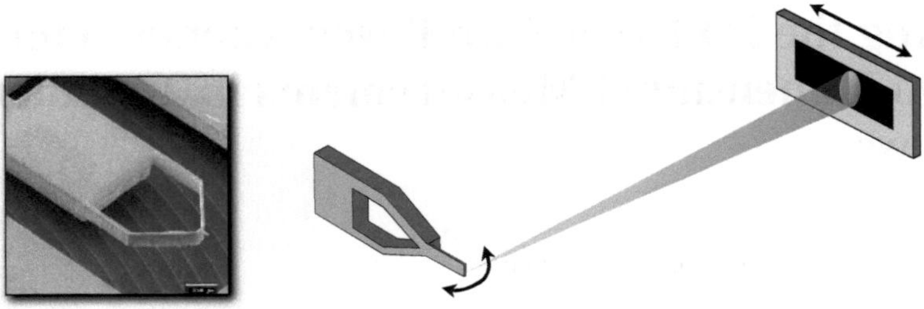

Fig. 1 **Left:** SEM picture of cantilever chip. **Right:** Sketch of the arrangement of cantilever and PSD element. A change in the signal of the PSD element corresponds to a change in the deflection of the cantilever.

2 Development of the 2D LCA

The Laser-Cantilever-Anemometer (LCA) developed at the University of Oldenburg was originally designed to measure only the longitudinal velocity component of a turbulent flow [1]. Therefore the signal of the LCA for inclined flow $Sig(\alpha)$ was expected to be the *cosine*−projection of the angle of attack α of the flow

$$Sig(\alpha) = cos(\alpha) \cdot Sig(\alpha = 0°). \tag{2}$$

Deviations from this behavior of the LCA for inclined flow led to the image of the bending and twisting cantilever for angular inflow and under turbulent flow, respectively. To verify this idea, the 1D PSD element was replaced by a 2D PSD element, which allows for detection of the center of gravity of the incident light in two dimensions. The orientation of the cantilever was chosen according to figure 2. A bending of the cantilever results in a change of the horizontal component of the 2D PSD signal, a twisting results in a change of the vertical component of the signal.

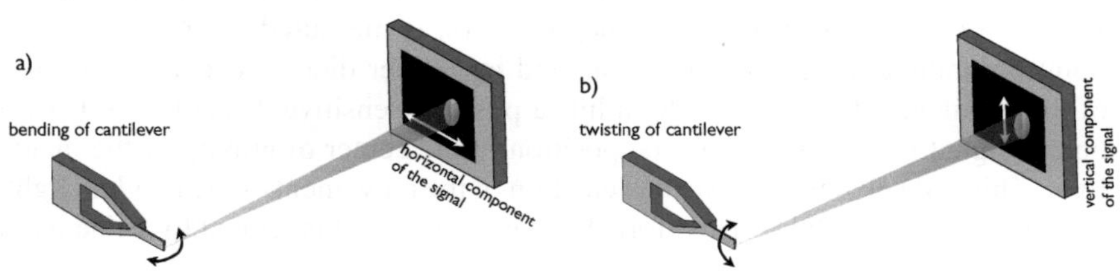

Fig. 2 **a)** Bending of the cantilever results in a change of the horizontal component of the 2D PSD signal. **b)** Twisting of the cantilever results in a change of the vertical component of the 2D PSD signal.

Figure 3 shows the vertical component of the 2D PSD signal Sig_{vert} over the horizontal component of the 2D PSD signal Sig_{hor} for velocities from $0\,m/s$ up to

$12m/s$ and angles of attack from $-40°$ to $40°$. It shows that each combination of Sig_{vert} and Sig_{hor} can clearly be assigned to one angle of attack and one velocity value. Based on such a calibration function the new 2D LCA can be used to perform two dimensional measurements in turbulent flows.

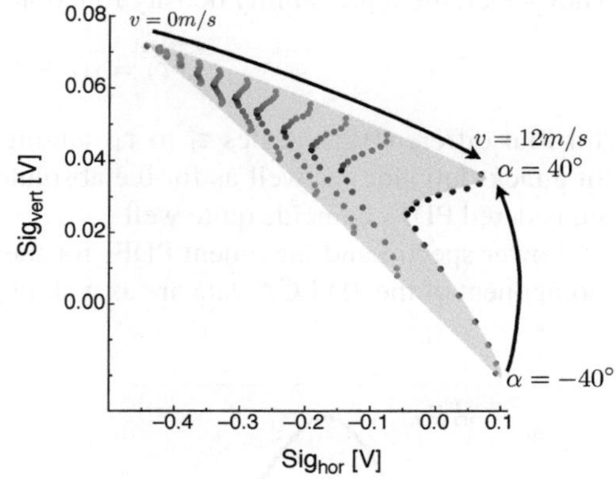

Fig. 3 Calibration function of 2D LCA for velocities from $0m/s$ to $12m/s$ and angles of attack from $-40°$ to $40°$. Each combination of Sig_{vert} and Sig_{hor} is clearly assigned to one angle of attack and one velocity value.

3 Measurements with 2D LCA and x-Wire

Measurement in the wake of a cylinder with a diameter of $D = 1cm$ were carried out with a 2D LCA and a commercial x-wire anemometer. The distance between the sensors and the cylinder was set to $68D$. Figure 4 shows the power spectra of

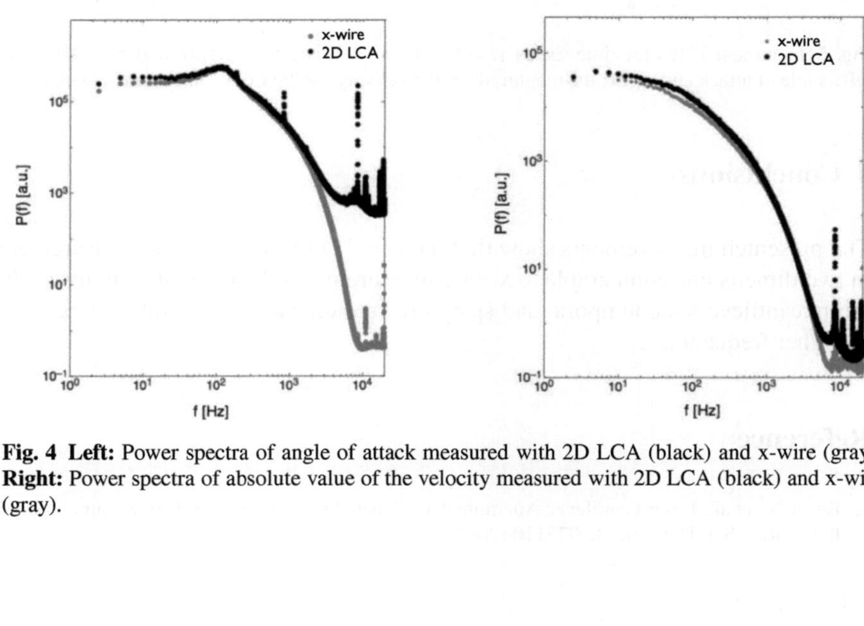

Fig. 4 Left: Power spectra of angle of attack measured with 2D LCA (black) and x-wire (gray). **Right:** Power spectra of absolute value of the velocity measured with 2D LCA (black) and x-wire (gray).

the measured time series for the angle of attack (left) and the absolute value of the velocity (right) for both sensors. It can clearly be seen that for the angle of attack both spectra fit quite well up to about $3kHz$, for the absolute value of the velocity this range extend up to about $5kHz$. Marginal differences can be allocated to a slightly off calibration function for the 2D LCA. For a more detailed comparison figure 5 shows increment probability density functions (PDFs) according to

$$x(\tau) = x(t+\tau) - x(t) \tag{3}$$

for four different time scales τ_1 to τ_4 ranging from $0.5ms$ to $100ms$. For the angle of attack (left side) as well as for the absolute value of the velocity (right side) all considered PDFs coincide quite well.

Power spectra and increment PDFs for the transversal and longitudinal velocity component of the 2D LCA data are as well in good agreement with the x-wire data.

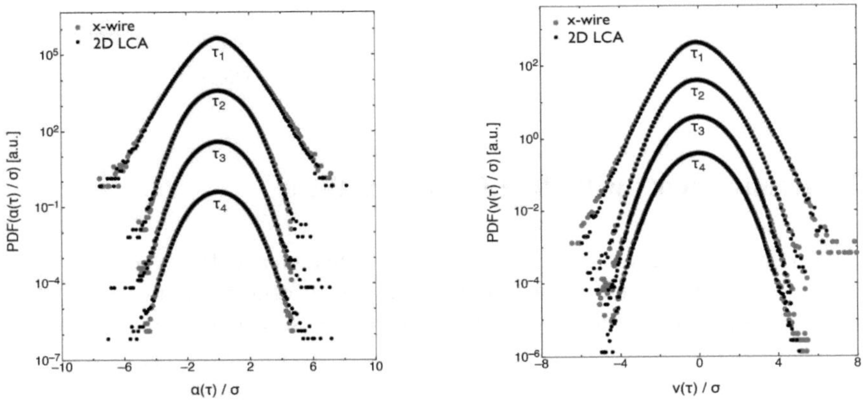

Fig. 5 Increment PDFs for time scales $\tau_1 = 0.5ms$, $\tau_2 = 10ms$, $\tau_3 = 2.5ms$, and $\tau_4 = 100ms$ for **left:** angle of attack and **right:** the magnitude of the velocity for 2D LCA (black) and x-wire (gray).

4 Conclusions

The presented measurements show that the new 2D LCA allows for measurements in two dimensions comparable to x-wire measurements. Note, by the choice of different cantilevers the temporal and spatial resolution can, in principle, be extended to higher frequencies.

Reference

1. Barth, S., et al.: Laser-Cantilever-Anemometer - A new high resolution sensor for air and liquid flows. Rev. Sci. Instrum. 76, 075110 (2005)

3D Structures from Stereoscopic PIV Measurements in a Turbulent Boundary Layer

David J.C. Dennis and Timothy B. Nickels

Abstract. Experiments using stereoscopic high-speed particle image velocimetry (PIV) to take measurements in a cross-stream (i.e. wall-normal/spanwise) plane in a turbulent boundary layer have been used to produce full 3D velocity fields. The 3D fields were constructed from planar 3C fields by using Taylor's hypothesis to create a pseudo-spatial x-dimension from the temporally resolved measurements. This has produced a 3D view of the elongated regions of high and low streamwise momentum found in the boundary layer, often referred to as 'long structures', and provided information on the arrangement, length and characteristic angle of these structures. Long structures are also seen to be associated with regions of high Reynolds stress. Vortical motions are visualised using swirling strength, and indicate that there is a prevalence for vortices to surround the low speed long structures. The vortices are characterised, and are found to resemble hairpin vortices in some respects.

1 Introduction

This experiment was performed to show the structures present in the turbulent boundary layer. Structures that are of interest are 'long structures'. These are elongated regions of high and low streamwise momentum that have previously been found in a turbulent boundary layer using hot-wires (Hutchins & Marusic[6]), and PIV measurements taken from a streamwise/spanwise plane in the log region (Ganapathisubramani et al.[4]). As well as PIV measurements in a streamwise/wall-normal plane in the outer region (Adrian et al.[2]), and Direct Numerical

David J.C. Dennis
Cambridge University Engineering Department, Trumpington Street, Cambridge, CB2 1PZ, UK
e-mail: djcd2@cam.ac.uk

Timothy B. Nickels
Cambridge University Engineering Department, Trumpington Street, Cambridge, CB2 1PZ, UK
e-mail: tbn22@cam.ac.uk

Simulations (Ringuette et al.[8]). These long structures are of interest because they have been found to contain a substantial portion of the Reynolds stress (Ganapathisubramani et al.[5]), and have an influence on the near-wall motions (Hutchins & Marusic[7]). Also of interest are hairpin vortices, vortex packets, and the relationship between long structures and the vortex arrangement in the boundary layer, see Adrian[1] for an overview of these.

2 Experiment

The experiments were performed using high-speed, stereoscopic PIV to take measurements in a cross-stream (i.e. wall-normal/spanwise) plane in the Cambridge University Engineering Department's turbulent boundary layer water tunnel research facility. This facility has a 0.9m $\times$ 0.5m $\times$ 8m long working section, and has been specially designed to produce thick turbulent boundary layers. The flow is tripped at inlet, and the measurement plane is 5m downstream. At this location the boundary layer thickness is $\delta = 90$mm, the freestream velocity is $U_\infty = 0.69$m/s, the Reynolds Number based on momentum thickness is $\mathrm{Re}_\theta = 4685$. The water is seeded with silver-coated hollow glass spheres with 10μm mean diameter.

The stereoscopic system consisted of a New Wave Pegasus-PIV laser, and two LaVision HighSpeedStar 4 CMOS cameras. It provided all three components (3C) of the velocity vector, and the use of a cross-stream measurement plane meant that the flow was advected through the measurement plane, and therefore full 3D velocity fields could be constructed from the planar 3C fields by using Taylors hypothesis to create a pseudo-spatial x-dimension from the highly resolved temporal measurements, which is shown to be valid for short distances in Dennis & Nickels [3]. A similar technique to this was used by vanDoorne [9] (on a smaller scale), in a pipe.

3 Results

Figure 1 presents a result showing a grey iso-surface representing a region where the streamwise velocity component is 90% of the local mean, (i.e. $u_{iso} = -0.1\overline{U}$), alongside a meshed iso-surface representing a region where the streamwise velocity component is 110% of the local mean, (i.e. $u_{iso} = 0.1\overline{U}$). These two iso-surfaces are good representations of low and high speed structures in the flow.

In this example the low speed structure is found to be 2.5δ in length, with the height of its upper surface increasing with downstream distance such that it is at an angle of $\approx 11^\circ$ to the wall. This is found to be fairly indicative of all the structures in the dataset as a whole. The length of structures reached a maximum of 8.4δ, but such a long structure was quite rare with almost half found to be between 2-3δ, and 97% below 7δ in length. This is supported by the two-point correlation of the streamwise velocity fluctuation (R_{uu}), which extends to about 5δ, and is strong for around 2δ. The angle to the wall varies considerably across structures, but there is a definite tendency for structures to be at a positive angle to the wall (73%). This and the streamwise velocity fluctuation correlation indicate that structures are at quite a shallow angle to the wall (4-17°). High speed structures are found to have similar characteristics in general, although it was *not* seen that a low-speed structure always had a matching high-speed partner as seen in figure 1.

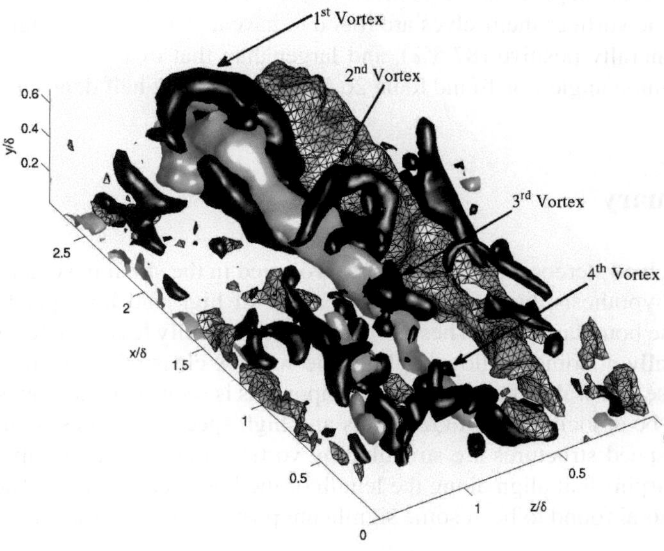

Fig. 1 Visualisation of vortices with high and low speed structures. Black iso-surface: $|\lambda_{ci}|_{iso} = 0.18|\lambda_{ci}|_{max}$. Grey iso-surface: $u_{iso} = -0.1\overline{U}$. Meshed iso-surface: $u_{iso} = 0.1\overline{U}$.

Study of the wall-normal velocity fluctuation (v) showed a relationship between it and the streamwise velocity fluctuation (u). An association was seen between regions of high $-u$ (low speed structures) and regions of high $+v$, a combination that represents a Q2 event (an ejection). Similarly regions of high $+u$ (high speed structures) and high $-v$, were found together, forming a Q4 event (a sweep). Both Q2 and Q4 events are notable because they increase $-uv$, which shows that events associated with long structures make a significant contribution to the Reynolds stress.

Vortical motions have been indicated using the black iso-surfaces in figure 1. These are based on the swirling strength criterion λ_{ci}, (the imaginary part of the complex-conjugate eigenvalue of the velocity gradient tensor). The figure shows a large vortex at the head of the low speed structure labelled "1st Vortex", that is almost an archetypal horseshoe/hairpin shape, and is the apparent cause of the negative streamwise velocity fluctuation. The long length of the low speed structure can be explained by the series of similar vortices behind the "1st Vortex", labelled "2nd",

"3rd" and "4th Vortex". These vortices are in line with the initial vortex and decrease in size and height as we progress upstream.

Vortices such as these are seen throughout the dataset, and they are seen to be concentrated around the low speed structures. By contrast there are very few vortices attached to high speed structures. The vortices share some of their properties with hairpin shapes although it is very rare to see an actual hairpin, in many cases only the leg, or only the head is seen, or perhaps one leg and a head (a 'cane' vortex). Whatever shape it takes it is invariably asymmetric and therefore not a perfect hairpin. The vortices themselves are found to have a characteristic angle to the wall that is generally positive (87.5%), and larger than that of the long structures, the most common angle was found to be 26.5° (to the nearest half degree).

4 Summary

Time-resolved stereoscopic PIV results, projected in the streamwise direction using Taylor's hypothesis, have shown the presence of high and low speed long structures in the boundary layer. These structures are generally less than 7δ in length and are normally at some shallow angle to the wall. A clear relationship between the streamwise and wall-normal velocity components is seen such that low speed structures can be associated with Q2 events and high speed structures with Q4 events. The low speed structures are straddled by vortices that resemble imperfect asymmetric hairpins that align along the length of the low speed regions. These vortices are in general found to be at some significant positive angle to the wall.

References

1. Adrian, R.: Hairpin vortex organization in wall turbulence. Phys. Fluids 19, 041301 (2007)
2. Adrian, R., Meinhart, C., Tomkins, C.: Vortex organisation in the outer region of the turbulent boundary layer. J. Fluid Mech. 422, 1–54 (2000)
3. Dennis, D.J.C., Nickels, T.B.: On the limitations of Taylor's hypothesis in constructing long structures in a turbulent boundary layer. J. Fluid Mech. 614, 197–206 (2008)
4. Ganapathisubramani, B., Clemens, N., Dolling, D.: Large-scale motions in a supersonic turbulent boundary layer. J. Fluid Mech. 556, 271–282 (2006)
5. Ganapathisubramani, B., Longmire, E.K., Marusic, I.: Characteristics of vortex packets in turbulent boundary layers. J. Fluid Mech. 478, 35–46 (2003)
6. Hutchins, N., Marusic, I.: Evidence of very long meandering features in the logarithmic region of the turbulent boundary layers. J. Fluid Mech. 579, 1–28 (2007)
7. Hutchins, N., Marusic, I.: Large-scale influences in near-wall turbulence. Phil. Trans. R. Soc. A. 365, 647–664 (2007)
8. Ringuette, M., Wu, M., Pino Martin, M.: Coherent structures in direct numerical simulation of turbulent boundary layers at Mach 3. J. Fluid Mech. 594, 59–69 (2008)
9. Van Doorne, C.: Stereoscopic PIV on transition in pipe flow. Ph.D. thesis, Delft University of Technology (2004)

The Sphere Anemometer – A Fast Alternative to Cup Anemometry

Hendrik Heißelmann, Michael Hölling, and Joachim Peinke

Abstract. The main problem of cup anemometry is the different response time for increasing and decreasing wind velocities due to its moment of inertia. This results in an overestimation of wind speed under turbulent wind conditions, the so-called over-speeding. Additionally, routine calibrations are necessary due to the wear of bearings. Motivated by these problems the sphere anemometer, a new simple and robust sensor for wind velocity measurements without moving parts, was developed at the University of Oldenburg. In contrast to other known thrust-based sensors, the sphere anemometer uses the light pointer principle to detect the deflection of a bending tube caused by the drag force acting on a sphere mounted at its top. This technique allows the simultaneous determination of wind speed and direction via a two-dimensional position sensitive detector.

The behaviour of the newly designed sensor under turbulent conditions was investigated by comparative measurements of gusty wind using the sphere anemometer, cup anemometer and a hot-wire as a reference. It turned out that the sphere anemometer provides a better temporal and spatial resolution than cup anemometers do.

Hendrik Heißelmann
ForWind - Center for wind energy research, University of Oldenburg, Marie-Curie-Str. 1, D-26129 Oldenburg
e-mail: hendrik@heisselmann.de

Michael Hölling
ForWind - Center for wind energy research, University of Oldenburg, Marie-Curie-Str. 1, D-26129 Oldenburg
e-mail: michael.hoelling@uni-oldenburg.de

Joachim Peinke
ForWind - Center for wind energy research, University of Oldenburg, Marie-Curie-Str. 1, D-26129 Oldenburg
e-mail: peinke@uni-oldenburg.de

1 Sphere Anemometer Principle and Setup

The developed sphere anemometer uses a bending tube to detect the wind velocity and direction simultaneously. The deflection s of the tube is proportional to the point force F acting on the tip only, as sketched in figure 1 (left). The deflection is given by

$$s = \frac{l^3}{3} \cdot \frac{F}{E \cdot J}, \tag{1}$$

with the length of the tube l, the elasticity modulus E and the second moment of area J. For a sphere with a much larger radius r than the radius of the tube, the force can be assumed to be acting on the tip only. With the drag force acting on a sphere

$$F = \frac{1}{2} \cdot c_d \cdot \rho \cdot \pi \cdot r^2 \cdot v^2 \tag{2}$$

and the second moment of area for a tube with inner radius R_i and outer radius R_o equation (1) transforms to

$$s = \frac{2}{3} \cdot \frac{l^3 \cdot \rho \cdot r^2}{E \cdot (R_o^4 - R_i^4)} \cdot c_d \cdot v^2, \tag{3}$$

where c_d is the drag coefficient of the sphere and ρ the density of air. Beside material constants, the displacement s depends only on the drag coefficient c_d times wind speed v squared. The drag coefficient can be considered constant in the Reynolds number range of about 800 up to 200,000 [1], corresponding to wind speeds from 0.2 m/s to 48 m/s for a sphere of 7 cm diameter. In this range, the deflection s of the tube is proportional to the wind speed squared, allowing an easy calibration of the sensor.

In order to match the assumption of a point force acting on the tip only, a sphere of 7 cm diameter is mounted on top of an acrylic glass tube. The sphere is made of styrofoam so that effects of inertia are minimized. The displacement of the sphere, typically in order of micrometers, is detected by means of a light pointer, which facilitates the construction compared to other thrust-based anemometers [2], [4].

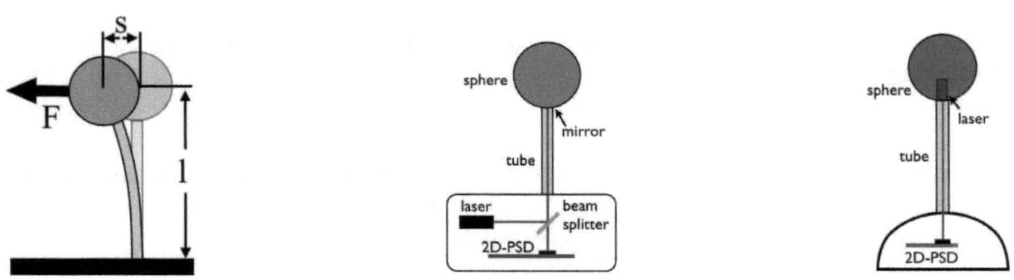

Fig. 1 *Left:* Bending tube due to point force acting on the tip; *Middle:* Schematic drawing of the first anemometer setup; *Right:* Schematic drawing of the new anemometer setup

In the first sphere anemometer setup a laser beam was aimed on a beam splitter and reflected onto a mirror at the top of the tube with the sphere. The reflected beam passed the beam splitter and hit a two-dimensional position sensitive detector (2D-PSD) (see fig. 1, middle). At large displacements of the sphere the beam was disturbed by reflections at the mount of the beam splitter and additionally the adjustment of this setup turned out to be difficult. Therefore an improved anemometer design was developed by replacing the mirror at the top of the tube by a laser diode (fig. 1, right). This supersedes the beam splitter and simplifies the adjustment of the anemometer.

2 Measurements

As a first step in the evaluation of the sphere anemometer, a one-dimensional calibration function was recorded in laminar flow. The inflow direction was chosen identically to the x-axis of the PSD. Figure 2 (left) shows the wind velocity plotted against the detected voltage output for the x-component of the 2D-PSD signal. The calibration data can be fitted with a square root function and is in accordance with the expectations stated in section 1. Furthermore, a two-dimensional calibration characteristic was obtained by varying the inflow direction from 0° to 350° by steps of 10°. In figure 2 the measured x- and y-components of the PSD output are plotted for every inflow angle. The PSD signals, corresponding to the position of the laser spot on the detector, are unambiguously linked to the inflow angle. This feature of the sphere anemometer permits the simultaneous determination of wind speed and direction.

Subsequently, measurements under gusty wind conditions were performed with the sphere anemometer in the wind tunnel. Sphere anemometer, cup anemometer and a hot-wire, which served as a reference due to its higher spatial and temporal resolution, were positioned behind a motor-driven gust generator to characterize the response behaviour to fast velocity changes. Wind gusts of approximately 1 Hz

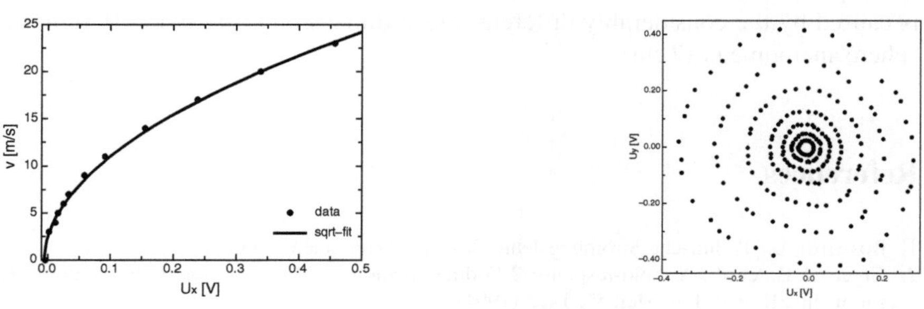

Fig. 2 *Left:* Calibration function for the U_x component; *Right:* U_y-component plotted against U_x-component of the 2D-PSD for inflow angles between 0° (flow from left side) to 350°

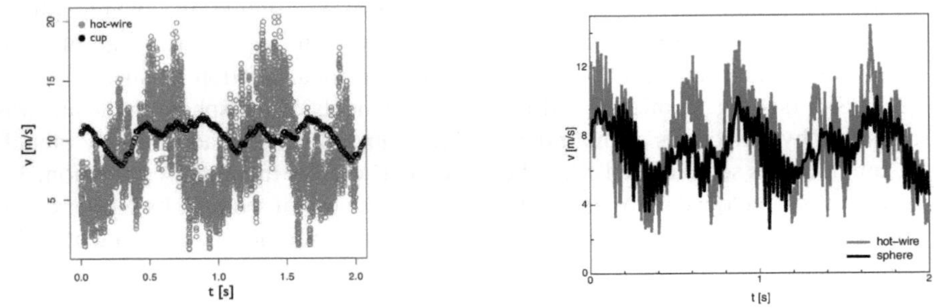

Fig. 3 Two seconds excerpts of a time series taken with *Left:* cup anemometer (black) and hot-wire (grey) [3]; *Right:* sphere anemometer (black) and hot-wire (grey) behind a motor-driven gust generator.

were produced and time series were recorded. Two seconds of the taken time series for cup anemometer and sphere anemometer are compared to the hot-wire data (fig. 3). Unlike cup anemometers, which exhibit the well-known effect of overspeeding under fast wind speed changes [3], the sphere anemometer is able to resolve fast wind speed fluctuations up to its natural frequency of 44 Hz, which is significantly higher than the resolution cup anemometers (about 1 Hz).

3 Conclusions

The sphere anemometer is a simple and robust sensor, capable of measuring the wind speed and direction simultaneously. Since it uses the light pointer principle known from atomic force microscopes, it has no wear parts that require maintenance. It provides a better temporal and spatial resolution than cup anemometers, which is only limited by its natural frequency. The data recorded with the sphere anemometer is in good agreement with the hot-wire data concerning its temporal response characteristics, while the measured maximum velocities differ. This is caused by the considerably different sensor dimension of hot-wire (0.5 cm) and sphere anemometer (7 cm).

References

1. Böswirth, L.: Technische Strömungslehre. Vieweg, Wiesbaden (2007)
2. Green, A.E., et al.: A rapid-response 2-D drag anemometer for atmospheric turbulence measurements. Bound.-Lay Met. 57, 1–15 (1991)
3. Hölzer, M.: Stochastische Beschreibung des Schalensternanemometers. Diplomarbeit, Oldenburg (2008)
4. Smith, S.D.: Evaluation of the mark 8 thrust anemometer-thermometer for measurement of boundary-layer turbulence. Bound.-Lay. Met. 19, 273–292 (1980)

CICLoPE – A Large Pipe Facility for Detailed Turbulence Measurements at High Reynolds Number

J.-D. Rüedi, A. Talamelli, H.M. Nagib, P.H. Alfredsson, and P.A. Monkewitz

Abstract. High Reynolds number turbulence is ubiquitous in a number of flow of practical interest and crucial to draw conclusions regarding the physics of turbulence. Although recent laboratory experiments, measurements in planetary boundary layer and direct numerical simulations provide a huge amount of information, none of these data sets provide high Reynolds number, high spatial resolution and well converged statistics at the same time. As a response to this problem, an international collaboration between a group of universities and research centers started some years ago to build large scale infrastructures for high Reynolds number experiments. The Center for International Cooperation in Long Pipe Experiments, CICLOPE (www.ciclope.unibo.it) at the University of Bologna, was created for this purpose and will be open to international scientists through different collaboration programs. The laboratory is currently under construction and the first facility, which will be installed there is a large pipe flow experiment that will allow fully resolved turbulence measurements even at high Reynolds number.

J.-D. Rüedi
II Facoltà di Ingegneria, Università di Bologna, Forlì, 47100, Italy
e-mail: ciclopeproject@mac.com

A. Talamelli
II Facoltà di Ingegneria, Università di Bologna, Forlì, 47100, Italy
e-mail: alessandro.talamelli@unibo.it

H.M. Nagib
IIT, Chicago, IL 60616, USA
e-mail: nagib@iit.edu

P.H. Alfredsson
Linné FLOW Centre, KTH Mechanics, Stockholm, 10044, Sweden
e-mail: hal@mech.kth.se

P.A. Monkewitz
Fluid Mechanics Laboratory, EPFL, Lausanne, 1015, Switzerland
e-mail: peter.monkewitz@epfl.ch

1 Fully Resolved High Reynolds Number Experiment

High Reynolds number turbulence is ubiquitous in aerospace engineering, ground transportation systems, flow machinery, energy production as well as in nature and is of prime interest to draw conclusions regarding the general physics of turbulence. The computing power available nowadays makes it possible to simulate and probe turbulence in great details but the Reynolds number attainable with fully resolved direct numerical simulations remains moderate and is not likely to reach the higher values of practical interest for decades to come. High Reynolds number can be obtained from experiments but requires the spatial size of the smallest scales to be sufficiently large to be resolvable by available measurement techniques. Such conditions can be obtained in nature but well converged data cannot be obtained easily, hence well controlled laboratory experiments remain the only alternative available at present. The existence of a logarithmic relationship between the mean streamwise velocity U and the distance from the wall y is well accepted for wall bounded turbulent flows. There is strong evidence that in boundary layers, the logarithmic region starts around $y^+ = 200$ and stretches out to approximately $0.15\ \delta/\ell_*$, where $y^+ = y/\ell_*$, δ is the boundary layer thickness, $\ell_* = \nu/u_\tau$, ν is the kinematic viscosity and u_τ the friction velocity, see Österlund et al. [1]. One would like the logarithmic region to stretch over at least one decade of y^+ in order to have sufficient scale separation, which means $200 < y^+ < 2000$. If one considers that the same is true for pipe flow, this leads directly to the conclusion that $R^+ = Re_\tau = R/\ell_*$, where R is the pipe radius, should be at least 13300 (= 2000/0.15). The Reynolds number range must be further expanded by a factor of at least 3 to investigate scaling behavior, hence the minimum Reynolds number required is $13300 < R^+ < 40000$.

In order to measure turbulence fluctuations with sufficient spatial resolution, the size of the sensing element must be small enough compared to the viscous length scale. For hot-wires it is accepted that a probe lengths larger than $10\ \ell_*$ may lead to spatial averaging [2]. Single hot-wire probes can be manufactured with a sensing length as small as 120 μm (with a wire diameter of 0.6 μm and a length-to-diameter ratio of the sensing element of 200 [3]), which sets a lower limit for the viscous length of about 12 μm. For pipe flow one can write

$$\ell_* = \frac{R}{Re_\tau} \tag{1}$$

which shows that in order to have large enough scales at high Re_τ, the Radius of the pipe, hence the overall dimension of the facility must be large. The Reynolds number and the viscous length defined above yields together with equation 1 a radius of 0.48 m, or a diameter of 0.96 m. Finally the length-to-diameter ratio (L/D) of the pipe itself must be long enough to ensure fully devloped flow. According to Zagarola & Smits [4] a length of the pipe of at least 100 D, or nearly 100 m in this case, is necessary to achieve a fully developed flow in terms of mean profile and higher statistical moments for this range of Reynolds numbers. The characteristics of various pipe experiments are plotted in figure 1 together with the planned

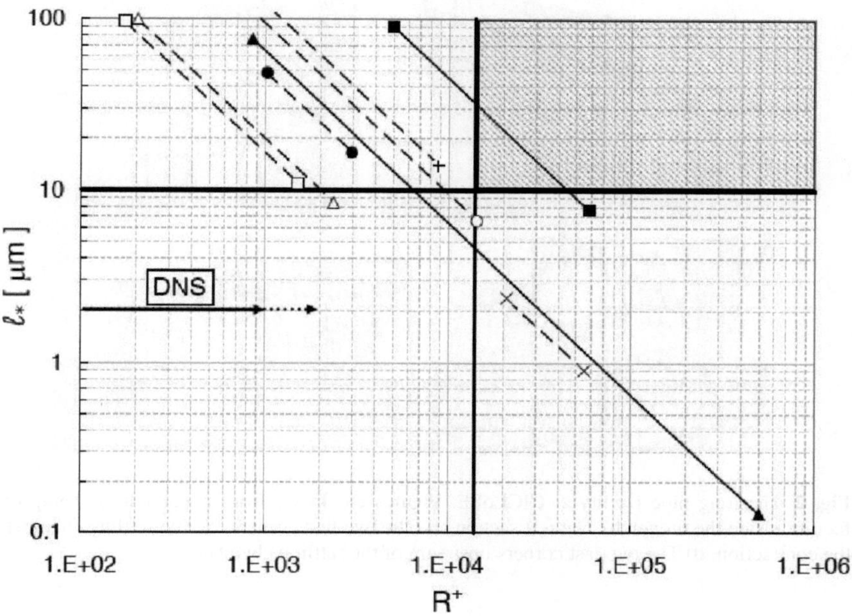

Fig. 1 Range of Reynolds number and viscous length scale of various pipe flow experiments. □: Wygnanski & Champagne (1973), air, $D = 0.033$ m; △: van Doorne & Westerweel (2007), water, $D = 0.040$ m; •: Monty (2005), air, $D = 0.0988$ m; ▲: Zagarola & Smits (1998), compressed air, $D = 0.129$ m; ×: Nikuradse (1932), water, $D = 0.10$ m; ○: University of Cottbus (under construction), air, $D = 0.19$ m; +: Laufer (1954), air, $D = 0.123$ m; ■: CICLoPE experiment, Talamelli (2009), air, $D = 0.90$ m. The highest Reynolds number for a Direct Numerical Simulation reported so far for turbulent pipe flow is the one by Satake et al. (2000) at R^+=1050. For turbulent channel flow Hoyas & Jiménez (2006) has reported a DNS at $h^+ = h/\ell_* = 2003$, where h is the mid height of the channel. The solid vertical line refers to the criterion of a well developed overlap region ($R^+ > 13300$). Horizontal line gives the limit for $\ell_* > 10$ μm which is the minimum for sufficient spatial resolution. Hatched region in upper right corner shows where the criteria for both spatial resolution and high enough Reynolds number are met.

CICLoPE pipe, in terms of Reynolds number (R^+) versus viscous length scale (ℓ_*). The experiments include air, compressed air and water as flow medium. As it can be seen, most experiments do not achieve very high Reynolds numbers and the large pipe at CICLoPE is the only one that passes through the range of interest defined above, i.e. $R^+ > 13300$ and $\ell_* > 10$ simultaneously [5].

2 The Long Pipe Facility at CICLoPE

The long pipe facility at CICLoPE [6], shown in figure 2, has a diameter of 0.9 m and a length of 115 m, hence a L/D of more than 126. The pipe will be made in carbon

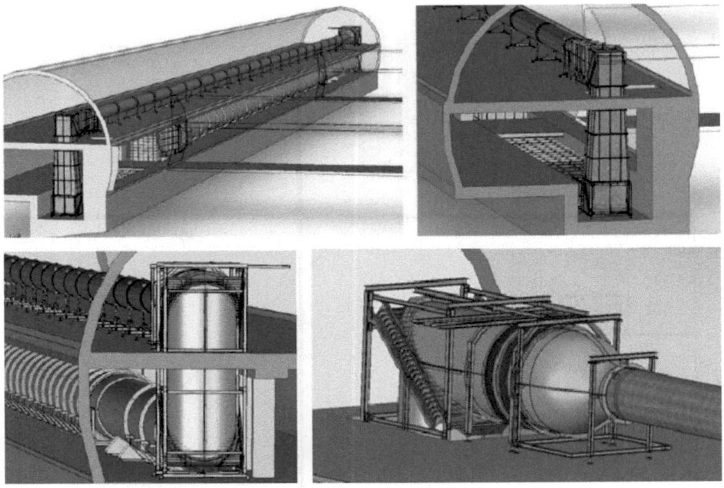

Fig. 2 The long pipe facility at CICLoPE. Figures clockwise from upper left. a) Setup of the facility inside the tunnel, b) The test section and the two first corners, c) The settling chamber and the contraction, d) The two last corners upstream of the settling chamber.

fiber with very tight specifications in terms of diameter and surface roughness to ensure optimum flow quality and smooth flow regime up to the highest Reynolds number. Access holes every 5 m along pipe will allow to place small measurements devices, e.g. traversing system along the whole length of the pipe. The pipe is built in a modular way and can accommodate custom test sections to match the different requirements of every users and measurement techniques. The facility has a closed loop circuit for accurate control of the velocity, temperature and humidity of the flow. The laboratory is located 60 m underground in a tunnel of the former *Caproni* aircraft industry in Predappio, which was excavated to shelter the production from bombing during world war 2. The complex consists in two 130 m long tunnels with a diameter of about 9 m each, linked together and to the exterior by two access galleries. The stability of the ambient conditions in the tunnel, the complete absence of perturbations (vibration, electro-magnetic field, etc.) provides ideal conditions to operate this facility and to perform high quality flow measurements.

3 The CICLoPE Laboratory

The CICLoPE laboratory is operated with the idea of gathering world-leading scientists in the field of turbulence research for a collaborative effort to make decisive breakthroughs in the fundamental issues of high Reynolds number turbulence. The centre is hosted by the university of Bologna and was founded in collaboration with a group of different Universities and research Centres around the world: the International Centre of Theoretical Physics in Trieste, the Royal Institute of Technology in Stockholm, Illinois Institute of Technology in Chicago, the Ecole Polytechnique Fédérale de Lausanne, La Sapienza in Rome and Princeton University. The center will be open for international collaborations through different research programs and allow visitors to have full access to the facility, the usage of standard measuring devices, office space and technical support.

References

1. Österlund, J.M., et al.: A note on the overlap region in turbulent boundary layers. Phys. Fluids 12, 1–4 (2000)
2. Johansson, A., Alfredsson, H.: Effect of imperfect spatial resolution on measurements of wall-bounded turbulent shear flows. J. Fluid Mech. 137, 409–421 (1983)
3. Bruun, H.H.: Hot-wire Anemometry Principles and Signal Analysis. Oxford science publications (1994)
4. Zagarola, M.V., Smits, A.J.: Mean-flow scaling of turbulent pipe flow. J. Fluid Mech. 373, 33–79 (1998)
5. Talamelli, A., et al.: CICLoPE - a response to the need for high Reynolds number experiments. Fluid Dynamics Research 41, 021407 (2009)
6. Talamelli, A., et al.: The new high Reynolds number pipe flow facility at CICLoPE. In: 26th AIAA Aerodynamic Measurement Technology and Ground Testing Conference, AIAA-2008-3966, Seattle (2008)

Aerodynamics of an Airfoil at Ultra-Low Reynolds Number

Md. Mahbub Alam, Y. Zhou, and H. Yang

Abstract. A comprehensive experimental study is conducted of the aerodynamic characteristics of an NACA0012 airfoil over a large range of angle (α) of attack and low- to ultra-low cord Reynolds numbers, 5.3×10^3 - 5.1×10^4, which is of both fundamental and practical importance. While the mean and fluctuating lift and drag coefficients were measured using a load cell, the detailed flow structure is documented using particle image velocimetry and laser-induced fluorescence flow visualization. C_D increases monotonically from $\alpha = 0°$ to $90°$, whilst C_L grows from 0 to its maximum at $\alpha = 45°$ and then drops. The near wake characteristics examined were found to be consistent with the force measurements, including the vortex formation length, wake width, spanwise vorticity, wake bubble size and wavelength of K-H vortices.

1 Introduction

The aerodynamic characteristics of airfoils at a chord Reynolds number ($Re_c = \frac{\rho c U_\infty}{\mu}$, where ρ and μ are the density and viscosity of the fluid, respectively, and c is the chord length of an aerofoil) of less than 5×10^5 are becoming increasingly important from both fundamental and industrial point of view [1]. To our knowledge, the lowest Re_c, at which airfoil lift and drag coefficient (C_L and C_D) data are available, is 1.0×10^4. General researches on airfoil aerodynamics have focused on conventional aircraft design with Re_c beyond 5.10^5 and the angle (α) of attack below stall. Mueller & DeLaurier [2] reviewed the available low Re_c studies, with almost all the measured Re_c higher than the wind turbine values quoted above. The flow physics of airfoil flow at low Re_c and high α is scarcely addressed. Aerodynamics of airfoils operated on the condition of a low $Re_c (< 10^5)$ has recently gained an increasing practical importance, with a variety of applications such as

Md. Mahbub Alam · Y. Zhou
Department of Mechanical Engineering, The Hong Kong Polytechnic University, Hong Kong

H. Yang
Department of Building Services Eng., The Hong Kong Polytechnic University, Hong Kong

micro air vehicles (MAVs), unmanned air vehicles (UAVs), small wind turbines, and lowspeed/high-altitude aircraft [2]. The work aims to determine the dependence on α of C_L and C_D of NACA 0012 airfoil at $Re_c = 5.3\times10^3 \sim 5.1\times10^4$.

2 Experimenral Details

Experiments were conducted in a closed-loop water tunnel, with a test section of $0.3m$ (width) $\times$ $0.6m$ (height) $\times$ $2.4m$ (length). The flow speed in the test section ranges from $0.05m/s$ to $4m/s$. NACA0012 airfoil was used as the test model with a chord length of $c = 0.1m$ and a span of $0.27m$. The tests were carried out at $Re_c = 5.3\times10^3 \sim 5.1\times10^4$, over which the free-stream turbulence was between 0.4% and 0.5%. α varied from 0° to 90° with an increment step of 10°. C_L and C_D were measured using a 3-component load cell (Kistler 9251A) and the detailed flow structure is documented using particle image velocimetry (PIV) and laser-induced fluorescence flow visualization.

3 Results and Discussion

Mean drag and lift. Figure 1 presents the load cell measured C_L and C_D. Note that the data at $Re_c = 5.1\times10^4$ is only shown for $\alpha \le 40°$ because the force at $\alpha > 40°$ is relatively large, exceeding the valid range of the load cell used. Evidently, C_L is dependent on Re_c for all α, growing with higher Re_c. The figure shows the occurrence of the stall at about 10° for $Re_c = 1.05\times10^4$ and 5.1×10^4. With an increment of $\Delta\alpha = 10°$, the stall should occur beyond 10° slightly. Interestingly, the stall is absent at $Re_c = 5.3\times10^3$, C_L climbing until $\alpha \approx 45°$, without any rapid drop as at other Re_c. Beyond the stall α, C_L displays a maximum at $\alpha \approx 45°$, regardless of Re_c. C_D increases monotonously with α and reaches maximum at $\alpha = 90°$. C_D is between 1.35 and 1.48, depending on Re_c, at $\alpha = 90°$ where the airfoil is like a flat plate normal to incident flow.

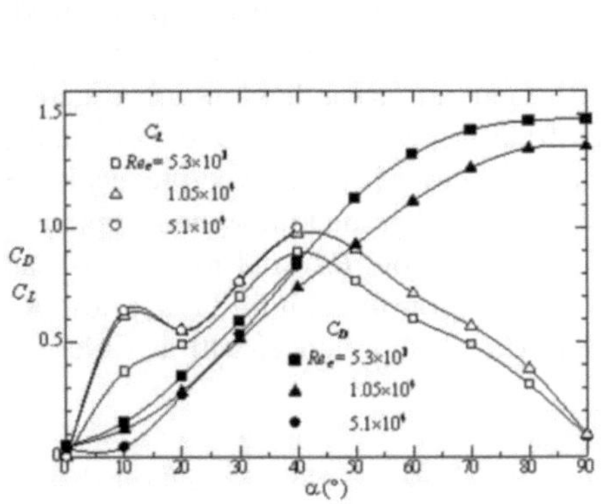

Fig. 1 Dependence of C_L and C_D on α

Flow structure. The flow structure behind an airfoil is dependent on both Re_c and α. The flow structure evolution with increasing α at a very low Re_c, where stall is absent, has not been thoroughly documented previously. Figure 2 presents typical photographs of the flow structure at $\alpha = 10° \sim 90°$ for $Re_c = 5.3 \times 10^3$. At very small α, flow was fully attached (not shown). With increasing α ($< 10°$), the suction side boundary layer separated initially near the trailing edge, and the separation point moved gradually toward the leading edge. At the same time, C_L grew significantly, so did C_D, though at a lower rate. At $\alpha = 10°$, the separated boundary layer was laminar for a rather long longitudinal distance, without reattachment (Fig. 2a), and with further increase in α remained separated (Fig. 2b). The separation bubble has not been observed, conforming to the absence of stall. The separation point is near the leading edge at $\alpha \geq 20°$.

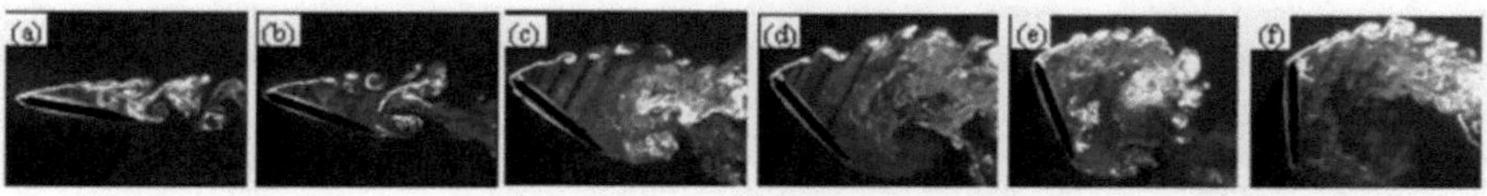

Fig. 2 Typical photographs from LIF flow visualization at (a) $\alpha = 10°$, (b) 20°, (c) 40°, (d) 50°, (e) 70°, (f) 90°. $Re_c = 5.3 \times 10^3$.

Velocity field. Figure 3 presents the contours of the mean streamwise velocity, $\overline{U}^*$, for different α ($Re_c = 5.3 \times 10^3$), where the superscript '*' stands for normalization by c and/or U_∞. The minimum $\overline{U}^*$ and the recirculation bubble size, enclosed by $\overline{U}^* = 0$, provide a measure for the strength of recirculation. The maximum velocities $\overline{U}^*_{max}$ in the leading- and trailing-edge shear layers and the minimum velocity $\overline{U}^*_{min}$ in the recirculation bubble are linked with C_D; an increase in the magnitude of $\overline{U}^*_{max}$ and/or $\overline{U}^*_{min}$ corresponds to an increased C_D. Figure 4 summarizes the dependence of $\overline{U}^*_{max}$, $\overline{U}^*_{min}$ and the maximum bubble width h^*_b on α. $\overline{U}^*_{max}$ in the leading-edge shear layer is higher than in the trailing edge. The recirculation size grows with increasing α, along with $\overline{U}^*_{max}$ and $\overline{U}^*_{min}$; their growth rate is higher at $\alpha < 45°$ than at $\alpha > 45°$. $\overline{U}^*_{max}$ at the trailing edge is lower than at the leading edge, consistent with the positive C_L. Fitting each set of data in figure 4 to a second-order polynomial curve and to a *sine* curve, respectively, using the least square technique yields empirical correlations given in the figure.

4 Conclusion

The stall of an airfoil characterized by a rapid drop in C_L occurs at about $\alpha \approx 10°$ at $Re_c = 1.05 \times 10^4$ and 5.1×10^4; the stall is however absent at $Re_c = 5.3 \times 10^3$, because the separated boundary layer remains laminar and separated for a sufficiently long downstream distance, without reattachment. C_D increases monotonically from $\alpha = 0°$ to 90°, whilst C_L grows from 0 to its maximum at $\alpha = 45°$ and then drops. With increasing α, (i) the mean vortex formation length $\overline{L}^*_f$ (not shown) and $\overline{U}^*_{min}$

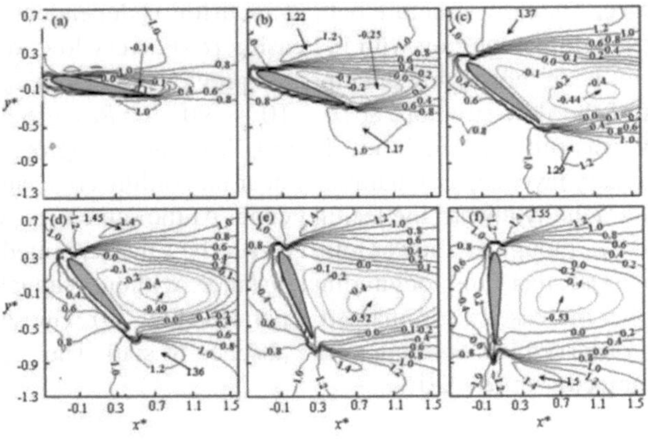

Fig. 3 Contours of the mean streamwise velocity, $\overline{U}^* = \frac{\overline{U}}{U_\infty}$(a) = 10°, (b) 20°, (c) 40°, (d) 50°, (e) 70°, (f) 90°. $Re_c = 5.3 \times 10^3$

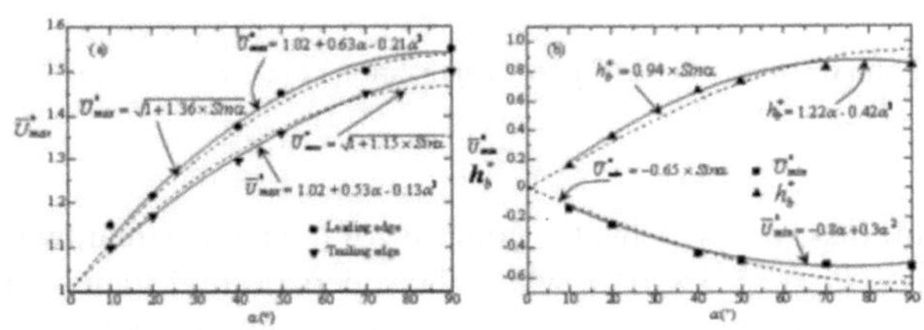

Fig. 4 Dependence on α of the maximum streamwise velocity $\overline{U}^*_{max}$ in the shear layers separated from the leading and trailing edge, the minimum streamwise velocity $\overline{U}^*_{min}$in the buble and the bubble width h^*_b. Solid and dashed lines represent the polynomial and sinusoidal fit to the measured data, respectively. In the equations, the unit of α is radian.

in the wake bubble diminish rapidly up to $\alpha = 45°$; (ii) the wake width d'^* , $\overline{U}^*_{max}$ minimum mean vorticity (not shown), wake bubble width h^*_b and flow velocity $\overline{U}^*_s$ (not shown) at separation all increase.

Acknowledgement. The work described in this paper was supported by a grant from The Hong Kong Polytechnic University (Project No. G-YD83). YZ wishes to acknowledge support given to him from Research Grants Council of Hong Kong Special Administrative Region through grant PolyU5334/06E.

References

1. Lin, J.C.M., Pauley, L.L.: Low-Reynolds-number separation on an airfoil. AIAA J. 34, 1570–1577 (1996)
2. Mueller, T.J., DeLaurier, J.D.: Aerodynamics of small vehicles. Annu. Rev. Fluid Mech. 35, 89–111 (2003)

Turbulence Energetics in Stably Stratified Atmospheric Flows

S.S. Zilitinkevich, T. Elperin, N. Kleeorin, V. L'vov, and I. Rogachevskii

Abstract. We propose a new turbulence closure model based on the budget equations for the key second moments: turbulent kinetic energy (TKE), turbulent potential energy (TPE) and vertical turbulent fluxes of momentum and buoyancy (proportional to potential temperature). Besides the concept of the turbulent total energy (TTE = TKE + TPE), we take into account the non-gradient correction to the traditional buoyancy flux formulation. The proposed model permits the existence of turbulence at any gradient Richardson number, Ri. For the stationary, homogeneous regime the turbulence closure model yields universal dependencies of the flux Richardson number, turbulent Prandtl number, anisotropy of turbulence, and normalized vertical fluxes of momentum and heat on the gradient Richardson number, Ri. We also take into account an additional vertical flux of momentum and additional productions of turbulent kinetic energy, turbulent potential energy and turbulent flux of potential temperature due to large-scale internal gravity waves (IGW). Accounting for the internal gravity waves, the Ri-dependencies of the flux Richardson number, turbulent Prandtl number, anisotropy of turbulence, vertical fluxes of momentum and heat lose their universality. In particular, with increasing wave energy, the maximal value of the flux Richardson number (attained at very large Ri) decreases. In contrast to the mean wind shear which generates only the horizontal TKE, IGW generate both horizontal and vertical TKE, and thus lead to a more isotropic turbulence at very large Ri. IGW also increase the share of TPE in the turbulent total energy. A well-known effect of IGW is their direct contribution to the vertical transport of momentum. Depending on the direction (downward or upward),

---

S.S. Zilitinkevich
Division of Atmospheric Sciences, University of Helsinki and Finnish Meteorological Institute, Helsinki, Finland

T. Elperin · N. Kleeorin · I. Rogachevskii
Department of Mechanical Engineering, Ben-Gurion University of the Negev, Beer-Sheva, Israel

V. L'vov
Department of Chemical Physics, Weizmann Institute of Science, Israel

it either strengthens of weakens the total vertical flux of momentum. Predictions from the proposed model are consistent with available data from atmospheric and laboratory experiments, direct numerical simulations (DNS) and large-eddy simulations (LES).

1 Energy and Flux Budget Turbulence Closure Model

Traditionally, turbulence energetics is characterized by turbulent kinetic energy (TKE), E_K, and modelled using solely the TKE budget equation. In stable stratification, the turbulent kinetic energy is generated by the velocity shear and expended through viscous dissipation and work against buoyancy forces. The effect of stratification is characterized by the ratio of the buoyancy gradient to squared shear, called gradient Richardson number, $\mathrm{Ri} = N^2/S^2$, where N is the Brunt-Väisälä frequency and S is mean shear. It is widely believed in meteorological community that at Ri exceeding a critical value, Ri_c, local shear cannot maintain turbulence, and the flow becomes laminar. However, experimental, observational, LES and DNS data (see, e.g., [1, 2]) demonstrate general existence of turbulence at very large Ri, up to $\mathrm{Ri} > 10^2$ exceeding its commonly accepted critical values by more than two orders of magnitude.

We revise the concept of the critical Richardson number by extending the energy analysis to turbulent potential energy (TPE) and turbulent total energy (TTE = TKE + TPE), consider their budget equations, and conclude that the turbulent total energy is conserved and maintained by shear for any stratification [1, 2]. This analysis does not support the existence of the critical Richardson number, in contrast to the hydrodynamic-instability threshold, Ri_c-instability, whose typical values vary from 0.25 to 1. We have demonstrated in [1, 2] that this interval, $0.25 < \mathrm{Ri} < 1$, separates two different turbulent regimes: strong mixing and weak mixing rather than the turbulent and the laminar regimes, as the classical concept states. The suggested theory explains persistent occurrence of turbulence in the free atmosphere and deep ocean at $\mathrm{Ri} \gg 1$, clarifies the principal difference between turbulent boundary layers and free flows, and provides a basis for improving operational turbulence closure models.

In the budget equation for the vertical flux of potential temperature we take into account a crucially important mechanism: generation of the counter-gradient flux due to the buoyancy effect of potential-temperature fluctuations, compensated, but only partially, by the correlation between the potential-temperature and the pressure-gradient fluctuations. We show that this is precisely the mechanism responsible for the principle difference between the heat and the momentum transfer [1]. We derive a simple algebraic version of an energetically consistent closure model for the steady-state, homogeneous regime, and verify it against available observational, experimental, LES and DNS data.

2 Effect of Large-Scale Internal Gravity Waves

Energy and flux budget turbulence closure model developed in [1] does not taken into account the vertical transports by large-scale internal gravity waves (IGW) which can be excited by strong wind shears, flows over topography and convection. Internal gravity waves transport the energy and momentum, contribute to the turbulence production and eventually enhance mixing. Our analysis focuses on an additional vertical flux of momentum and additional production of turbulent kinetic energy, turbulent potential energy and turbulent flux of potential temperature due to large-scale internal gravity waves in the context of the energy and flux budget turbulence closure model for the stably stratified atmospheric flows [1]. We consider separately the large-scale and small-scale internal gravity waves. The small-scale internal gravity waves are considered as a part of turbulent flow, while the large-scale IGW are treated as a mean flow with random phases.

We generalize the energy and flux budget turbulence closure model derived in [1] considering the budget equations for the turbulent kinetic and potential energies and the vertical turbulent fluxes of momentum and buoyancy taking into account for the effects of large-scale internal gravity waves. This turbulent closure model allows us to determine the following non-dimensional functions: the turbulent Prandtl number, Pr_T, the vertical anisotropy parameter, $A_z = E_z/E_K$ (where E_z is the vertical TKE), the dimensionless turbulent fluxes of momentum and heat, and the ratio of the turbulent potential energy to the total turbulent energy.

These non-dimensional functions become dependent on the three governing parameters: the gradient Richardson number Ri, the wave-energy parameter that is proportional to the normalized kinetic energy of large-scale internal gravity waves, and the dimensionless lapse rate $Q = N(z)/N(z_0)$ that determines the propagation of large-scale internal gravity waves in the stably stratified atmospheric boundary layer. When the sources of IGW are located at the lower boundary ($z_0 = 0$), large-scale internal gravity waves are generated by the flow interaction with mountains or hills. In the energy and flux budget turbulence closure model developed in [1] whereby the effects of large-scale internal gravity waves have not been taken account, these non-dimensional functions depend only on Ri.

Consider the main effects of large-scale internal gravity waves on stably stratified turbulence. IGW result in two contributions to the production of the turbulent flux of potential temperature. The indirect effect of the production of the turbulent flux of potential temperature is due to the production of the potential temperature fluctuations by large-scale internal gravity waves. The production of the potential temperature fluctuations changes the turbulent flux of potential temperature. The direct effect of the production of the turbulent flux of potential temperature is caused by the IGW source term in the equation for this flux. As the result of the competition between the direct and indirect effects of large-scale internal gravity waves on the turbulent flux of potential temperature, the maximum value of the flux Richardson number, $\mathrm{Ri}_f = \mathrm{Ri}/\mathrm{Pr}_T$, decreases with increase of the energy of IGW. The turbulent Prandtl number Pr_T increases when the wave power of IGW increases. Large-scale internal gravity waves decrease the dimensionless turbulent fluxes of momentum and heat, because IGW increase the turbulent kinetic and potential energies.

Large-scale internal gravity waves increase the vertical anisotropy parameter A_z. The reason is that the large-scale waves produce horizontal and vertical components of TKE, whereas large-scale shear produces only horizontal component of TKE. Therefore, the parameter A_z in the presence of IGW is larger than that without waves. Large-scale internal gravity waves increase the ratio of the turbulent potential energy to the turbulent total energy, i.e., they increase the share of turbulent potential energy in the turbulent total energy.

3 Conclusions

In the present study we investigate the effect of large-scale internal gravity waves on the stably stratified sheared turbulence that contribute to the production of turbulent kinetic and potential energies and generation of turbulent flux of potential temperature. Since large-scale internal gravity waves produce the potential temperature fluctuations, there is an indirect effect of large-scale internal gravity waves on the turbulent flux of potential temperature. The direct and indirect productions of the turbulent flux of potential temperature result in decrease of the maximum value of the flux Richardson number with increase of the energy of large-scale internal gravity waves. The large-scale velocity shear generates only horizontal components of TKE, while the large-scale internal gravity waves generate both, horizontal and vertical components of TKE. This is the reason why large-scale internal gravity waves increase isotropy of turbulence even for large gradient Richardson numbers, while the large-scale shear increases anisotropy of turbulence.

Acknowledgement. This work has been supported by the EC FP7 projects ERC PBL-PMES and MEGAPOLI, and the Israel Science Foundation governed by the Israel Academy of Sciences.

References

1. Zilitinkevich, S., Elperin, T., Kleeorin, N., Rogachevskii, I.: Boundary-Layer Meteorol 125, 167–192 (2007)
2. Zilitinkevich, S., Elperin, T., Kleeorin, N., Rogachevskii, I., Esau, I., Mauritsen, T., Miles, M.: Quart. J. Royal Meteorol. Soc. 134, 793–799 (2008)
3. Mauritsen, T., Svensson, G.: J. Atmos Sci. 64, 645–655 (2007)
4. Canuto, V.M., Cheng, Y., Howard, A.M., Esau, I.N.: J. Atmos Sci. 65, 2437–2447 (2008)

Measurements of the Flow Upstream a Rotating Wind Turbine Model

Davide Medici, Jan-Åke Dahlberg, and P. Henrik Alfredsson

Abstract. Wind turbines are commonly placed in complex terrain meaning that they may work under conditions of e.g. high turbulence level. This and other effects may give rise to that the turbine works under yawed inflow with respect to the turbine disc plane during a large fraction of its operational time. In this paper we show how the inflow towards the rotor disc is affected by yaw through PIV measurements.

1 Introduction

Wind energy is one of the fastest growing renewable energy sources and it becomes more and more important to accurately predict the power output from single turbines as well as from wind turbine parks. Most wind turbine studies made in laboratories or through computational analysis assume ideal conditions, as for instance the flow is approaching the turbine normal to the disc plane, the turbulence level is low, there is no mean wind gradient, the ground effect can be neglected etc. However, in a real installation this is not the case and all the above complexities arise to a certain extent.

One specific issue is that turbines may operate under yawed conditions. Such an inclination may stem from four different sources, a) the wind direction is not known accurately which will result in an inaccurate positioning of the turbine relative to the wind (yaw misalignment), b) active wake control (AWC) is used to increase the output from downstream turbines by yawing an upstream turbine (see ref. [1]),

Davide Medici
Garrad Hassan Italia, I–40026 Imola, Italy
e-mail: `davide.medici@garradhassan.com`

Jan-Åke Dahlberg
Vattenfall Power Consultant, S-162 16 Stockholm
e-mail: `Jan-ake.Dahlberg@vattenfall.com`

P. Henrik Alfredsson
Linné FLOW Centre, KTH Mechanics, Royal Institute of Technology, S–100 44 Stockholm, Sweden
e-mail: `hal@mech.kth.se`

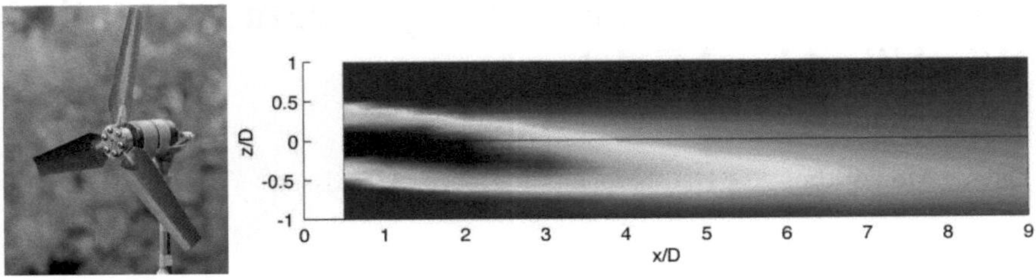

Fig. 1 Turbine model used in the experiments and "visualization" of the wake region at 20° yaw from hot wire measurements. The side force on the turbine is directed upwards in the photograph, making the wake move downwards. Colours indicate streamwise velocity, blue is low, red is high.

c) the turbine is located on a ridge so the flow is approaching from below, d) the turbulence level is high and the scales large so the turbine is working in a slowly, but turbulent, varying wind direction.

A wind turbine operating under yawed conditions will be affected by a side force in addition to the thrust force which is directed in the wind direction. The wind will be affected by the same forces but in opposite direction, according to Newton's second law. This will make the wake deflect as is illustrated in the "flow visualization" in figure 1. Also the flow upstream the turbine will be affected and in this paper we will describe experimental results which give information of that flow, both under normal inflow and at yaw misalignment.

2 Experimental Conditions

The experiments were performed in the MTL wind tunnel with an energy extracting three bladed model with a diameter of 0.18 m (for more information see ref. [3]). The PIV images were produced by a horizontal laser sheet at the model hub-height and the camera was mounted below the test section floor. The image pairs give the streamwise and radial velocity components. The seeding was introduced downstream of the test section and recirculated around the closed circuit wind tunnel giving an even seeding in the test section. Averaging was done using approximately 1600 stochastically independent images. The flow upstream of the turbine was studied for three different yaw angles (γ) according to table 1.

Table 1 Experimental conditions for the three cases studied. λ is the tip speed ratio C_P the power coefficient, and C_T the thrust coefficient (only measured at zero yaw).

Yaw angle	U_∞[m/s]	λ	C_P	C_T
0°	7.8	2.803	0.308	0.834
10°	7.9	2.802	0.295	n.a.
20°	7.8	2.613	0.242	n.a.

3 Results

Figure 2 shows the upstream conditions at zero yaw. Due to the expected symmetry around z =0 only one side of the turbine plane was measured. In the figure the upstream blade passage position is outlined as a thick line. The left hand figure gives the velocity component normal to the turbine disc whereas the right hand figure shows the radial velocity component. For the normal component the figure shows the expected behaviour, a gradual decrease of the the velocity towards the disc, where the decrease is stretching furthest out in the hub region. By integrating the velocity defect near the disc across the area swept by the blades it is possible to obtain a value of the axial interference factor in Betz' theory. This value was estimated to a =0.305 and the thrust coefficient is related to the axial interference factor as

$$C_T = 4a(1-a)$$

With a =0.305 we obtain $C_T = 0.848$ which is close to the directly measured value of C_T=0.834.

The radial component on the other hand increases in the radial direction and reaches values of the order of 25% at the tip of the turbine.

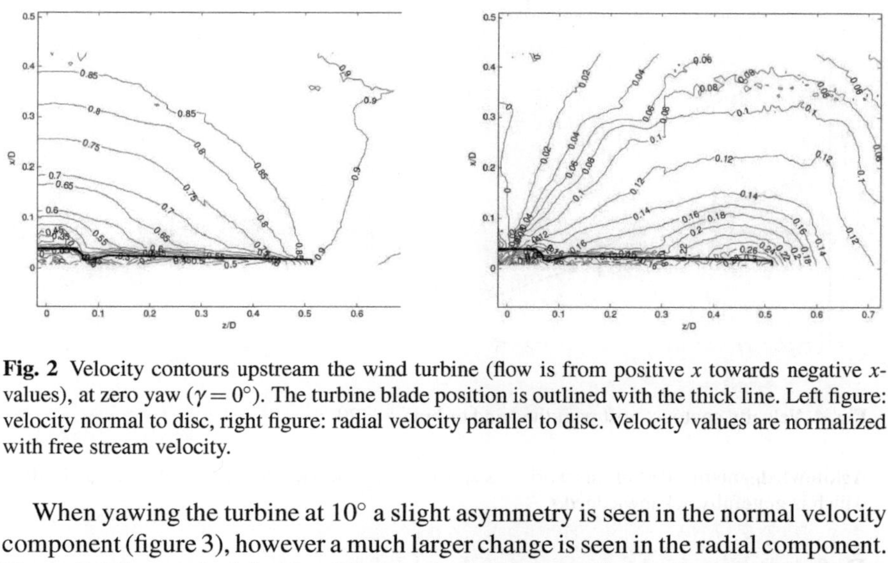

Fig. 2 Velocity contours upstream the wind turbine (flow is from positive x towards negative x-values), at zero yaw ($\gamma = 0°$). The turbine blade position is outlined with the thick line. Left figure: velocity normal to disc, right figure: radial velocity parallel to disc. Velocity values are normalized with free stream velocity.

When yawing the turbine at 10° a slight asymmetry is seen in the normal velocity component (figure 3), however a much larger change is seen in the radial component. The radial flow at the left side of the disc is almost zero whereas the radial flow at the right hand side has increased to at least 0.42. which is almost double that of the zero yaw case. In the final case with $\gamma = 20°$ (figure 4) the flow along the full diameter of the turbine is towards the right side and the radial velocity reaches values larger than 50% of the free stream velocity. Both figures 3 and 4 are composed of three different image planes in order to be able to cover the full flow field in front of the

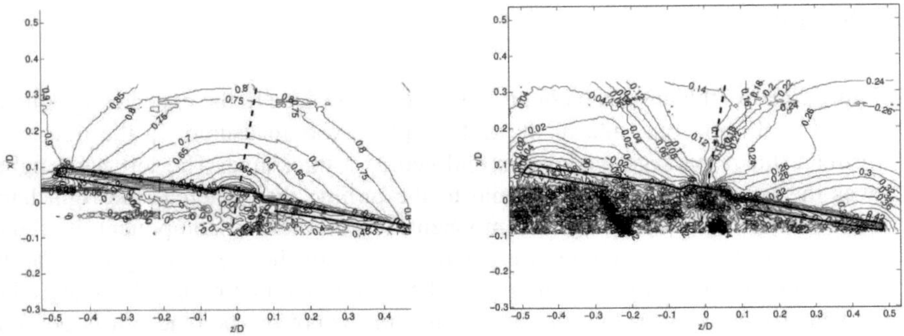

Fig. 3 Velocity contours upstream the wind turbine at $\gamma = 10°$

turbine. Note that the contour lines downstream of the turbine plane should not be considered.

As a final comment the present experiments show that wind turbine performance under yawed conditions will be affected by the change in the inflow towards the turbine. Such changes may occur also under steady wind conditions, through the presence of large scale turbulence structures (see e.g. ref. [2]).

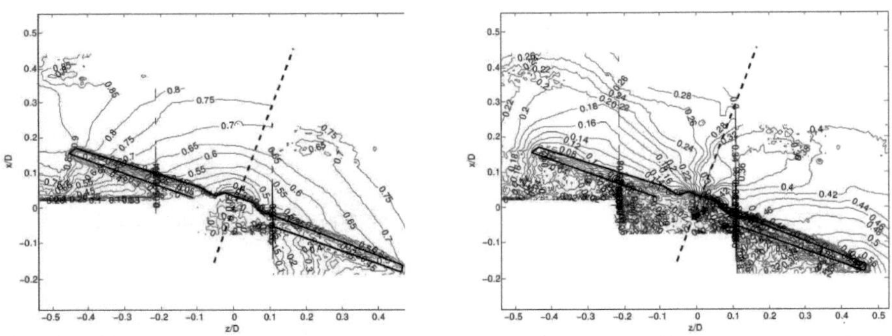

Fig. 4 Velocity contours upstream the wind turbine $\gamma = 20°$

Acknowledgement. Part of this work was sponsored by the Swedish Energy Agency and Elforsk, which is gratefully acknowledged.

References

1. Dahlberg, J.Å., Medici, D.: Potential improvement of wind turbine array efficiency by active wake control (AWC). In: Proc. European Wind Energy Conf., Madrid (2003)
2. Hutchins, N., Marusic, I.: Evidence of very long meandering features in the logarithmic region of turbulent boundary layers. J. Fluid Mech. 579, 1–28 (2007)
3. Medici, D., Alfredsson, P.H.: Measurements behind model wind turbines: further evidence of wake meandering. Wind Energy 11, 211–217 (2008)

Is the Meandering of a Wind Turbine Wake Due to Atmospheric Length Scales?

Guillaume Espana, Sandrine Aubrun, and Philippe Devinant

Abstract. This study is about the meandering phenomenon of a wind turbine wake. The atmospheric boundary layer is reproduced in a wind tunnel as well as wind turbines using porous discs. Hot wire anemometry is used to carry out time investigations through the study of space-time correlations. Particle Image Velocimetry is also used to observe the spatial development of the wake and especially the horizontal oscillations characterising the meandering.

1 Introduction

At the present time, in numerical and experimental approaches, the wake is often modelled as a steady phenomenon: the main studied items are the wake deficit (power loss) and the production of turbulence in the wake (fatigue of a downstream wind turbine). However, field observations showed an unsteady behaviour of the wake called "meandering". The whole wake is seen to oscillate randomly. The reasons are not yet well known: is the wake meandering due to the intrinsic instabilities of the wake (as for a bluff body)? Is it due to the tip vortices? To the characteristic length scales of the turbulence contained in the Atmospheric Boundary Layer ABL (as observed for atmospheric flows or plume dispersion)?

The following approach, based on physical modelling in wind tunnels, dissociates the different possible sources of meandering, focusing in the present study on the influence of ABL turbulence length scales.

Guillaume Espana
PRISME Institute, Orleans, France
e-mail: guillaume.espana@univ-orleans.fr

Sandrine Aubrun
PRISME Institute, Orleans, France
e-mail: sandrine.aubrun@univ-orleans.fr

Philippe Devinant
PRISME Institute, Orleans, France
e-mail: philippe.devinant@univ-orleans.fr

2 Experimental Set Up

In this study, the return test section (5m high, 5m wide and 20m long) of the PRISME Institute wind tunnel was used to reproduce the properties of an ABL at a geometric scale of 1 : 400. Many previous studies enable the choice of the proper parameters to model a moderately rough ABL (Roughness length $z_0 = 6cm$, power law coefficient $\alpha = 0.15$). At $z = 40m$ in full scale, the upstream mean velocity was $U_o \approx 3m/s$ and the upstream turbulence intensity was $Iu_0 \approx 17\%$. The wind turbines were modelled according to the "Actuator Disc" concept. Porous discs made of metallic mesh were used to reproduce the velocity deficit between the upstream and the downstream of a wind turbine. Several previous measurements have also proved the validity of this type of modelling [1]. The 2 porous discs used for this study had a diameter $D = 100mm$ ($40m$ in Full Scale), a mast height $H_{hub} = D = 100mm$ ($40m$ in F.S) and power coefficients equal to $C_p = 0.50$ and $C_p = 0.37$ respectively. A solid disc was also used for bluff-body wake comparisons. This paper only exposes the comparisons between the solid disc and the porous disc $C_p = 0.50$.

The first part of this study tried to observe the time evolution of the wakes using hot wire anemometry. More precisely, the goal was to process space-time correlations of turbulent velocities to bring to the foreground the global oscillation of the wake. 2 hot wire probes were used : the first one was placed at hub height and at a fixed transversal position in the shear layer of the wake whereas the second one was fixed to the traversing system and could move transversally (at hub height). Data was acquired simultaneously by the 2 probes for about 1 minute at $f_{acquisition} = 2kHz$ ($f_c = 1kHz$). The measurements were repeated at several locations downstream of the discs ($x/D = 3,4,6$).

The second part of the study was dedicated to a spatial analyse of the meandering. Horizontal PIV measurements were then carried out at several locations downstream of the different types of discs used. A CCD Camera ($2048x2048 pixels$) and a double pulsed Nd:Yag laser ($2x200mJ$ at $532nm$) were used. For each configuration, 400 image pairs measuring $200x200mm$ were acquired at a frequency of $7.5Hz$. As the disc diameter measured $100mm$, this window size allowed the visualisation of the wanted oscillations.

3 Analysis

According to [2], if the disc is porous with a solidity (defined as the ratio of solid to total frontal area) lower than 60%, the wake is devoid of the typical eddy detachment observed for bluff bodies. In our cases, the solidities were 35% and 45% for the disc $C_p = 0.37$ and $C_p = 0.50$ respectively. As a consequence, no eddy detachment should be observed and if wake oscillations do exist for the porous discs, another explanation must be found.

3.1 Temporal Study

According to [3], the well-known periodic vortex shedding behind a solid disc is characterised by a Strouhal number $St_D = \frac{f.D}{U_0} = 0.135$ (defined by the disc diameter D and the undisturbed incoming flow velocity U_0). Contrary to the solid disc measurements, where $0.15 < St_D < 0.17$ was found, the spectral signature downstream of the porous disc showed no obvious spectral peak, proving that no periodic phenomenon seems to exist downstream of this porous disc (not shown here).

Space-time correlations were also performed between the velocity signals registered by the 2 hot wire probes. Figure 1 shows the space-time correlations, $3D$ downstream of the solid disc (left figure) and $3D$ downstream of the porous disc (right figure). Concerning the solid disc measurements, alternating regions of positive correlations and of negative correlations are seen. However, because of the high level of ambient turbulence, it is difficult to observe more than 2 or 3 periods. Concerning the porous disc measurements, no alternation can be seen. Nevertheless, the wanted oscillations are clearly observed (but without any periodicity) thanks to the high level of negative correlations at $\tau = 0s$.

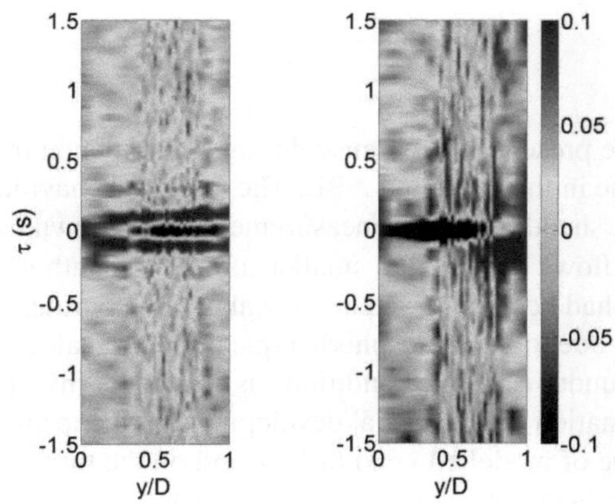

Fig. 1 Space-time correlations at $x/D = 3$, for the solid disc (left figure) and the porous disc (right figure)

The same type of measurements had previously been carried out by Medici and Alfredsson ([4]) behind a rotating wind turbine model in a wind tunnel. Their correlations behaviour was the same as this of the solid disc (periodic oscillations at $St_D = 0.17$), so that a phenomenon of vortex shedding existed on their wind turbine model.

3.2 Spatial study

The PIV system was used to collect steady information but also, and especially, to statistically study the unsteadiness of the wakes. In this way, a special image processing was applied to the 400 images pairs for each configuration. It consisted of finding the contours of the wake on images of the instantaneous velocity field.

Figure 2 is an example of image of instantaneous velocity and figure 3 shows the result of this treatment.

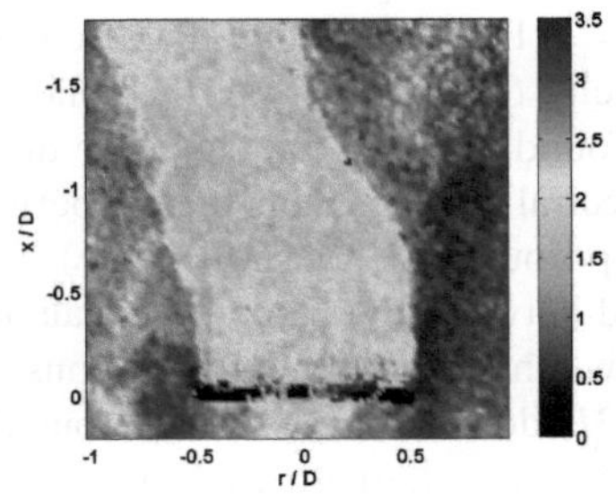

Fig. 2 Image of an instantaneous velocity field

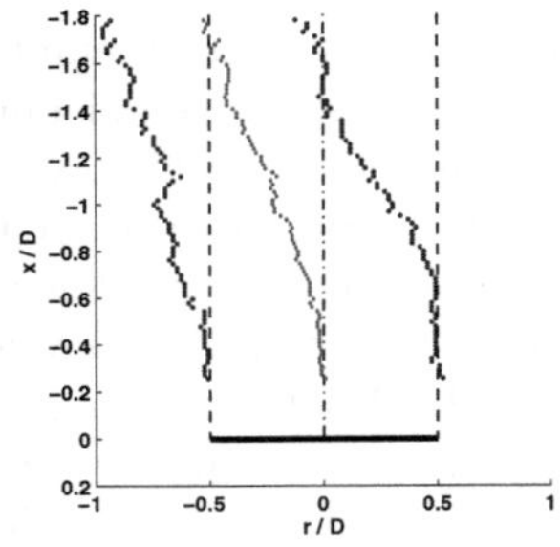

Fig. 3 Wake reconstructed after image processing

When every 400 images of a configuration are treated, a statistical approach can be done on the wake width, on its evolution with the downstream location, and on the trajectory distances from the middle of the axis of symmetry.

4 Conclusion

The temporal study underlines the presence of the meandering phenomenon in the wake of the modelled wind turbine in the modelled ABL. The random behaviour is different from the periodic vortex shedding. Other measurements (not shown here) had been carried out in isotropic flows (with much smaller turbulent length scales than in the ABL) and no flapping had been observed in the wake of the porous disc. This typical flapping may then be due to the atmospheric typical length scales.

The spatial study, in ABL and under isotropic conditions, is still under investigation but should give further information on the spatial development of the meandering. Parametric studies on the type of modelled wind turbine and on the turbulence intensity level should also give further explanations on the phenomenon.

References

1. Aubrun, S.: Modelling wind turbine wakes with a porosity concept. In: Peinke, J., Schaumann, P., Barth, S. (eds.) Proceedings of the Euromech Colloquium 464b, Wind Energy, pp. 265–270 (2005)
2. Cannon, S., Champagne, F., Glezer, A.: Observations of large-scale structures in wakes behind axisymmetric bodies. Experiments in Fluids 14, 447–450 (1993)
3. Fuchs, H.V., Mercker, E., Michel, U.: Large-scale coherent structures in the wake of axisymmetric bodies. Journal of Fluid Mechanics 93, 185–207 (1979)
4. Medici, D., Alfredsson, P.H.: Measurements on a wind turbine wake: 3D effects and bluff body vortex shedding. The science of making torque from wind. In: Special Topic Conference, Delft, April 19-21, 2004, pp. 155–165 (2005)

Impact of Atmospheric Turbulence on the Power Output of Wind Turbines

Julia Gottschall and Joachim Peinke

Abstract. In this contribution, we focus on the relevance of effects of atmospheric turbulence on the power output of a wind energy converter system. In particular, we introduce and critically discuss a dynamical approach for an appropriate mapping of the power conversion process, which provides a basis for modelling the transfer of turbulent structures from the wind to the corresponding output of electrical power.

Introduction

The objective of this work is to discuss the relevance of turbulence for the power performance of a wind turbine or wind energy converter system (WECS). The term power performance covers the general relation between a representative wind velocity acting on the turbine and the corresponding power output. A power chracteristic that quantifies this relation can be understood as a summarizing characteristic for the entire WECS displaying the effective control behaviour of the considered complex system. Common methods for the estimation of power characteristics are based on the evaluation of averaged values of wind speed and power output, typical averaging periods are thereby 1 or 10 min (cf. e.g. [1]). Such an approach does not allow to study short-time dynamics. At the same time, it is known that a respective power characteristic is sytematically effected by local turbulence effects. Likewise, non-Gaussian small-scale statistics have been found to be relevant for the power output of a WECS that is exposed to a turbulent wind field.

In the folllowing, we study high-frequency power performance data and propose a phenomenological model for the corresponding relation between wind speed and output of electrical power and, in particular, deduce hereof the definition of a dynamical power characteristic that may be used to characterize, monitor or eventually certify the power performance of the analyzed system. In accordance with the

Julia Gottschall · Joachim Peinke
ForWind, University of Oldenburg, 26111 Oldenburg, Germany
e-mail: julia.gottschall@uni-oldenburg.de, peinke@uni-oldenburg.de

international standard IEC 61400-12-1 [1] for power performance testing, we consider as representative wind velocity the horizontal wind speed measured at the hub height of the turbine and in a distance of 2–3 times the rotor diameter with a meteorological mast.[1]

Characterization of Turbulent Structures

Following [2], atmospheric turbulence is suitably described and particularly distinguished from laboratory, i.e. homogeneous, isotropic and stationary turbulence, by considering the distribution of respective velocity increments for different time separations. For both cases, turbulent wind velocities are characterized by an intermittent shape of the probability density functions (pdfs) of their increments on small scales. But while for increasing time separations the pdfs for typical laboratory data approach more and more to a Gaussian one, this characteristic change of shape is missing for the atmospheric data. Figure 1(a) indicates this behaviour for the atmospheric wind speed increments $u_\tau(t) := u(t+\tau) - u(t)$, where $u(t)$ is the measured time series of horizontal wind speed and τ the respective time separation.

In an analogous manner, we analyzed the power increments $P_\tau(t) := P(t+\tau) - P(t)$ for a measured time series of power output $P(t)$ from a WECS. As shown in figure 1(b), the intermittent wind structures are almost directly transferred to the power output, resulting in respective small-scale fluctuations of electrical power. Note that the secondary peaks besides the maximum in the lowest pdf are due to sampling limitations of the data and not due to the conversion dynamics of the WECS. To investigate the transfer of turbulent structures in more detail, a model for the power conversion dynamics is needed.

In [1], the corresponding mapping between wind speed and power output is given by the power curve. which defines the steady state of power output for a fixed value of wind speed, and is derived on the basis of the 10-min averages of the measured data and by applying a bin averaging for a fixed wind speed binning. By definition, the standard power curve (due to IEC 61400-12-1) does not give any information about the actual short-term dynamics of power conversion.

Stochastic Relaxation Model

According to [3], we describe the dynamics of a wind turbine's power output with a stochastic relaxation model. We propose a set of univariate Langevin processes that are defined by the first-order stochastic differential equations

$$\dot{P}(t) = h_i(P,t) + \sqrt{g_i(P,t)}\ \Gamma_i(t) \tag{1}$$

[1] The data, referred to in this contribution, was measured and processed within the scope of the joint project mentioned below. The measurement was performed in a wind park situated in a rather complex terrain. The tested wind turbine is a commercial MW-class turbine with a hub height of 98 m. Wind speed was measured with an ultrasonic anemometer.

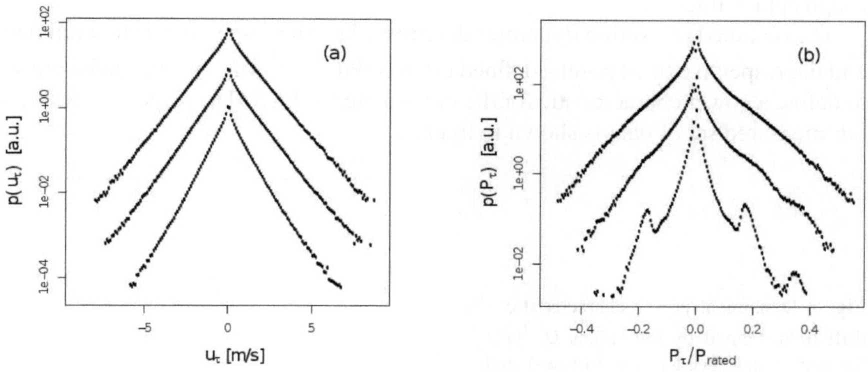

Fig. 1 Pdfs of (a) wind speed increments u_τ and (b) power output increments P_τ for different time separations τ – from the bottom up $\tau = 5$ s, $\tau = 15$ s and $\tau = 30$ s plotted half-logarithmically and shifted against each other for clarity of presentation.

for fixed values of wind speed u_i. The terms h_i and g_i are arbitrary functions of the power output P at time t, and $\Gamma_i(t)$ is a noise process. The wind speed is a parameter, as indicated by the index i, but does not act as a second variable, so that this model corresponds to a quasi-onedimensional approach. The conditioning on the wind speed is realized by applying a wind speed binning with a fixed bin width, as it is introduced in the standard IEC 61400-12-1 [1]. The deterministic part of the dynamic equation (1) describes the relaxation behaviour of the system with respect to one or more steady states that are also referred to as fixed points. The wind speed fluctuations act as a noisy driving force in this model, and induce a kind of switch-over between the single wind speed bins and the respective one-dimensional Langevin models.

Due to the assumption that (1) describes a Langevin process and $\Gamma(t)$ accordingly corresponds to a Langevin force defined as Gaussian distributed white noise with $\langle \Gamma_i(t) \rangle = 0$ and $\langle \Gamma_i(t_1)\Gamma_j(t_2) \rangle = 2\delta_{ij}\delta(t_1 - t_2)$, the functions $h_i(P,t)$ and $g_i(P,t)$ equate drift and diffusion coefficients, i.e. $h_i(P,t) \equiv D_i^{(1)}(P,t)$ and $g_i(P,t) \equiv D_i^{(2)}(P,t)$, and can be reconstructed directly from a corresponding set of data according to

$$D_i^{(n)}(P) = \frac{1}{n!} \lim_{\tau \to 0} \frac{1}{\tau} \left\langle [P(t+\tau) - P(t)]^n | P(t) = P_j, u(t) = u_i \right\rangle . \qquad (2)$$

For further details on the explicit procedure and specific technical problems with real data sets see [5].

We applied this reconstruction scheme to derive the coefficients $D_i^{(1)}(P)$ and $D_i^{(2)}(P)$ for our measured set of data, and to simulate power increments on the basis of a measured wind speed time series – cf. [4]. The results for the reconstructed increment pdfs, compared to the measured values, suggest that the proposed model describes the power conversion on specific scales quite well but is in total rather a rough approximation.

The obtained relaxation dynamics determined by the respective drift coefficients and the respective fixed points, defined by the relation $D_i^{(1)}(P_{FP}) \equiv 0$, can be utilized to define a power characteristic for the investigated WECS. The respective result for our measured set of data is shown in figure 2.

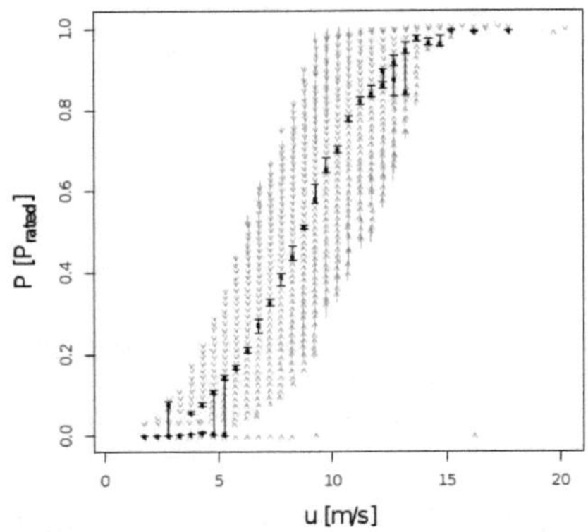

Fig. 2 Dynamical power characteristic – drift field defined by the values $D_i^{(1)}(P_j)$ for each wind speed bin u_i (arrows) and stable fixed points of relaxation dynamics (dots).

Discussion and Concluding Remarks

Self-consistency tests indicate that the investigated power conversion dynamics follow not the proposed Langevin model in a strict sense. Reducing the relevance of the reconstruction to the deterministic drift part, and furthermore to the fixed points as characteristic states, we can disregard this shortcoming. In a similar manner, we can argue for the robustness of the reconstructed fixed points against different external disturbances (cf. [5]). The drift part in the introduced relaxation model corresponds to an effective relaxation that describes not only the internal dynamical behaviour of the WECS but also incorporates e.g. delay effects that are caused by the spatial distance between wind turbine and wind measurement. The presented approach provides a flexible way to handle the impact of turbulent wind structures on the power output of a WECS. The resulting dynamical power characteristic is a promising monitoring tool to detect systematic changes in the response behaviour of the investigated system.

This contribution has been part of the joint project "Wind turbulences and their impact on the utilization of wind energy" funded by the Federal Ministry of Education and Research (BMBF) under the code number 03SF0314A.

References

1. International standard IEC 61400-12-1, 1st edn. (December 2005)
2. Böttcher, F., Barth, S., Peinke, J.: Stoch. Environ. Res. Ris. Assess. 21, 299–308 (2007)
3. Anahua, E., Barth, S., Peinke, J.: Wind Energy 11, 219–232 (2007)
4. Gottschall, J., Peinke, J.: J. Phys. Conf. Ser. 75, 012345 (2008)
5. Gottschall, J., Peinke, J.: New Journal of Physics 10, 083034 (2008)

About First Order Geometric Auto Regressive Processes for Boundary Layer Wind Speed Simulation

Thomas Laubrich and Holger Kantz

Abstract. Under certain conditions the first order geometric auto regressive process has statistical properties similar to atmospheric boundary layer wind speed. In this contribution, we investigate this similarity and analyse the extent to which this stochastic process is a suitable model for wind speed simulation.

1 Motivation

Ref. [1] introduces a method to analyse the fluctuations of atmospheric boundary layer (ABL) wind speed at time t around the wind speed V averaged over the time period Δt around time t. Recordings at a flat terrain at the Lammefjord site (55° 47' 41" N, 11° 26' 52" E) were used to show that the distribution of the fluctuation conditioned on a mean wind speed V agrees well with a symmetric Gaussian distribution with a standard deviation linear in V.

The fluctuation of the first order geometric auto regressive (AR) process

$$\frac{u_{n+1}}{U} = \left(\frac{u_n}{U}\right)^{\alpha} e^{b\xi_n} \tag{1}$$

where $n \geq 1$ has the same statistical property if $\alpha = 1 - \varepsilon$ with $\varepsilon \ll 1$ and $b \ll 1$. U denotes the unit of u_n having the dimension of speed. ξ_n stands for an independent delta-correlated normally distributed random variable with zero mean and unit variance. Since $\ln(u_n/U)$ corresponds to a stationary AR(1) process, u_n is of a log-normal distribution with median U and mean $U\omega$ where $\omega^2 = \exp[b^2/(1-\alpha^2)]$. Hence, the particular choice of the initial value u_1 is irrelevant.

Thomas Laubrich · Holger Kantz
Max Planck Institute for the Physics of Complex Systems,
Noethnitzer Str. 38, 01187 Dresden, Germany
e-mail: laubrich@pks.mpg.de

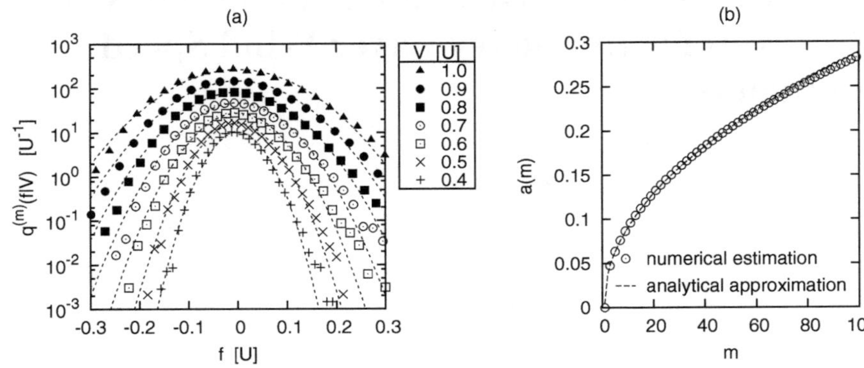

Fig. 1 **(a)** Fluctuation distributions conditioned on the $m = 11$ window mean V for a geometric AR(1) process with $\alpha = 0.99$ and $b = 0.1$. Note that the distributions are shifted against one another for clarity improvement. **(b)** The proportionality factor $a(m)$ as a function of the window size m.

2 Fluctuation Statistics

The fluctuation series $f_n^{(m)} = u_n - \bar{u}_n^{(m)}$ with $\bar{u}_n^{(m)} = \sum_{k=-\tilde{m}}^{\tilde{m}} u_{n+k}/m$ and $m = 2\tilde{m}+1$ for $\tilde{m} = 0,1,2,\ldots$ can be computed approximately for $\alpha \lesssim 1$ such that $\alpha^{\pm s} \sim 1$ for $1 \le s \le \tilde{m}$. We arrive at $\bar{u}_n^{(m)} \sim u_n\{1+\sum_{s=1}^{\tilde{m}}[\varpi_s^- + \varpi_s^+]\}/m$ where $\varpi_s^- = \prod_{l=1}^{s} e^{-b\xi_{n-l}}$ and $\varpi_s^+ = \prod_{l=0}^{s-1} e^{b\xi_{n+l}}$ so that expanding $\varpi_s^\pm$ up to the first order in b yields

$$f_n^{(m)} \sim -\bar{u}_n^{(m)} \frac{b}{m} \sum_{l=0}^{\tilde{m}-1} (\tilde{m}-l)(\xi_{n+l} - \xi_{n-l-1}). \tag{2}$$

The fluctuation conditioned on $\bar{u}_n^{(m)} = V$ corresponds to a sum of independent Gaussian variables. Consequently, it is of a symmetric normal distribution of the form

$$q^{(m)}(f|V) \sim \frac{1}{a(m)V\sqrt{2\pi}} \exp\left[-\frac{f^2}{2a^2(m)V^2}\right] \tag{3}$$

with a standard deviation being proportional to V. The proportionality factor is

$$a(m) \sim \frac{b}{2}\sqrt{\frac{m^2-1}{3m}}. \tag{4}$$

We check this result numerically by generating 10^6 points using (1) with $\alpha = 0.99$ and $b = 0.1$. The conditioned histograms of the fluctuation for $m = 11$ are shown in Fig. 1 (a). The good agreement with Gaussian shaped curves according to (3) and (4) is evident. Fig. 1 (b) displays the window size dependence of the proportionality factor $a(m)$, demonstrating good agreement between the analytical approximation and the numerical estimation through the standard deviation of the set $\{u_n/\bar{u}_n^{(m)}\}$.

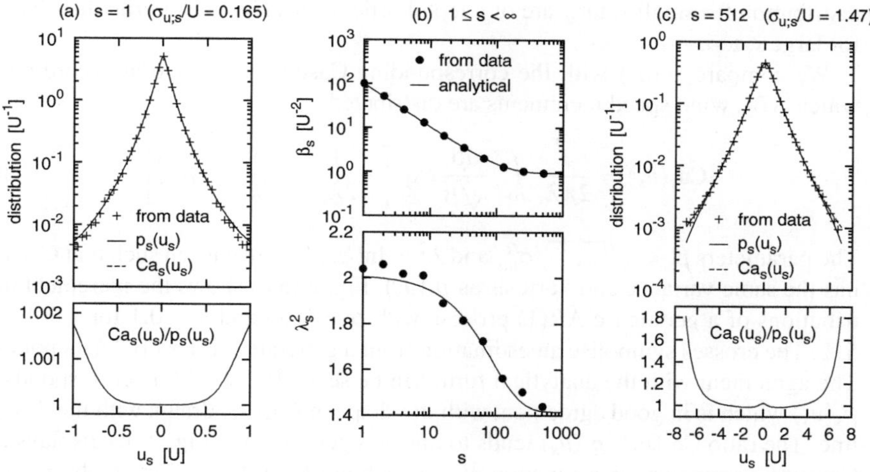

Fig. 2 **(a)** and **(c)** Increment distributions of a geometric AR(1) process with $\alpha = 0.99$ and $b = 0.1$ for $s = 1$ and $s = 512$. **(b)** The position and shape parameter of the Castaing distribution as a function of s.

3 Increment Statistics

B. Castaing et al. [2] deduced an analytical expression for the marginal increment distribution of ABL wind speed from the assumption of a log-normally distributed energy transfer rate in turbulence. Castaing's hypothesis assumes that on a short time scale the wind speed increment is of a Gaussian distribution with zero mean and variance $1/\beta$ which varies on a large time scale according to a log-normal distribution. As a result, the increment distribution of ABL wind speed is symmetric and leptokurtic for small increment lengths s. In Refs. [1, 3, 4] it is shown that the kurtosis of ABL wind speed increments decreases with increasing s and approaches three for sufficiently large s.

The distribution of the increment series $u_{s;n} = u_{n+s} - u_n$ is given by an integral of u_n over the joint distribution of u_n and u_{n+s} with $u_{n+s} = u_n + u_{s;n}$. It can be written as

$$p_s(u_s) = \frac{1}{UN(s)} \int_{|u_s/U|}^{\infty} \frac{\mathrm{d}y}{y_+ y_-} \exp\left\{ -\frac{\ln^2 y_+ + \ln^2 y_- - 2\alpha^s \ln y_+ \ln y_-}{4(1-\alpha^{2s}) \ln \omega} \right\} \tag{5}$$

with $y_\pm = (y \pm u_s/U)/2$ and the normalisation constant $N(s) = 8\pi\sqrt{1-\alpha^{2s}} \ln \omega$. Defining $z(s)$ as $1-\alpha^s$, the variance reads $\sigma^2_{u;s} = 2U^2\omega^4(1-\omega^{-2z(s)})$ and the kurtosis $k_{u;s} = 3\omega^8(1+2\omega^{-2z(s)}+3\omega^{-4z(s)})/6$ is larger than three and tends to three if ω approaches one. $k_{u;s}$ decreases with increasing increment length. However, if the increments are—depending on the value of ω—of a leptokurtic or (nearly) mesokurtic distribution for small s, they are of a leptokurtic or (nearly) mesokurtic distribution for large s, too.

We compare $p_s(u_s)$ with the corresponding Castaing distribution according to which ABL wind speed increments are distributed:

$$\mathrm{Ca}_s(u_s) = \frac{1}{2\pi\lambda_s}\int_0^\infty \frac{\mathrm{d}\beta}{\sqrt{\beta}} \exp\left[-\frac{1}{2\lambda_s^2}\ln^2\frac{\beta}{\beta_s} - \frac{1}{2}\beta u_s^2\right]. \tag{6}$$

The parameters $\beta_s = \sqrt{k_{u;s}/3}/\sigma_{u;s}^2$ and $\lambda_s^2 = \ln(k_{u;s}/3)$ are chosen such that $\mathrm{Ca}_s(u_s)$ has the same variance and kurtosis as $p_s(u_s)$. Fig. 2 (a,c) shows the increment distributions of a geometric AR(1) process with $\alpha = 0.99$ and $b = 0.1$ for $s = 1$ and 512. The crosses symbolise an estimation from a generated series (10^6 data points). The agreement with the analytical form can be seen. The solid line corresponds to $p_s(u_s)$ which is in good agreement with the shape of $\mathrm{Ca}_s(u_s)$ drawn with the dashed line. The ratio $\mathrm{Ca}_s(u_s)/p_s(u_s)$ tends to one as s gets smaller. Fig. 2 (b) displays the Castaing parameters as a function of s, which are both decreasing as observed in ABL wind speed increments.

4 Summary

We showed that if the exponent $\alpha \lesssim 1$ and the noise strength $b \ll 1$, the geometric AR(1) process (Eq. (1)) has the same fluctuation distribution conditioned on the moving window mean V as observed in ABL wind speed data, namely a symmetric Gaussian with a standard deviation being proportional to V. The degree of proportionality increases as α approaches one.

The increment distribution of the geometric AR(1) process approximately has the shape of a Castaing distribution. Hence, it has the same increment statistics as observed in ABL wind speed data. The kurtosis decreases as the increment length s becomes larger. However, this model does not provide for a clear cross-over between a leptokurtic distribution for small s and a Gaussian distribution for large s.

The present study was supported financially by Germany's Federal Ministry for Education and Research under grant number 03SF0314. It is part of the joint project "Statistical analysis and stochastic modelling of turbulent gusts in surface wind".

References

1. Laubrich, T., Ghasemi, F., Peinke, J., Kantz, H.: (2008) (to be published)
2. Castaing, B., Gagne, Y., Hopfinger, E.J.: Physica D 46, 177 (1990)
3. Boettcher, F., Renner, C., Waldl, H.P., Peinke, J.: Bound.-Layer Meteor. 108, 163 (2003)
4. Boettcher, F., Barth, S., Peinke, J.: Stoch. Environ. Res. Risk Assess 21, 299 (2007)

Unsteady Numerical Simulation of the Turbulent Flow around a Wind Turbine

Markus Rütten, Julien Penneçot, and Claus Wagner

Abstract. A study of the turbulent air flow around a wind turbine based on unsteady Reynolds-averared Navier-Stokes (URANS) simulation in order to determine the induced loads on its blades for off-design flow cases is performed. In particular the impact of applied turbulence models on the flow simulation results will be discussed.

1 Introduction

A modern wind turbine consists of three major parts. One is the tower which is typically made from tubular steel, concrete, or steel lattice. The second one is the rotor which consists usually of two or three blades and the hub. The third part is the nacelle which sits atop the tower and contains the gear box, speed shafts, generator, controller and brake. The wind blowing over the blades generates lift and, therefore, rotation. The wind speed is directly correlated with the power output. However, the drag of the earth surface, higher air boundary layers and upwind obstacles such as other hills, trees or buildings slows down the wind at ground level. Thus, in order to increase the production of electricity higher towers and turbines with longer blades are built, since wind speed increases with the seventh root of the altitude according to the wind profile power law. At higher altitudes however, there the wind is

Markus Rütten
Institute of Aerodynamics and Flow Technology, German Aerospace Center, Bunsenstrasse 10, 37073 Göttingen, Germany
e-mail: markus.ruetten@dlr.de

Julien Penneçot
CIMNE Barcelona, Edifici C-1, Campus Nord UPC, Gran Capitá, s/n, 08034 Barcelona, Spain
e-mail: julien.pennecot@dlr.de

Claus Wagner
Institute of Aerodynamics and Flow Technology, German Aerospace Center, Bunsenstrasse 10, 37073 Göttingen, Germany
e-mail: claus.wagner@dlr.de

more gusty. Such time-dependent far field flows lead to a more disturbed flow field in the vicinity of the blades. This enhances flow separation and the generation of vortices, which in turn affect the efficiency, durability, noise emissions and environmental impact of the wind turbine [4]. Hence, in order to evaluate the impact of such inhomogeneous onflow conditions numerical simulation techniques can help to investigate the sensitivity of wind turbines against off-design conditions.

2 Configuration and Investigated Flow Case

The US National Renewable Energy Laboratory (NREL) has conducted the Phase IV test campaign [7] for which a two blade wind turbine has been designed especially for the world's largest wind tunnel at NASA Ames. We chose this configuration as our validation experiment since they conducted detailed surface pressure measurements at different locations. For the simulation a constant wind speed of 10 m/s has been specified at the inflow plane of the computational domain. This is the lower limit of the wind speed for the onset of stall whereas vortices can emerge at the blades. Bearing in mind a rotational speed of 72 rpm and a blade length from of 1.257 m to 5 m the local Reynolds number varies spanwise from 5.9e+5 to 8.0e+5, due to the twist the angle of attack is nearly constant with 16 degrees. For this Reynold number range and angle of attack a laminar flow separation but a turbulent flow attachment has to be expected.

3 Unsteady Numerical Simulation of the Blade Flow

The challenge in the conducted numerical study has been that all parts of the wind turbine including the ground have been considered. Since rotating and fixed turbine parts had to be simulated simultaneously the chimera technique has been applied dividing the CFD grid into several components with some of them overlapping. Unsteady numerical simulations have been performed on a multi processors cluster system using the DLR standard second order finite volume flow solver TAU. The CFD code has been developed for aerodynamic flow simulations around aircraft and flight vehicles and it solves the compressible, unsteady Reynolds-averaged Navier Stokes equations (URANS) [2]. Since the wind speeds around the wind turbine are relatively slow, the flow is incompressible, and, therefore, preconditioning techniques have to be applied. In order to minimize the computational effort for selecting a suitable turbulence models we conducted simulations of the flow around a S809 airfoil profile on a two-dimensional mesh. The hybrid mesh consists of 22922 points, 5760 quadrilaterials and 33844 triangles, the boundary layer was resolved by 20 quadrilaterial layers. Due to y+ adaptation an averaged y+ of 1 could be garanteed. In this study results generated with the modified Spalart-Almaras one-equation turbulence model (SAM) [5] and the Wilcox k-ω two-equation model [6] are compared. For a single-time snapshot a comparison of the distribution of the turbulent kinetic energy and the turbulent viscosity, ν_t, are presented in the Figures 1 and 2. The Wilcox k-ω model is more sensitive against flow separation conditions. Therefore, it overpredicts the production of turbulent kinetic energy and enforces unsteady and fluctuating flow separation leading to a more complex flow topology than predicted by the SAM model.

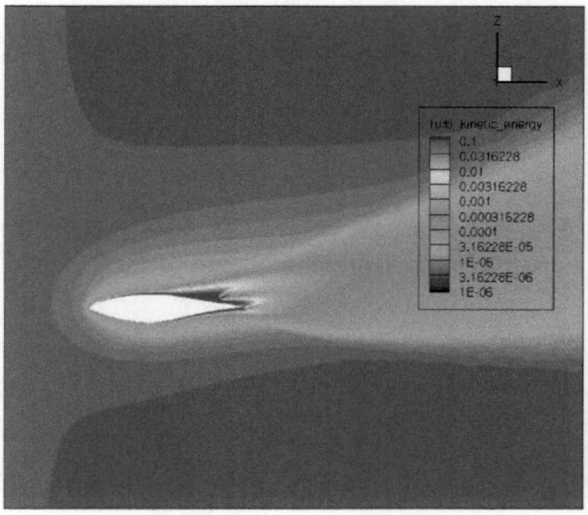

Fig. 1 Two-dimensional flow around the S809 airfoil, Wilcox k-ωturbulence model,contour plot of turbulent kinetic energy

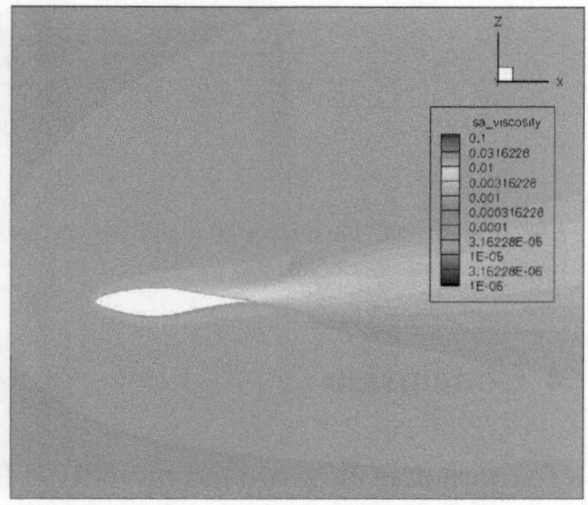

Fig. 2 Two-dimensional flow around the S809 airfoil, SAM turbulence model, contour plot of turbulent viscosity

The three-dimensional (3-D) flow characteristics of rotating blades are an essential feature of any wind turbine, since the 3-D flow over rotating blades is significantly different to the flow over wings. Rotating blades have significant radial (or spanwise) flow, and the blade speed varies linearly from root to tip. In addition, the three-dimensional wake of a rotating blade remains in close proximity to the blade for a long period of time compared to the wake of a wing [3]. For such a complicated 3-D flow a large computational CFD mesh has been generated which resolves the blade surface in detail and the comprehension of the whole machine including a large surrounding area. The mesh has 9 millions vertices, it is unstructured and non regular, consisting of 18 millions volume cells, tetrahedra and non-convex prisms for resolving the boundary layer. The cell sizes range from 0.51 mm at the turbine wall to 6.29 m in the farfield. The y+ is in average close to 1. Due to the computational effort only one-third of the revolution has been simulated so far. An impression of the complex flow field is depicted in the Figures 3 and 4. The tip vortices are following the rotation of the blades. As discussed above the flow around the blades is characterized by unsteady laminar separation and turbulent attachment of the flow at the trailing edge of the blade.

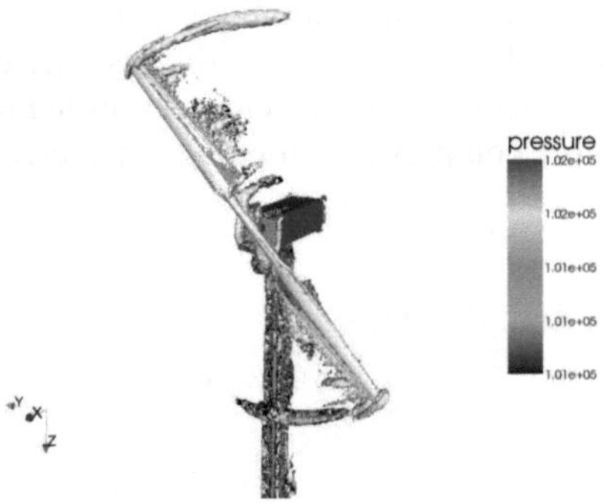

Fig. 3 Lambda2 isosurface to visualize vortices, pressure contour on surface

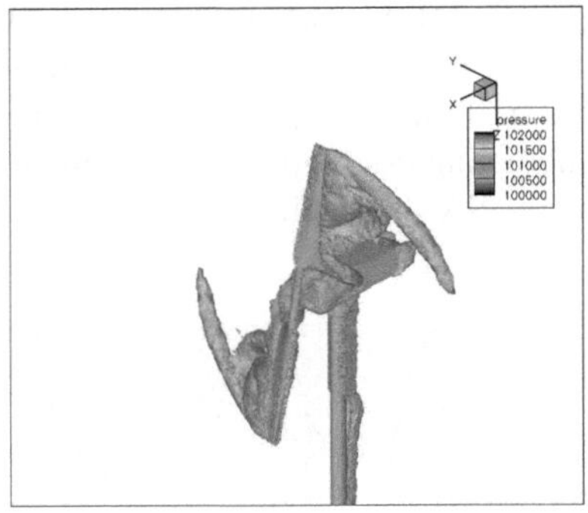

Fig. 4 Isosurface of vortcity to visualize flow structures, pressure contour on surface

4 Conclusion

The turbulent flow around the NREL Phase VI wind turbine rotor has been computed using the 3-D, unsteady, parallel, finite volume flow solver TAU. It was shown that even with one-equation turbulence models reasonable predictions are possible for the case of a wind turbine. In agreement with measurements the numerical simulations predict that the onset of stall begins at a wind speed of 10 m/s and that the flow massively separates over the entire blade in the tail section. Considerable spanwise pressure variations, in addition to the chordwise variations, were be observed. These simulations will provide an initial solution for the investigation of more complex flow cases, for example the interaction of atmospheric gusts, turbulence and atmospheric shear layers with the wind turbine.

References

1. Benjanirat, S., Sankar, L.N., Xu, G.: Evaluation of Turbulence Models for the Prediction of Wind Turbine Aerodynamics. In: AIAA 2003-517, 41st AIAA Aerospace Sciences Meeting and Exhibit, Reno, Nevada, January 6-9 (2003)
2. Gerhold, T., Evans, J.: Efficient Computation of 3D-Flows for Complex Configurations with the DLR Tau-Code Using Automatic Adaptation. Notes on Numerical Fluid Mechanics, vol. 72. Vieweg, Wiesbaden (1999)
3. Leishman, J.G.: Challenges in Modeling the Unsteady Aerodynamics of Wind Turbines. In: AIAA- 2002-0037, 21st ASME Wind Energy Symposium, Reno, NV, January 14-17 (2002)
4. Muljadi, E., Butterfield, C.P., Buhl, M.L.: Effects of Turbulence on Power Peneration for Variable-Speed Wind Turbines. In: Proceedings of the 1997 ASME Wind Energy Symposium, Reno, Nevada, January 6-9 (1997)
5. Spalart, P.R., Allmaras, S.R.: A One-Equation Turbulence Model for Aerodynamic Flows, AIAA-1992-0439 (1992)
6. Wilcox, D.C.: Turbulence Modeling for CFD, 2bd ed. (1998), DCW Industries, La Cañada, California (1998)
7. Schreck, S.: The NREL Full-Scale Wind Tunnel Experiment. Journal of Wind Energy 5, 77–84 (2002)

A New Non-gaussian Turbulent Wind Field Generator to Estimate Design-Loads of Wind-Turbines

A.P. Schaffarczyk, H. Gontier, D. Kleinhans, and R. Friedrich

Abstract. Climate change and finite fossil fuel resources make it urgent to turn into electricity generation from mostly renewable energies. One major part will play wind-energy supplied by wind-turbines of rated power up to 10 MW. For their design and development wind field models have to be used. The standard models are based on the empirical spectra, for example by von Karman or Kaimal. From investigation of measured data it is clear that gusts are underrepresented in such models. Based on some fundamental discoveries of the nature of turbulence by Friedrich [1] derived from the Navier-Stokes equation directly, we used the concept of Continuous Time Random Walks to construct three dimensional wind fields obeying non-Gaussian statistics. These wind fields were used to estimate critical fatigue loads necessary within the certification process. Calculations are carried out with an implementation of a beam-model (FLEX5) for two types of state-of-the-art wind turbines The authors considered the edgewise and flapwise blade-root bending moments as well as tilt moment at tower top due to the standard wind field models and our new non-Gaussian wind field model. Clear differences in the loads were found.

A.P. Schaffarczyk
CEwind and University of Applied Scienced Kiel, Grenzstrasse 3, D-24149 Kiel, Germany
e-mail: Alois.Schaffarczyk@FH-Kiel.de

H. Gontier
WINDnovation Engineering Solutions GmbH, D-10243 Berlin, Germany

D. Kleinhans
Mathematical Sciences, Chalmers University of Technology, SE-412 96 Göteborg, Sweden

R. Friedrich
Institut für Theoretische Physik, Westphälische Wilhelms Universität Münster,
Wilhelms-Klemm-Strasse 9, D-48291 Münster, Germany

1 Introduction

Wind Energy has been very successful worldwide in the last ten years. Reasonable estimates give numbers like 35 G Euros for future turnover per year. Therefore design standards comparable to other established branches (aerospace, civil engineering) have to be applied. Unfortunately wind turbines operate continuously within the lower atmosphere, thereby being exposed to several G fatigue cycles during their life-time. In close connection to this more and more sophisticated *engineering* wind-field models [3, 5] have been developed however, omitting the progress made in understanding turbulence within the framework of Statistical Physics [1, 8]. Further constraints are given by the very slowly advancing international standardization [2, 4].

In this paper, a comparison is undertaken between typical fatigue[1] load ranges resulting from different turbulence models, i.e. between the usual models as used in the standards and a new model designed by Friedrich and Kleinhans.

2 Model Description

2.1 Wind-Field Models

The international most important standard formulating the procedure to define design load cases (DLCs) is called IEC 61400-1. As important part of this load definition the NTM (Normal Turbulence Model) has to be used. The standard, edition 2004 proposes the Kaimal and von Karman models, in the edition of 2006, the von Karman model has been withdrawn replaced by the Mann model [4]. The Mann model is described in detail in [5] and was discussed recently by Veltcamp [6].

All these models rely at some stage to Gaussian stochastic processes, meaning that only their power spectra and coherences have to be specified.

Additionally parameters intervening in the defining equations are (coherence) lengths, also precisely defined in the standards. In our case, the hub height of the turbine) is 100m, which means integral length scales for the Kaimal model of L_u = 170.1m, L_v = 56.7m, L_w = 13.86m and L_c = 73.5m. In the edition of 2006 of the IEC guidelines, those lengths are doubled as the parameter is 42m instead of 21m for a hub height higher as 60m. The length for the von Karman model is taken equal to 73.5m.

In addition we have for the Kaimal model the following ratios between the standard deviations of the various wind components, also defined in the standards: $\sigma_2 = 0.8\sigma_1$ and $\sigma_3 = 0.5\sigma_1$. Being isotropic, the von Karman model used equal ratios: $\sigma_1 = \sigma_2 = \sigma_3$.

[1] Fatigue as an engineering term is used to failure by application of **frequent low loads** in contast to static failure given by only **one** load exceeding the ultimate strength of a material.

Those models have been integrated in Windsim7, FLEX5's wind generator.[2]

Based on work of Friedrich [1] Kleinhans [7] has formulated a CTRW model to generate wind-fields including intermittent and non-Gaussian statistical properties.

2.2 *Wind Turbine Model*

Two state-of-the-art wind turbines of 2 and 5 MW, manufactured by DeWind and REpower resp. were modeled completely and typical fatigue load spectra were investigated. However, main design data of these engines are confidential, meaning that only gross informations are available elsewhere.

3 Result: Blade-Root Bending Moment

For evaluation the resulting load data are represented by cumulated distributions of frequency occurrences.

The values for the Friedrich-Kleinhans model are higher for small wind speed and decrease to get lower for the higher wind speeds. For 5m/s, we have a mean of 107.6 % and for 18m/s, 94.7 %.

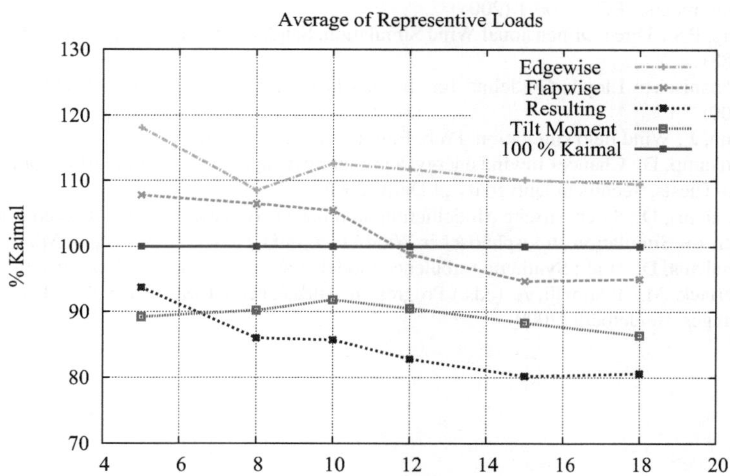

Fig. 1 Average of representative Loads

[2] FLEX5 is a semi-commercial computer code to model and simulate a complete wind turbine within a so-called simple elastic beam-theory of Gallilei, J. Bernoulli and L. Euler.

4 Summary and Conclusions

A recently developed Non-Gaussian wind-field model was applied to state-of-the-art MultiMegawatt wind turbines and compared to results from wind-field models prescribed in the international standards. Main results are:

- Results differ significantly compared to Gaussian models but with no clear trend
- Two types of wind-turbines may not be sufficient

As a conclusion to be drawn it seems to be necessary to investigate specific loads (blade-root bending moment, tower tilt-moment) in more detail. Other types of wind turbines should be included.

Acknowledgement. The authors wish to thank for many helpful discussions: Dr. D. Veltkamp, Vestas, The Neterhlands and TU Delft, Dipl.-Ing. B. Hillmer, DeWind, Lübeck, Germany and D. Steudel, REpower AG, Rendsburg, Germany. The funding by Bundesministerium für Bildung und Forschung, Germany under grant number 03SF0314E is gratefully acknowledged.

References

1. Friedrich, R.: Statistics of Lagrangian Velocities in Turbulent Flow. Phys. Rev. Lett. 90, 084501 (2003)
2. International Electrotechnical Commission: Wind turbine generator systems Part 1: Safety requirements. IEC 61400-1 (2004)
3. Veers, P.S.: Three-Dimensional Wind Simulation. Sandia National Laboratories SAND88-0152 (1988)
4. Germanischer Lloyd: Guideline for the certification of wind turbines. Hamburg, Germany (2000)
5. Mann, J.: Wind field simulation. Prob. Eng. Mech. 13(4) (1998)
6. Veldkamp, D.: Chances inwind energy, a probabilistic approach to wind turbine fatigue design. PhD-Thesis, Technical University of Delft (2006)
7. Kleinhans, D.: Stochastische Modellierung komplexer Systeme: Von den theoretischen Grundlagen zur Simulation atmosphärischer Windfelder. PhD-Thesis, University of Münster (2008)
8. Kleinhans, D., et al.: Synthetic turbulence models for wind turbine applications. In: Peinke, J., Oberlack, M., Talamelli, A. (eds.) Progress in Turbulence III. SPPHY, vol. 131, pp. 110–113. Springer, Heidelberg (2009)

Synthetic Turbulence Models for Wind Turbine Applications

D. Kleinhans, R. Friedrich, A.P. Schaffarczyk, and J. Peinke

Abstract. Wind energy converters such as wind turbines permanently work in the atmospheric boundary layer. For the modelling of the dynamics and for the optimisation of design and material of wind turbines synthetic models for atmospheric turbulence are applied already for a long time. The main purpose of these models is to provide fast and efficient methods for numerical simulation of random fields, that show some characteristic features of atmospheric turbulence. Typically they only have a partial connection to the fundamental equations of fluid dynamics. After a short overview summarizing widespread models by Veers and Mann, that are based on the simulation of random fields in the Fourier domain, advanced models for the simulation of velocity fields are discussed.

1 Introduction

For engineering applications in the field of wind energy conversion enormous amounts of wind field data is needed, that neither is available from measurements nor from direct numerical simulation of Navier-Stokes equations at reasonable costs. Therefore, turbulence models reproducing *some* properties of atmospheric boundary layer turbulence are of great interest for this purpose. A general demand on such models is the spatio-temporal simulation of the wind speed vector in the rotor plane.

D. Kleinhans
Mathematical Sciences, Chalmers University of Technology and the University of Gothenburg, SE-412 96 Göteborg, Sweden
e-mail: davidkl@chalmers.se

D. Kleinhans · R. Friedrich
Institut für Theoretische Physik, Universität Münster, D-48145 Münster, Germany

A.P. Schaffarczyk
Fachhochschule Kiel, Grenzstraße 3, D-24149 Kiel, Germany

Joachim Peinke
Institut für Physik & ForWind, Universität Oldenburg, D-26111 Oldenburg, Germany

2 Overview over Existing Models

Within the scope of this work recent approaches are discussed, that are categorized in table 1. In the following we refer to the categories introduced in this table.

Table 1 Categorization of recent models for the simulation of atmospheric wind field data[a]. For the sake of completeness categories VI to VIII have been included, although models in these categories are not yet applicable for the simulation of three dimensional velocity *fields*.

Category	Year	Author	Ref	Category	Year	Author	Ref
Ia	1988	Veers	[15]	IV	1992	Fung et al.	[3]
Ib	1998	Mann	[8]	Va	2006	Kleinhans et al.	[7]
Ic	2001	Frehlich et al.	[2]	Vb	2007	Gontier et al.	[4]
II	2004	Bierbooms	[1]	VI	2005	Schmiegel et al.	[13]
IIIa	1997	Gurley et al.	[5]	VII	2003	Kitagawa et al.	[6]
IIIb	2004	Nielsen et al.	[11]	VIII	2006	Nawroth et al.	[10]
IIIc	2007	Nielsen et al.	[12]				

[a]Please mind that this listing makes no claim to be complete.

2.1 Spectral Models (Category I)

Already for a long time, simulation algorithms are dominated by so-called *spectral simulations*, that rely on Gaussian stochastic processes in the Fourier domain [9]. Using spectral simulations, the component j of the wind velocity u_i^j at grid point i is generated by means of trigonometric series of the form

$$u_i^j(t) = \sum_{k=1}^{m} W_{i,k}^j \cos\left(\omega_k t - \phi_{i,k}^j\right) \quad , \tag{1}$$

where the phases $\phi_{i,k}^j$ are random. The second moments of the amplitudes $W_{i,k}^j$ are directly connected to the desired spectral properties of the wind field. The very first model applied by wind engineers has been described in great detail by Veers in the 80th [15]. Later on, Mann developed a more detailed algorithm, that includes spatio-temporal correlations between the three components of wind velocity [8].

As a common feature the spectral properties of resulting wind fields can be controlled very well. Spectral simulation algorithms, however, only rely on second order statistics as Veers already pointed out in an earlier publication [14]. Hence, the non-Gaussian and intermittent properties of atmospheric turbulence cannot be captured by these approaches.

2.2 Improvements of Spectral Models (Categories II-IV)

Categories II to IV denote several enhancements of spectral models. Bierbooms et al. extend the simulation algorithm such that measurements available for single grid points can be included [1]. Works of category III address the question, how deviations of the stationary distributions from a Gaussian shape can be considered. Finally, Fung formulates a quite powerful (and more involved) combination of spectral simulations and large eddy simulations (LESs), that are used for the simulation of the small and large scale, respectively [3]. The application of these so-called *Kinematic Simulations* for wind energy purposes seems to be promising.

2.3 Probabilistic CTRW-Model (Category V)

A model based on continuous time random walks (CTRWs), that has been introduced by some of the authors of this contribution, follows a different pathway [4, 7]. Simulations are not performed in the Fourier domain but in real space. Spatio-temporal correlations are included by means of drift-diffusion-processes in connection with a transform of the system's time. As a main feature, this approach naturally includes intermittent velocity increments in time as also observed in atmospheric turbulence. Significant effects on the resulting loads on wind turbines become evident from validations with respect to the spectral models Ia and Ib [4].

2.4 Modelling of Energy Cascade (Categories VI and VII)

Some other groups, that perhaps follow the possibly most physical approach, model the turbulent energy cascade. Here, especially Schmiegel et al. (see [13] and references therein) should be emphasised. Velocity time series then can be generated with help of appropriate integration kernels. Until now this ansatz, however, has not been used for simulation of three dimensional velocity fields.

The latter is also the case for an approach by Kitagawa et al., that models the energy transfer through a cascade of wavelets on different scales [6].

2.5 Multi-scale Reconstruction of Time Series (Category VIII)

Measurements on turbulent velocity fields typically show significant contribution over broad ranges of scales. A recent approach by Nawroth et al. is based on the simulation of (yet one dimensional) velocity time series *in the scale* and the consecutive transformation to real space [10]. By this means, higher order moments naturally are included.

3 Summary

Summarising, in the last decades a multiplicity of models has been introduced, that go far beyond the capabilities of old-fashioned simple *spectral simulation* codes. Some of the new approaches already are applicable for the simulation of three dimensional wind *fields*. First comparisons show distinct differences especially in some of the resulting loads on wind turbines [4]. The estimation of these loads forms a very critical point during design of wind energy converters. Therefore, the application of more sophisticated models urgently has to be advanced.

The authors kindly acknowledge financial support by the *Bundesministerium für Bildung und Forschung* (BMBF) within the project *Windturbulenzen und deren Bedeutung für die Windenergie*.

References

1. Bierbooms, W.: Simulation of stochastic wind fields which encompass measured wind speed series – enabling time domain comparison of simulated and measured wind turbine loads. In: European Wind Energy Conference Proceedings (2004)
2. Frehlich, R., et al.: Simulation of three-dimensional turbulent velocity fields. Journal of Applied Meteorology 40, 246–258 (2001)
3. Fung, J.C.H., et al.: Kinematic simulation of homogeneous turbulence by unsteady random fourier modes. Journal of Fluid Mechanics 236, 281–318 (1992)
4. Gontier, H., et al.: A comparison of fatigue loads of wind turbine resulting from a non-Gaussian turbulence model vs. standard ones. Journal of Physics: Conference Series 75, 012070 (2007)
5. Gurley, K.R., et al.: Analysis and simulation tools for wind engineering. Probabilistic Engineering Mechanics 12, 9–31 (1997)
6. Kitagawa, T., Nomura, T.: A wavelet based method to generate artificial wind fluctuation data. Journal of Wind Engineering and Industrial Aerodynamics 9, 943–964 (2003)
7. Kleinhans, D., et al.: Simulation of intermittent wind fields: A new approach. In: Proceedings of DEWEK 2006 (2006)
8. Mann, J.: Wind field simulation. Probab. Eng. Mech. 13(4), 269–282 (1998)
9. Monin, A., Yaglom, A.: Statistical fluid mechanic: Mechanics of turbulence, vol. 2. MIT Press, Cambridge (1975)
10. Nawroth, A., Peinke, J.: Multiscale reconstruction of time series (2006)
11. Nielsen, M., et al.: Wind simulation for extreme and fatigue loads. Technical report, Risø National Laboratory, Roskilde, Denmark (2004)
12. Nielsen, M., et al.: Simulation of inhomogenous, non-stationary and non-Gaussian turbulent fields. Journal of Physics: Conference Series 75, 012060 (2007)
13. Schmiegel, J., et al.: Stochastic energy-cascade model for $(1+1)$-dimensional fully developed turbulence. Physics Letters A 320, 247–253 (2005)
14. Veers, P.S.: Modeling stochastic wind loads on vertical axis wind turbines. Technical report, Sandia National Labs., Albuquerque, USA (1984)
15. Veers, P.S.: Three dimensional wind field simulation. Technical report, Sandia National Labs., Albuquerque, USA (1988)

Numerical Simulation of the Flow around a Tall Finite Cylinder Using LES and PANS

Siniša Krajnović and Branislav Basara

Abstract. Two unsteady numerical techniques, LES and PANS, with difference computer requirements, were used for prediction of the flow around a tall finite cylinder. The well resolved LES was found to predict the flow in agreement with previous experimental observations, while PANS was found to suffer from the combination of $k-\varepsilon$ model in conjuction with wall function close to the wall of the cylinder and too coarse resolution.

1 Description of the Set-Up and Numerical Methods

The finite cylinder studied in this paper is that used in the experiments of Park and Lee [5]. The cylinder was mounted vertically on a flat plate. A free-stream inlet velocity of $U_o = 10$ m/s and a diameter, D, equal to 0.03 m give a Reynolds number of approximately 2×10^4. A test section of 24D×20D×28D ($W \times H \times L$) was used in simulations (Fig. 1a). The inlet and the outlet in the numerical wind tunnel are placed 8D and 19D from the cylinder, respectively. The velocity profile at the inlet in the simulations was constant in time and produced a boundary layer thickness of 2 m at the location of the cylinder (x=0), similar to the experimental flow condition. No-slip boundary conditions are used at the surface of the cylinder in the LES, and the instantaneous wall functions based on the log-law were applied on the channel floor, the ceiling and the cylinder wall in the PANS. The convective and the homogeneous Neumann boundary conditions were used in LES and PANS, respectively, at the outlet (Fig. 1a). The lateral surfaces were treated as slip surfaces using symmetry conditions. LES simulations used the Smagorinsky sub-grid scale model with the van Driest damping function and the Smagorinsky constant of $C_S = 0.1$. The Partially-Averaged Navier-Stokes (PANS) approach recently proposed by Girimaji

Siniša Krajnović
Chalmers, Dept of Applied Mechanics, Göteborg, Sweden
e-mail: sinisa@chalmers.se

Branislav Basara
AVL List GmbH, Advanced Simulation Technologies, Graz, Austria

et al. [2] was a second method applied to the cylinder flow and is presented in this paper. PANS changes seamlessly from RANS to the direct numerical simulation (DNS). The PANS model used in the present work is based on the $k-\varepsilon$ turbulence model, where the unresolved kinetic energy and dissipation equations are systematically derived from the $k-\varepsilon$ turbulence model. The parameter that determines the unresolved-to-total kinetic energy ratio, i.e. $f_k = k_u/k$, is defined on the basis of the grid spacing as $f_k = 1/\sqrt{C_\mu}(\Delta/\Lambda)^{2/3}$, where Δ is the grid cell dimension and Λ is the Taylor scale of turbulence. The ratio of the corresponding dissipations, f_ε, was taken to be 1. Parameter f_k in the present work was implemented in the computational procedure as a dynamic parameter, changing at each point and every time step, and is then used as a fixed value for the same location during the next time step. The computational grids contained approximately 21×10^6 and 3.8×10^6 computational cells in the LES and the PANS, respectively.

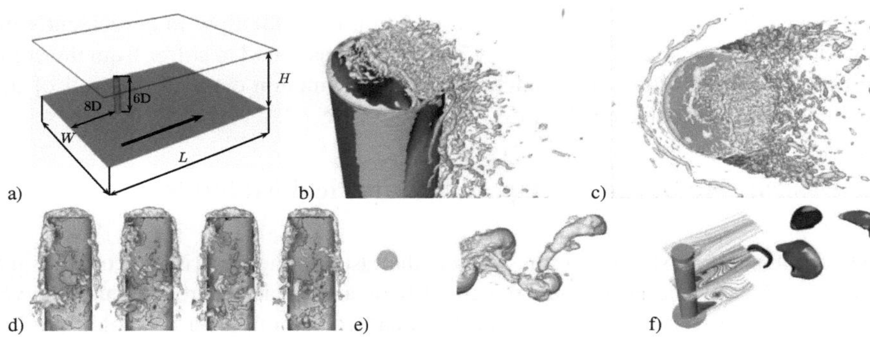

Fig. 1 a) Computational domain. b) - c) An iso-surface of the second invariant of the velocity gradient, Q=130 000, from the LES. d) - e) Iso-surface of static pressure p=-0.7 from the LES. (Time difference between two pictures in d) is $\Delta t_* = \Delta t U_o/D = 4.6$). f) Instantaneous flow from PANS. Streamlines at three different planes, $z/L = 0.05$, $z/L = 0.5$ and $z/L = 0.94$. Iso-surface of pressure $p=-2$, colored with vorticity component ω_z

2 Results

Before presenting the results of the simulations, it is valuable to say something about the flow studied here. At a Reynolds number of $Re = 2\times10^4$, the flow around the middle part of the cylinder is in so called subcritical state, where the transition occurs in the shear layer. This flow belongs to the second phase of the transition along the free-shear layer characterized by the formation of transition eddies ($1000 < Re < 40000$). Thus, the simulation is supposed to resolve the laminar boundary layer on the front side of the cylinder, the formation of the transition eddies in the shear layer and the large-scale coherent structure dynamics in the wake. Further, the junction flow at the position at which the cylinder is mounted on the floor will produce a thin (due to the thin boundary layer) horseshoe vortex, and the sharp-edged free end of the cylinder will result in a massively separated flow. Such a variety in flow regimes and the wide spectrum of flow scales makes simulations of this flow extremely challenging.

The roll-up of the transition waves into the discrete eddies along the shear layer before they became turbulent is predicted in the LES, as shown in Fig. 1b. After the separation over the free end of the cylinder, the formation of fully turbulent Kelvin-Helmholtz-like vortices is observed in the LES (Fig. 1b). As these vortices move downstream, they form hairpin vortices and, once they reach the rear of the free end surface, they decline and interact with the vortices coming from the transition in the shear layer along the cylinder. This leads to a downwash process that can be followed in Fig. 1d. Here we can follow coherent structures formed from the interaction between the flow structures coming from above the free end and the shed of the vortices from the lateral sides of the cylinder. The horseshoe vortex around the junction of the cylinder and the flat ground plate is well represented in the LES (Fig. 1c), while the PANS fails to represent this structure. The PANS simulation does not predict the above mentioned fine scale structures near the cylinder observed in the LES. Although the vortex shedding process in the near-wake is predicted in the PANS, only the large flow structures are predicted (Fig. 1f). A similar difference between simulations is observed in the far wake (compare Figs. 1e and f). While the LES predicts complex structure of alternate eddies, the half-arches predicted by PANS are very smooth and regular (Fig. 1f).

The time-averaged flow structures are displayed in Figs. 2a-b. Although both simulations predict the counter-rotating vortices formed close to the lateral edges of the free end and their descent towards the middle section of the cylinder behind the cylinder,a difference is seen here between the LES and the PANS. This is similarly valid for the vortices shed from the sides of the cylinder (Figs. 2a-b). Figure 2c-f shows a comparison of the streamlines projected onto the symmetry plane and on the top of the cylinder. These figures can also be compared with the LES presented by Afgan *et al.* [1], as the resolution in their LES and in the present PANS is similar. Both the present PANS and the LES from [1] predict very short recirculating region compared with the present well resolved LES. Besides, the present well-resolved LES shows much more complex flow on the free top of the cylinder (Fig. 2e) as compared with that found in the PANS (Fig. 2f) or [1] (both the present PANS and the LES reported in [1] display only one reattachment curve on the top of the cylinder at an almost identical location). This indicates that, beside the influence of the modeling (PANS vs. LES), the coarse grid used in the present PANS and previous LES [1] plays a significant role in the prediction. Surface pressure coefficients at four positions along the cylinder are compared in the present paper with the experimental data in [5] and [4] in Figs. 2g-h. Note that the reference pressure is chosen to produce $C_p = 1$ at angle$= 0^o$ at all positions z/L. The LES produced pressure coefficient in agreement with the experimental data, while the PANS resulted in too late a separation and a C_p coefficient that was too low. Such a poor prediction of the pressure coefficient in the present PANS is a combination of too coarse a computational grid and the $k-\varepsilon$ turbulence model in conjuction with a wall function that is not capable of capturing the correct separation position and the resulting near wake. Similar results for the C_p were observed in previous PANS of an infinite cylinder [3]. Despite the superior

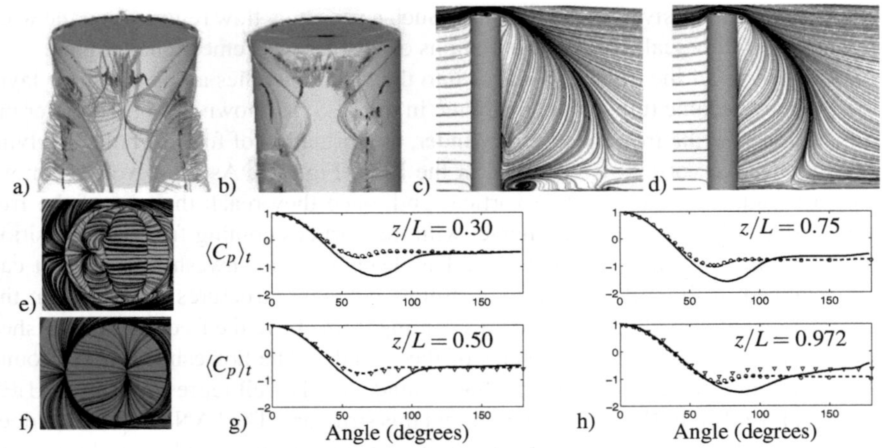

Fig. 2 Streamlines produced from the vortex on the top and behind the cylinder from LES (a) and PANS (b); streamlines projected onto the symmetry plane and the top of the cylinder from LES (c and e) and PANS (d and f); g-h) surface pressure coefficient at different locations. Solid line: PANS ; dashed line: LES ; triangles: experiments by [5]; circles: experiments by [4].

accuracy of the present well-resolved LES as compared to the PANS, an extrapolation of the LES resolution towards a higher Reynolds number is hardly feasible. An improvement of PANS's prediction of the present flow can be obtained by using better resolution and introducing a more accurate turbulence model and a more advanced near-wall model that are capable of capturing separations on smooth surfaces, junction flow at the bottom of the cylinder and free-end separation.

Acknowledgement. This work was supported by Banverket (the Swedish National Rail Administration) and AVL List GmbH. Computer time at SNIC (Swedish National Infrastructure for Computing) resources at the Center for Scientific Computing at Chalmers (C3SE) is gratefully acknowledged.

References

1. Afgan, I., Moulinec, I., Prosser, R., Laurence, D.: Large eddy simulation of turbulent flow for wall mounted cantilever cylinder of aspect ratio 6 and 10. Int. J. Heat and Fluid Flow 28, 561–574 (2007)
2. Girimaji, S., Srinivasan, R., Jeong, E.: PANS Turbulence Models For Seamless Transition Between RANS and LES: Fixed Point Analyses and Preliminary Results. In: ASME paper FEDSM 2003-4336 (2003)
3. Lakshmipathy, S.: Pans method of turbulence: simulation of high and low Reynolds number flows past a circular cylinder. Thesis for Master of Science, Texas A & M University (December 2004)
4. Luo, S.C., Li, L.L., Shah, D.A.: Aerodynamic stability of the downstream of two tandem square-section sylinders. Journal of Wind Engineering and Industrial Aerodynamics 79, 79–103 (1999)
5. Park, C.-W., Lee, S.-J.: Flow structure around a finite circular cylinder embedded in various atmospheric boundary layers. Fluid Dynamics Research 30, 197–215 (2002)

Large Eddy Simulation of Turbulent Flows around a Rotor Blade Segment Using a Spectral Element Method

A. Shishkin and C. Wagner

Abstract. Large Eddy Simulations of turbulent flows around a segment of the FX-79-W151 rotor blade have been performed for the Reynolds numbers $Re = 5 \cdot 10^3$ and $Re = 5 \cdot 10^4$ and the angle of attack 12° using a spectral element method with 8th order polynomials. The turbulence statistics obtained in the LES reveal regions of laminar and turbulent flow separation for the lower and higher Reynolds numbers, respectively, which lead to different loads on the blade.

1 Introduction

The presented work was part of a research project aiming at future improvements of wind turbines and particularly at investigating the effect of wind gusts on the rotor blades. The prediction of extreme loads caused by wind gusts is very important for the design of wind turbine blades, but this flow problem can not be accurately simulated based on the Reynolds-Averaged Navier-Stokes (RANS) equations or related numerical methods. In this respect, the Large Eddy Simulation (LES) is a promising technique. Here we present LES results of turbulent flows around a segment of the FX-79-W151 rotor blade with an angle of attack $\alpha = 12°$ for freestream Reynolds numbers $Re = 5 \cdot 10^3$ and $Re = 5 \cdot 10^4$.

The computational domain is shown in Fig.1 (left) and the hybrid structured/unstructured mesh consists of 2116 2D-elements (see Fig.1, right) with 64 Fourier planes in spanwise direction. The boundary conditions are time-independent laminar inflow and periodicity in spanwise direction. Further, the outflow boundary is realized with a sponge zone to damp the vortical structures before they leave the domain.

The used SEM is based on polynomial representation of the solution combined with the Fourier extension in homogenous spanwise direction. The temporal

A. Shishkin · C. Wagner
Institute for Aerodynamics and Flow Technology, DLR – German Aerospace Center, Bunsenstr. 10, 37073 Göttingen, Germany
e-mail: Andrei.Shishkin@dlr.deandClaus.Wagner@dlr.de

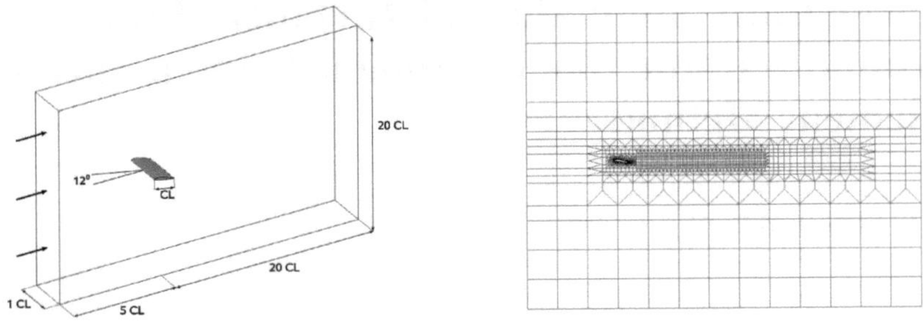

Fig. 1 Schematic view of the computational domain (left) and unstructured 2D mesh (right)

discretization is achieved by means of a higher than second order accurate splitting scheme by Karniadakis *et al.* [1]. For the dimensionless Navier-Stokes equations

$$\nabla\cdot\mathbf{u}=0,\quad \frac{\partial\mathbf{u}}{\partial t}+N(\mathbf{u})=-\nabla p+L(\mathbf{u}),$$

where $L(\mathbf{u})$ and $N(\mathbf{u})$ denote the linear and non-linear parts of momentum equation, respectively, the following three-step algorithm is used. First, the nonlinear term with the time derivative $\frac{\mathbf{u}^*-\mathbf{u}^n}{\Delta t}=N(\mathbf{u}^n)$ is treated explicitly, then the pressure equation with enforcing the continuity condition $\nabla^2 p^{n+1}=\nabla\cdot\frac{\mathbf{u}^*}{\Delta t}$ is solved, and in the last step the diffusive term $\frac{\mathbf{u}^{n+1}-\mathbf{u}^*}{\Delta t}=-\nabla p^{n+1}+L(\mathbf{u}^{n+1})$ is treated implicitly. Additionly, Adams-Bashforth/Adams-Moulton schemes are used. Further details on the splitting scheme can be found in [1, 2]. We use also the SEM-adapted Smagorinsky subgrid scale model as proposed by Karamanos [3].

2 Results of LES for $Re = 5\cdot 10^3$ and $Re = 5\cdot 10^4$

The LES were performed with polynomials of order $N = 8$ (spatial discretization) and a time step $\Delta t = 5\cdot 10^{-6}$. The predicted flow fields were averaged over 2.5 time units and over the spanwise length to provide the turbulence statistics.

The mean streamwise velocity components depicted in Fig. 2 reflect the flow acceleration over the leading edge and the backflow regions (flow separation) on the suction side (highlighted with white curves) for both Reynolds numbers. Considering the Turbulent Kinetic Energy (TKE) (Fig. 3) it is observed that the area of higher TKE is located downstream the trailing edge for $Re = 5\cdot 10^3$ (Fig.3, left), while the higher TKE values are located in the suction side area close to the trailing edge for the higher Reynolds number (Fig. 3, right). The mean pressure coefficients are shown in Fig. 4. The local minimum of the pressure coefficient on the suction side close to the trailing edge observed for the higher Reynolds number (Fig. 4, right) indicates a turbulent flow, while the separation region is laminar for the lower Reynolds number (Fig. 4, left). Finally, the spatial energy spectras at three locations over the suction side marked in Fig.3 (right) are presented in Fig. 5. Energy spectras taken in the region with higher TKE values (Fig. 5, middle and right top and

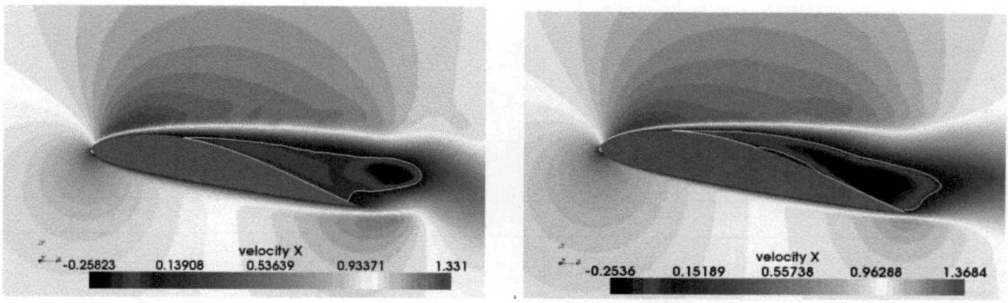

Fig. 2 The mean streamwise velocity components as obtained in the LES for $Re = 5 \cdot 10^3$ (left) and for $Re = 5 \cdot 10^4$ (right). The regions of negative mean velocity (backflow) are outlined.

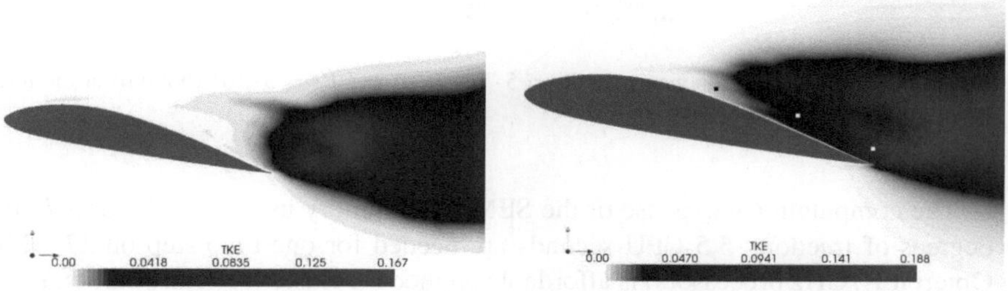

Fig. 3 Distribution of the turbulent kinetic energy obtained in the LES for $Re = 5 \cdot 10^3$ (left) and $Re = 5 \cdot 10^4$ (right)

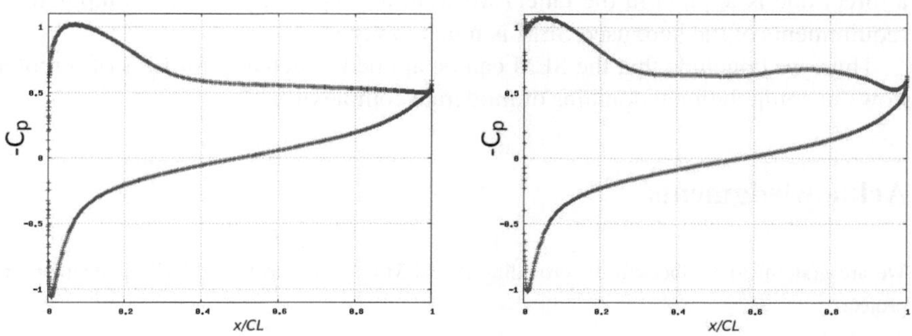

Fig. 4 Mean pressure coefficient obtained in the LES for $Re = 5 \cdot 10^3$ (left) and $Re = 5 \cdot 10^4$ (right)

right bottom) reflect a decay which partly agrees with the "-5/3" law of isotropic turbulence, while the spectras taken at the other locations are characterized by low energy values and faster decay for all velocity components.

The LES results reflect flow separation and complex flow structures in the vicinity of a rotor blade. The analysis of the flow fields also shows that turbulent separation is observed for the higher Reynolds number while the separation is laminar for the lower one.

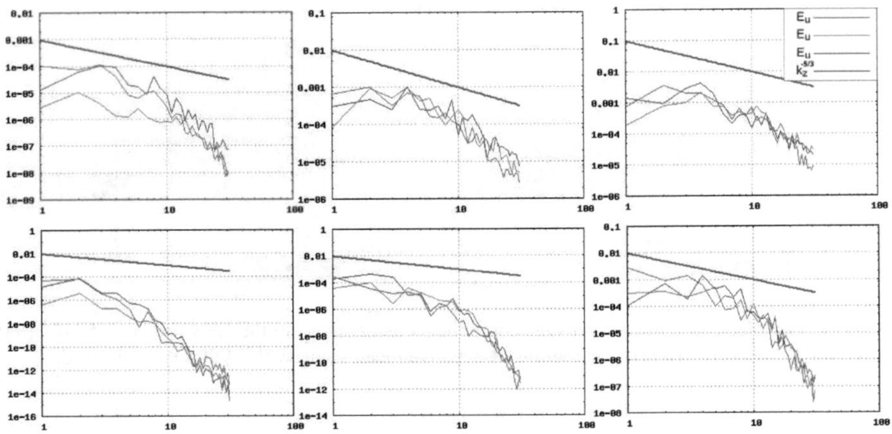

Fig. 5 Energy spectra components for $Re = 5 \cdot 10^4$ (top) and $Re = 5 \cdot 10^3$ (bottom) at the three positions above the trailing edge in Fig. 3

The computational expense of the SEM (the memory usage is 16G for $5.7 \cdot 10^6$ degrees of freedom; 5.5 CPU seconds are needed for one time step on 32 AMD Opteron 1.7GHz processors) is affordable on modern cluster systems. Earlier, in [4] we used a "full 3D" SEM to conduct Direct Numerical Simulations of turbulent pipe flow. Although a direct comparison of the SEM performance in [4] to the here used "2D+Fourier" version is not possible (the former uses an iterative solver, while a direct one is applied in the latter) we want to emphasize that the computational requirements of the here used SEM is much lower.

Thus, we conclude that the SEM can be applied effectively for LES of turbulent flows in computational domains of moderate complexity.

Acknowledgments

We are grateful to S. Sherwin for providing the SEM code and to the BMBF for financing the project.

References

1. Karniadakis, G.E., Israeli, M., Orszag, S.A.: High-order splitting methods for incompressible Navier-Stokes equations. J. Comp. Phys. 97, 414–443 (1991)
2. Karniadakis, G.E., Sherwin, S.: Spectral/HP Element Methods for CFD. Oxford University Press, Oxford (1999)
3. Karamanos, G.-S.: Large Eddy Simulation Using Unstructured Spectral/hp Elements, PhD Thesis, Imperial College (1999)
4. Shishkin, A., Wagner, C.: Direct Numerical Simulation of a turbulent flow using a spectral/hp element methode. Notes on Numer. Fluid Mech. 92, 405–412 (2006)

Vorticity and Helicity in Swirling Pipe Flow

Frode Nygård and Helge I. Andersson

1 Introduction

Direct Numerical Simulations (DNS) of pipe flow with and without swirl are carried out and compared. Swirling pipe flow due to pipe rotation has been reported numerous times in the past, e.g. by DNS in [5] or by experiments in [2]. In the present case, a pressure gradient in the azimuthal direction is setting up a swirl in the near wall region. In comparison to the rotating pipe, the swirl is to a greater extent concentrated close to the pipe wall. However, while the rotating pipe has been investigated experimentally several times, e.g. in [2], the present case is not realizable in practice. A 25R long pipeline with diameter $D = 2R$ is considered. The pipe is divided into about 6.3 million grid cells in a cylindrical coordinate system (129, 97, and 512 in θ, r, and z, respectively). The Navier-Stokes equations are solved by a finite-difference code, developed by Orlandi [4], on a staggered grid, non-uniform in the radial direction. In addition, the flow is driven by an axial pressure gradient sufficient to keep a constant bulk Reynolds number $Re_b = U_b D/\nu = 4900$. Here, U_b is the bulk velocity and ν is the kinematic viscosity. $U_p (= 2U_b)$ is the centerline velocity in the laminar Poiseuille profile used as the initial start up profile. Cyclic boundary conditions are used in the axial- and the azimuthal direction.

2 Results

The frictional Reynolds number, Re_τ, is defined as $u_\tau D/\nu$, with u_τ being the friction velocity. Re_τ is 332 and 337 for the flow with and without swirl, respectively, indicating a small drag reduction. The axial and azimuthal velocity fluctuations in inner variables (symbolized with $^+$) are shown in Fig. 1. The velocities are normalized with u_τ and the normalized wall distance $y = 1 - \frac{r}{R}$, is scaled with $\nu/(Ru_\tau)$. The <> around a variable indicates that it is averaged. The results without swirl are compared with the results by Eggels et. al. [1] with satisfactorily accordance. The

Frode Nygård
e-mail: fnygaard@ntnu.no

Helge I. Andersson
Fluids Engineering Division, Department of Energy and Process Engineering,
Norwegian University of Science and Technology (NTNU), Trondheim, Norway
e-mail: helge.i.andersson@ntnu.no

small deviation is probably due to the Reynolds number difference ($Re_b = 5300$ and $Re_\tau = 360$ in [1]). The reduction of axial velocity fluctuations in the buffer layer ($\sim 4 < y^+ <\sim 40$), when swirl is induced, indicates a reduction of the axial component of kinetic energy production in this region. However, an increase of the azimuthal velocity fluctuations due to the induced swirl indicates an increase of the azimuthal component of the kinetic energy production. As seen in Fig. 2, the vorticity fluctuations are reduced close to the wall ($y^+ < 20$) in the radial and azimuthal direction. The radial vorticity is assumed to be correlated to the high- and low-speed streak intensities [3]. Therefore, a reduction of the radial vorticity may indicate a potential drag reducing effect. As a result of the induced swirl, the axial vorticity is considerably increased in the near wall region.

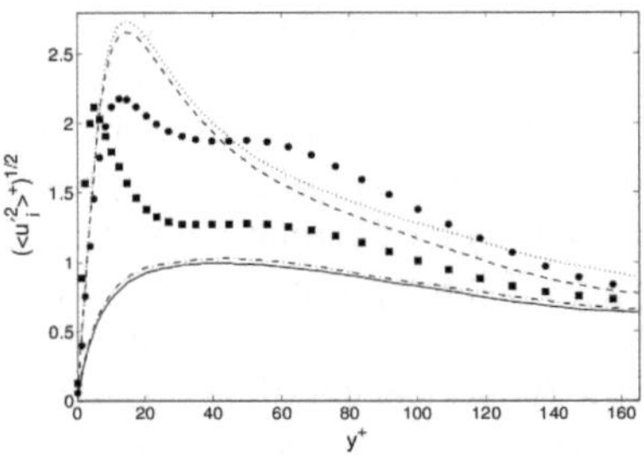

Fig. 1 Fluctuating velocity profiles in inner variables. Axial: ——, no swirl; ···, Eggels; •, induced swirl. Tangential: —, no swirl; – · –, Eggels; ▪, induced swirl.

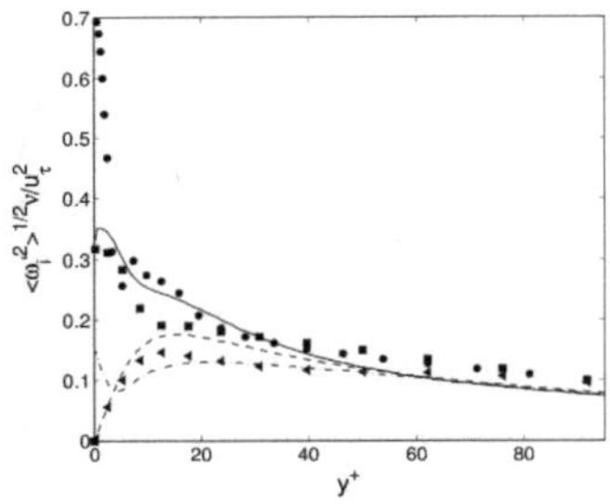

Fig. 2 Fluctuating vorticity profiles in inner variables. Lines are without swirl and symbols are with swirl. — and ▪, θ; —— and ◂, r; – · – and •, z.

Fluctuating helicity is defined as $< h_i > = < u_i' \omega_i' >$ and zero fluctuating helicity indicates an even distribution between right and left handed vortices [3]. Accordingly this is the case for the simulated flow without induced swirl as seen in Fig. 3 a). The induced swirl disrupts the even distribution, resulting in non-zero helicity. The pdf of the helicity density fluctuations, $P(h')$, visualizes the alignment between the fluctuating vorticity and the fluctuating velocity vector [3]. Here, h' is defined as $(\mathbf{u}'\omega')/(\sqrt{|\mathbf{u}|^2 |\omega|^2}) = \mathbf{cos}\phi$, with ϕ being the angle between $\mathbf{u}'$ and ω'. Close to the wall ($y^+ \leq 30$) Fig. 3 b)-d) show that without swirl, the alignment between the velocity and the vorticity fluctuations tend to be perpendicular. The reason for this is found in Fig. 1 and 2 where it can be seen that u_z' and ω_θ' dominate this region. The results for $y^+ \geq 50$ (not shown here) show an even distribution for the angle between the velocity and vorticity fluctuations, which is also reported in [6]. The even distribution seen in the log-law region confirms the motion towards isotropic flow away from the wall. Furthermore, the swirl reduces the axial velocity fluctuations together with the fluctuating azimuthal vorticity close to the wall ($y^+ \leq 30$). Consequently, a reduction of the misalignment between the fluctuating velocity and the fluctuating vorticity vector is observed at $y^+ = 2$ in Fig. 3 b). In addition, for $y^+ = 10$ seen in Fig. 3 c), it becomes most likely to have either aligned or counter-aligned fluctuating velocity and vorticity vectors. However, $P(h')$ is not far from being evenly distributed at $y^+ = 10$ ($y \approx 0.05$) which is in accordance with where the axial and the azimuthal fluctuating helicity terms are crossing the axis in Fig. 3 a). Fig. 3 d) shows that for $y^+ = 30$ ($y \approx 0.2$) it is most likely to have a counter-alignment between the fluctuating velocity and the fluctuating vorticity vector. Fig. 3 a) shows that the axial and azimuthal helicity fluctuations are negative and approximately equal, while the radial helicity fluctuation is almost zero at $y \approx 0.2$. This corresponds with the fluctuating velocity and the fluctuating vorticity vector tending to be counter aligned in the pdf-plot.

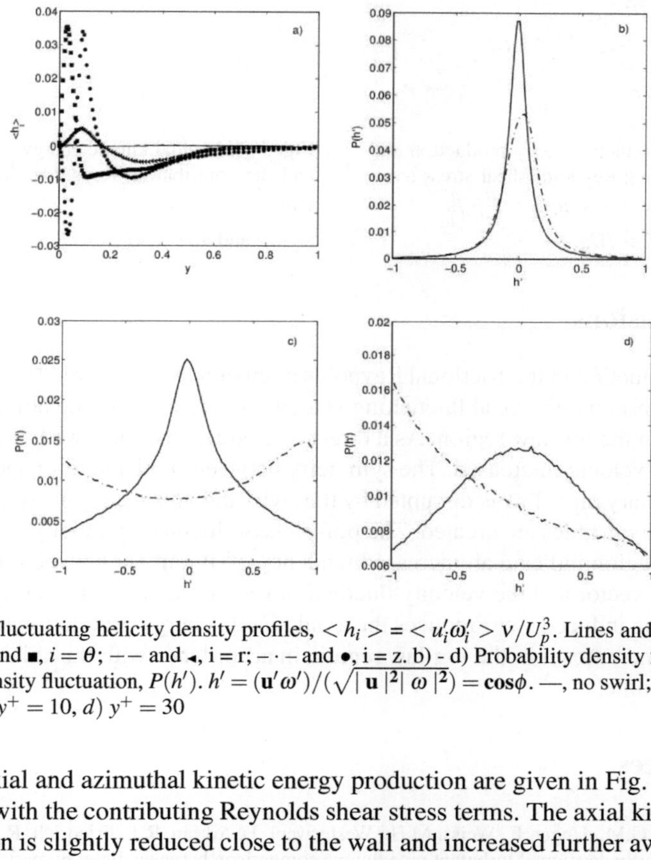

Fig. 3 a) Fluctuating helicity density profiles, $< h_i > = < u_i'\omega_i' > \nu/U_p^3$. Lines and symbols as in Fig. 2. — and ■, $i = \theta$; – – and ◂, i = r; – · – and •, i = z. b) - d) Probability density function of the helicity density fluctuation, $P(h')$. $h' = (\mathbf{u}'\omega')/(\sqrt{|\,\mathbf{u}\,|^2|\,\omega\,|^2}) = \cos\phi$. —, no swirl; – · –, swirl. *b)* $y^+ = 2$, *c)* $y^+ = 10$, *d)* $y^+ = 30$

The axial and azimuthal kinetic energy production are given in Fig. 4 and Fig. 5 together with the contributing Reynolds shear stress terms. The axial kinetic energy production is slightly reduced close to the wall and increased further away from the wall when a swirl is induced. This pattern is in accordance with the u_z'-profile given in Fig. 1 and it can be seen that the two curves (with and without swirl) are crossing at $y^+ \approx 43$ or $y \approx 0.25$ which corresponds with the crossing of the production curves in Fig. 4. Without swirl, the azimuthal production term and the $r\theta$-Reynolds shear stress term are zero. However, the induced swirl represents an azimuthal velocity with radial variation in the near wall region. As a consequence, a considerable increase in the azimuthal kinetic energy production occurs in this region. With the maximum azimuthal kinetic energy production being about 7 times the maximum axial production, an even bigger effect on the azimuthal velocity fluctuation than observed in Fig. 1 might be expected. However, as seen in Fig. 5, the peak of the azimuthal energy production occurs in the viscous sub layer where a considerable amount is damped by the viscous dissipation.

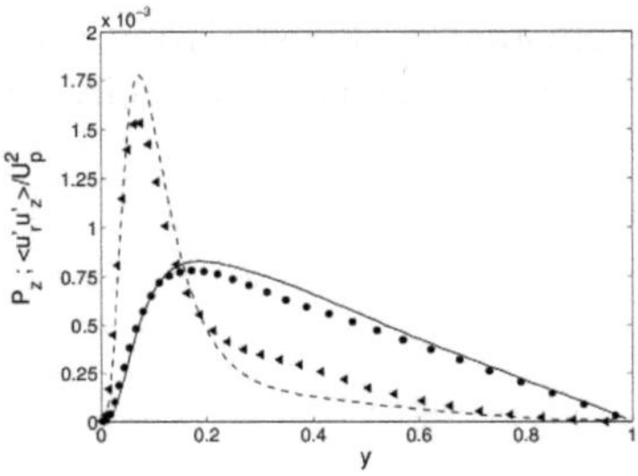

Fig. 4 Axial kinetic energy production and the contributing Reynolds shear stress term. —— and ◂, $P_z = - < u_r' u_z' > \frac{dU_z}{dr} \frac{R}{U_p^3}$; – – – and •, $< u_r' u_z' > / U_p^2$.

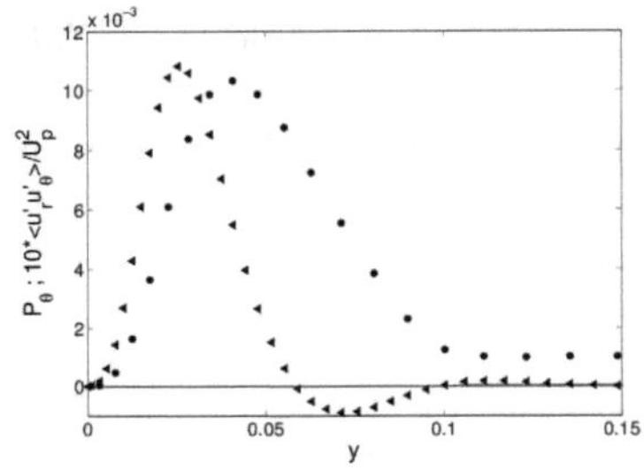

Fig. 5 Azimuthal kinetic energy production and the contributing Reynolds shear stress term. —— and ◂, $P_\theta = - < u_r' u_\theta' > \frac{dU_\theta}{dr} \frac{R}{U_p^3}$; – – – and •, $< u_r' u_\theta' > / U_p^2$.

3 Conclusions

A slight reduction in the frictional Reynolds number was discovered for the swirling flow. Furthermore, the axial fluctuating velocity was reduced in the buffer layer and increased in the log-law region. As a direct consequence of the swirl, the azimuthal fluctuating velocity increased. The symmetry between right and left handed vorticities in ordinary pipe flow is disrupted by the swirl and non-negligible fluctuating helicity density profiles are created. The pdf-plots of fluctuating helicity density in the near wall region indicate an increased tendency of alignment between the vorticity fluctuation vector and the velocity fluctuation vector when a swirl is introduced. In addition, the induced swirl reduces the axial kinetic energy production in the buffer layer and increases it in the log-law region, in accordance with the axial fluctuating velocity profile.

References

1. Eggels, J.G.M., Unger, F., Weiss, M.H., Westerweel, J., Adrian, R.J., Friedrich, R., Nieuwstadt, F.T.M.: Fully developed turbulent pipe flow: a comparison between direct numerical simulation and experiment. J. Fluid Mech. 286, 175–209 (1994)
2. Imao, S., Itoh, M., Harada, T.: Turbulent characteristics of the flow in an axially rotating pipe. Int. J. Heat Fluid Flow 17, 444–451 (1996)
3. Orlandi, P.: Helicity fluctuations and turbulent energy production in rotating and non-rotating pipes. Phys. Fluids 9, 2045–2056 (1997)
4. Orlandi, P.: Fluid Flow Phenomena, A Numerical Toolkit. Kluwer, Dordrecht (2000)
5. Orlandi, P., Fatica, M.: Direct simulations of turbulent flow in a pipe rotating about its axis. J. Fluid Mech. 343, 43–72 (1997)
6. Rogers, M.M., Moin, P.: Helicity fluctuations in incompressible turbulent flows. Phys. Fluids 30, 2662–2671 (1987)

Explicit Algebraic Subgrid Models for Large Eddy Simulation

Linus Marstorp, Geert Brethouwer, and Arne V. Johansson

Abstract. The objective of this study is to develop models for the subgrid-scale (SGS) stress and the SGS scalar flux by applying the same kind of methodology that leads to the explicit algebraic Reynolds stress model, EARSM [2], and the explicit algebraic scalar flux model, EASFM [3], for RANS. The idea is that these new models can improve the description of the anisotropy compared to eddy viscosity models. Since the new models can include the effect of system rotation in a natural way they have a particular potential for rotating flows.

1 Model Description

The explicit algebraic SGS stress model is based on a modelled transport equation involving an equilibrium assumption and reads

$$\tau_{ij} = K_{SGS}\left(\frac{2}{3}\delta_{ij} + \beta_1\tau^*\tilde{S}_{ij} + \beta_4\tau^{*2}(\tilde{S}_{ik}\tilde{\Omega}_{kj} - \tilde{\Omega}_{ik}\tilde{S}_{kj})\right) , \tag{1}$$

where K_{SGS} is the SGS kinetic energy, $\tilde{S}_{ij} = 0.5(\partial\tilde{u}_i/\partial x_j + \partial\tilde{u}_j/\partial x_i)$ is the resolved rate of strain, $\tilde{\Omega}_{ij} = 0.5(\partial\tilde{u}_i/\partial x_j - \partial\tilde{u}_j/\partial x_i)$ is the resolved rotation rate tensor and τ^* is the time scale of the subgrid scales. The first term on the right hand side in (1) is the isotropic part, the second term is an eddy viscosity part, and the third term is a nonlinear tensor that creates a realistic anisotropy of the SGS stress. β_1 and β_4 are coefficients depending on the resolved gradients and the model parameters. Two

Linus Marstorp
Linné Flow Centre, KTH Mechanics, SE-100 44 Stockholm, Sweden

Geert Brethouwer
Linné Flow Centre, KTH Mechanics, SE-100 44 Stockholm, Sweden
e-mail: geert@mech.kth.se

Arne V. Johansson
Linné Flow Centre, KTH Mechanics, SE-100 44 Stockholm, Sweden

versions of the model have been developed: (i) A dynamic model with a dynamic determination of K_{SGS}. (ii) A non-dynamic model with a non-dynamic determination of K_{SGS}. The explicit algebraic SGS scalar flux model is also based on a modelled transport equation involving an equilibrium assumption, and reads

$$q_i = -\tau^* c_{amp} A_{ij}^{-1} \tau_{jk} \frac{\partial \tilde{\theta}}{\partial x_k} , \quad (2)$$

where A_{ij} depends on the resolved velocity gradients and the model parameters, c_{amp} is a model parameter, and $\tilde{\theta}$ is the resolved scalar.

2 Validation

To validate the models, LES of turbulent channel flow with and without rotation about the spanwise direction at wall friction Reynolds number $Re_\tau = 180$ are carried out and compared to DNS data. Fig. 1 shows the mean bulk velocity at various rotation numbers. All simulations have the same mean pressure drop. According

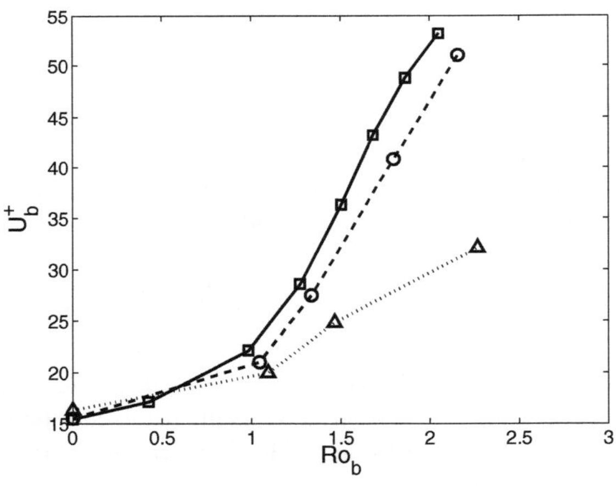

Fig. 1 Mean bulk velocity at various rotation numbers. DNS, solid line; new explicit non-dynamic model, dashed line; Smagorinsky model, dotted line.

to the DNS by [1] the mean bulk velocity increases with rotation and approaches the laminar value at high rotation speeds. Because of the included mean gradients a standard Smagorinsky model predicts a non-zero eddy viscosity even if the flow is laminar. Therefore, this model cannot handle the laminarisation and strongly underpredicts the bulk mean velocity. The new non-dynamic model performs much better in that respect because it can predict the laminarisation at the destabilised side of the channel to some extent. Especially, at high rotation rates this model benefits from the asymptotic behaviour $\beta_1,\ \beta_4 \to 0$. Additional simulations of $Re_\tau = 950$ turbulent channel flow shows that the explicit model provides for a better and less filter scale dependent description of the Reynolds stresses near the wall.

Fig. 2 shows the root-mean-square of the scalar fluctuations in non-rotating $Re_\tau = 180$ channel flow using the explicit algebraic SGS flux model and the eddy diffusivity model with a constant $Pr_T = 0.4$. A constant and uniform temperature

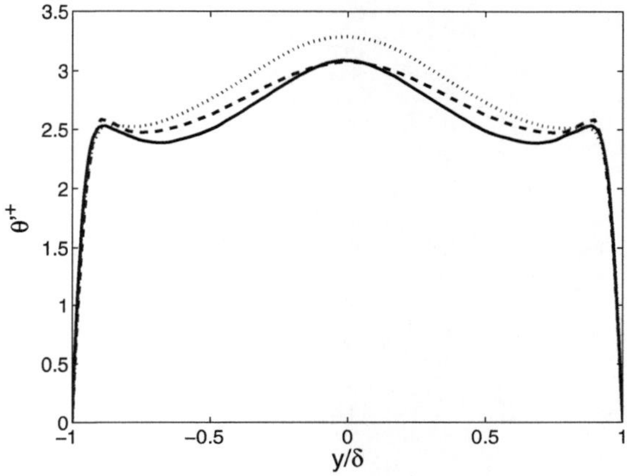

Fig. 2 Rms of the scalar fluctuations. Filtered DNS data, solid line; new explicit model, dashed line; eddy diffusivity model, dotted line.

difference between the walls is imposed. As a model for the SGS stress we apply the dynamic version of the new explicit model and dynamic Smagorinsky model respectively. The explicit SGS scalar flux model provides for a good description of the scalar fluctuations showing that the eddy diffusivity part of the model produces a realistic amount of mean SGS scalar dissipation. Additional simulations show that the explicit SGS scalar flux model provides for a better description of the scalar fluxes in rotating channel flow compared to the eddy diffusivity model.

References

1. Grundestam, O., Wallin, S., Johansson, A.V.: Direct numerical simulations of rotating turbulent channel flow. J. Fluid Mech. 598, 177–199 (2008)
2. Wallin, S., Johansson, A.V.: An explicit algebraic Reynolds stress model for incompressible and compressible turbulent flows. J. Fluid Mech. 403, 89–132 (2000)
3. Wikström, P.M., Wallin, S., Johansson, A.V.: Derivation and investigation of a new algebraic model for the passive scalar flux. Phys. Fluids 12, 688–702 (2000)

Direct Numerical Simulation of a Turbulent Flow with Pressure Gradients

Liang Wei and Andrew Pollard

1 Introduction

Flows with favorable pressure gradient (FPG) and adverse pressure gradient (APG) are of great importance practically and theoretically. In practice, many industrial applications, especially aerodynamics, involve flows with pressure gradients and separation. In theory, wall shear stress does not dominate this type of flow[8]. Much remains not understood.

Many experimental studies and numerical simulations have been done in the past. Spalart and Watmuff [8] performed both experiment and direct numerical simulation (DNS) of a turbulent boundary layer with pressure gradients. The DNS used a prescribed velocity to generate the pressure gradients of their experiment. Good agreement between the experimental results and simulations was found for mean wall-pressure coefficients, displacement and momentum thicknesses. However, they found some discrepancy between their experiment and DNS, especially structure functions and root-mean-square profiles. Nagib et al.[5] carried out experiments of high Reynolds number turbulent boundary layers with adverse, zero and favourable pressure gradients. The log-law parameter K was found to vary considerably for the non-equilibrium boundary layers under the various pressure gradients. Na and Moin [4] conducted an incompressible DNS of a separated turbulent boundary layer and employed a suction-blowing velocity profile at the top boundary. They found that locations of instantaneous spanwise-averaged detachment and reattachment points both fluctuate significantly in the streamwise direction and the maximum turbulent intensity occurred above the detachment region. Skote et al.[7] performed an incompressible DNS of self-similar APG turbulent boundary layers. Comparison of turbulence statistics from the zero pressure gradient and the two APG cases showed

Liang Wei, PhD Candidate
Queen's University, Kingston, Canada
e-mail: liang@me.queensu.ca

Andrew Pollard, Professor
Queen's University, Kingston, Canada
e-mail: pollard@me.queensu.ca

that the development of a second peak in turbulence energy was in agreement with experiment. Marquillie et al.[3] performed an incompressible DNS of a separated channel flow. The pressure gradients were obtained by employing a wall curvature through a mathematical mapping from physical coordinates to Cartesian ones. The inlet condition was taken from a highly resolved LES of plane channel flow. A thin separation bubble was found on the curved wall.

A compressible DNS of transition to turbulence with physical boundaries that prescribe favourable and adverse pressure gradients is more challenging. In the current simulation, the compressible Navier-Stokes equations were solved by using a discontinuous Galerkin method. In order to compare with incompressible results, the Mach number is 0.2 based on the inlet freestream velocity. The pressure gradients were obtained by using a top curved wall, similar to Spalart & Watmuff[8].

2 Computational Details

In the current simulation, all the length units are non-dimensionalized by the streamwise length of the plate ($L_x = 0.1m$). The computational domain is shown in the figure 1, where the periodic domain in streamwise direction (X) is [0.1,1.2]. At the computational inlet ($X = 0.1$), $Re_x \approx 42800$ based on the freestream velocity and the inlet x ($Re_\theta \approx 160$ based on the inlet momentum thickness), which matches Spalart and Watmuff's experiment [8] at $X = 0.1m$. It is then followed by a FPG region $[0.2, 0.6]$ which is long enough for turbulence to develop and lose memory of inflow conditions[8]. The flow enters the APG region from $X = 0.6$ to the end of the plate ($X = 1.0$). The region $[1.0, 1.2]$ is taken as fringe region, where a fringe region technique[7][1] was employed to drive the out-flow back to the inflow. It is implemented by the addition of a volume force function $F = \lambda(X) * (\widetilde{U_i} - U_i)$ to the Navier-Stokes equations, where $\widetilde{U_i}$ is the inflow velocity profile that velocities in the fringe region U_i are forced to and $\lambda(X)$ is the forcing strength, nonzero only in the fringe region. Periodic boundary conditions were applied to the streamwise direction. The flow is assumed homogeneous in the spanwise (Z) direction. In the wall normal (Y) direction no-slip wall boundary conditions were applied to the bottom plane wall and the top curved wall. There are 48x32x4 body fitted non-orthogonal grid elements in the streamwise, wall normal, span wise direction respectively. 10th-order refinement in each element makes a total of ~ 6 million modes. Over-integration was employed to eliminate de-aliasing errors.

3 Results and Discussion

The wall pressure coefficients of the current simulation are shown in the figure 2, which agrees well with Spalart and Watmuff's experiment and simulation data[8]. It can be seen that the lowest pressure coefficient happens at near X=0.6.

Iso-contours of the streamwise velocity fluctuations close to the bottom plane wall are shown in the figure 3. In the upstream FPG region where the flow is accelerated, the streaks are elongated, as is also observed by Piomelli et al.[6]. In the downstream APG region where the flow is decelerated, however, the streaks are shortened. At a certain point, they are totally disturbed. In other words, it looks more "turbulent" in the APG region than the FPG region. Acceleration would cause flow relaminarization if the nondimensional pressure gradient parameter $K = \frac{\nu}{U_\infty^2}\frac{dU_\infty}{dx} \geq 3\times10^{-6}$. In this simulation the maximum K in the FPG region is 1.4×10^{-6}.

Turbulent coherent structures shown in the figure 4 are iso-surfaces of the second invariant of the velocity gradient tensor $Q = 1/2(\omega^2 - S^2)$ in the FPG and APG regions. It can be seen that the structures are more densely distributed in the downstream APG region than the upstream FPG region. The start position of the structures at the bottom plane wall in the figure 4 is similar as the start position of disturbed low speed streaks close to the bottom wall shown in the figure 3.

Future work will include a detailed analysis of turbulence statistics and structure functions.

Acknowledgement. The authors would like to thank Dr. Karniadakis & his CRUNCH group at Brown University and Dr. Kirby at University of Utah for providing the original discontinuous Galerkin code and the related helpful email discussions. The research was funded through grants from NSERC Canada.

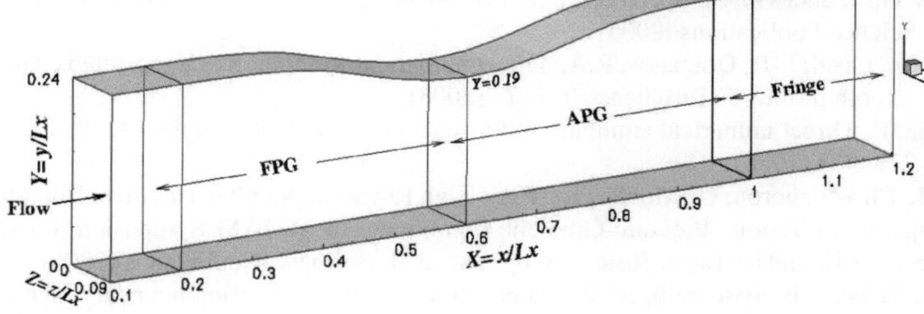

Fig. 1 Computational domain including fringe region

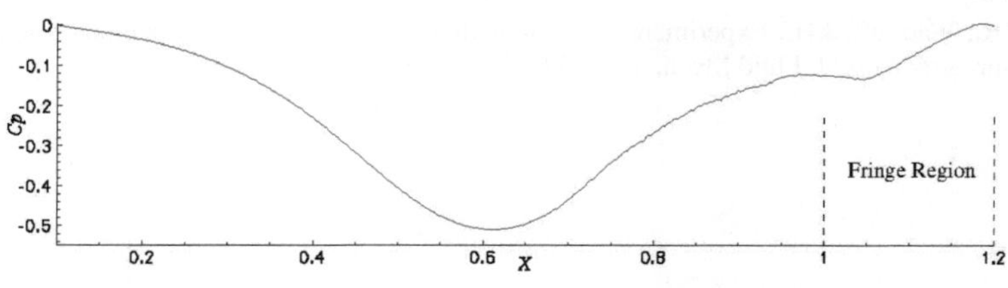

Fig. 2 Mean pressure coefficient at the bottom plane wall, based on the inlet core velocity

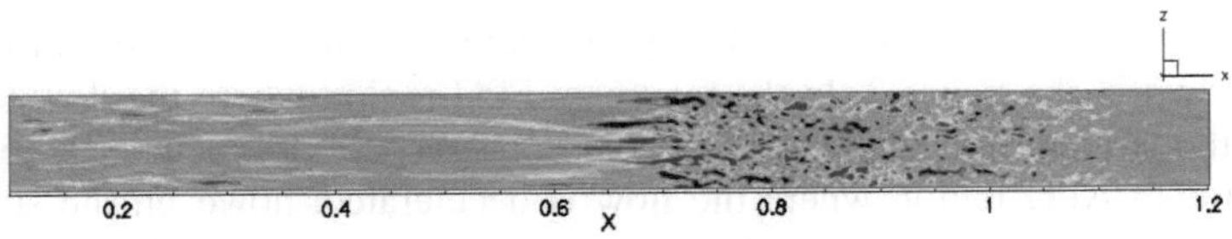

Fig. 3 Iso-contours of the streamwise velocity fluctuations close to the bottom plane wall

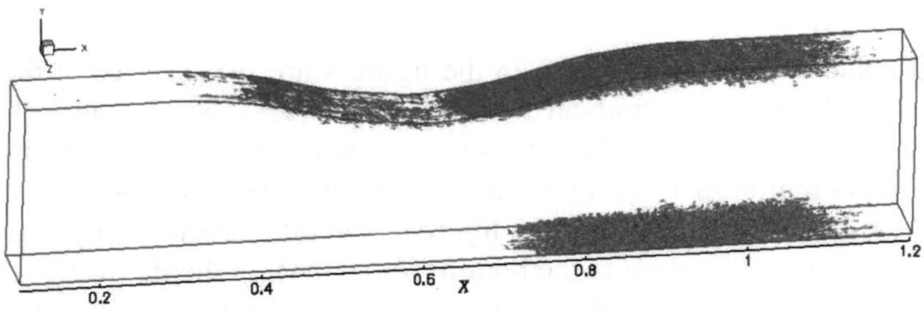

Fig. 4 Iso-surfaces of the second invariant of the velocity gradient tensor (Q) close to the bottom plane wall and the top curved wall, which creates the left FPG region and the right APG region

References

1. Herbst, A.H., Henningson, D.S.: The influence of periodic excitation on a turbulent separation bubble. Flow, Turbulence and Combustion 76, 1–21 (2006)
2. Karniadakis, G.E., Sherwin, S.: Spectral/hp Element Methods for Computational Fluid Dynamics. Oxford Science Publications (2005)
3. Marquillie, M., Laval, J.-P., Dolganov, R.A.: Direct numerical simulation of a separated channel flow with a smooth profile. J. Turbulence 9, 1–23 (2008)
4. Na, Y., Moin, P.: Direct numerical simulation of a separated turbulent boundary layer. J. Fluid Mech. 374, 379–405 (1998)
5. Nagib, H.M., Christophorou, C., Monkewitz, P.A.: High Reynolds Number Turbulent Boundary Layers Subjected to Various Pressure-Gradient Conditions. In: IUTAM Symposium on One Hundred Years of Boundary Layer Research, pp. 383–394. Springer, Netherlands (2006)
6. Piomelli, U., Balaras, E., Pascarelli, A.: Turbulent structures in accelerating boundary layers. J. Turbulence 1, 1–16 (2000)
7. Skote, M., Henningson, D.S., Henkes, R.A.W.M.: Direct numerical simulation of self-similar turbulent boundary layers in adverse pressure gradients. Flow, Turbulence and Combustion 60, 47–85 (1998)
8. Spalart, P.R., Watmuff, J.H.: Experimental and numerical study of a turbulent boundary layer with pressure gradients. J. Fluid Mech. 249, 337–371 (1993)

An Invariant Nonlinear Eddy Viscosity Model Based on a Consistent 4D Modelling Approach

Michael Frewer

1 Introduction

When developing turbulence modelling from scratch certain questions arise which inevitably turn into methodological problems regarding this topic

- What makes Euclidean transformations in classical continuum mechanics, in particular turbulence modelling, so special ?
- Why is frame-dependency in all unclosed terms of existing algebraic models, e.g. in the Reynolds-stress tensor, always only modelled by the mean objective intrinsic spin tensor, i.e. the mean vorticity tensor measured in a rotating frame relative to an inertial frame: $\langle W_{ij}\rangle = \langle \omega_{ij}\rangle + \varepsilon_{ijk}\Omega_k$?
- Why is the mean pressure or one of its gradients never taken along as a closure variable ?
- Why does there still not exist a clear-cut mathematical formulation of the material frame-indifference (MFI)-principle in general continuum mechanics (if it applies as a physical approximation for reducing constitutive equations) ? What consequences does a proper mathematical formulation have for modelling turbulence in the limit of a 2D flow state ?

Answers to these questions are given herein, except for the last question which is beyond the scope of this article [1]. From the outset it is clear, that in order to give an unambiguous answer to these interlinked questions one needs a new mathematical framework, or more precisely, a setting of universal form-invariance (UFI) which extends the classical framework of an Euclidean geometry being used so far [1].

To allow for UFI within Newtonian physics the usual 3D space manifold has to be extended by a true 4D space-time manifold, which, conceptually as well as mathematically, will be the *classical* limit of the manifold used in Einstein's general

---

Michael Frewer
Institute of Fluid Dynamics, Technische Universität Darmstadt
e-mail: frewer@fdy.tu-darmstadt.de

theory of relativity. However, the real advantage of using a different geometrical representation eventually lies in the analytical theory of modelling physical laws, irrespective of whether constitutive or dynamical equations are considered. Such a geometrical reformulation of classical continuum mechanics, without changing the physical content of the theory, automatically brings along consistent and structured modelling arguments in the most natural way whenever they are based on invariant principles. The claim here is that within Newtonian physics *(3+1)D modelling is not equivalent to 4D modelling*. The clear superiority of a true 4D approach, e.g. in modelling turbulence, can be manifestly fixed by the following arguments:

- The variables of space and time are fully independent.
- Physical quantities as velocities or stresses always transform as tensors, irrespective of whether they are objective (frame-independent) or not.
- Frame accelerations or inertial forces of any kind are automatically included and universally described by the underlying geometry, i.e. by the affine connection $\Gamma^{\rho}_{\mu\nu}$ of the 4D space-time manifold.
- The special space-time structure of the 4D-manifold allows for additional modelling constraints, which are absent in the usual (3+1)D geometry.

2 Construction of a Newtonian Space-Time Manifold

The minimal requirements for having UFI in Newtonian mechanics are met [1] by constructing a true 4D space-time manifold $\mathscr{N}$ i) with zero curvature, and ii) which emerges from Einsteinian mechanics in the classical limit $c \to \infty$. The result is a space-time manifold $\mathscr{N}$ which

- is non-Riemannian with a non-unique and singular metrical connection. Next to the affine connection $\Gamma^{\rho}_{\mu\nu}$ its geometry is described by four singular tensors

$$h^{\alpha\beta} = \begin{pmatrix} 0 & 0 \\ 0 & \delta^{ij} \end{pmatrix}, \; k_{\alpha\beta} = \begin{pmatrix} \|\mathbf{u}\|^2 & u^j \\ u^i & \delta_{ij} \end{pmatrix} ; \; g_{\alpha\beta} = \begin{pmatrix} 1 & 0 \\ 0 & 0 \end{pmatrix}, \; m^{\alpha\beta} = \begin{pmatrix} 1 & u^j \\ u^i & u^i u^j \end{pmatrix},$$

- only allows for space-time coordinate transformations in which the time coordinate transforms as an absolute quantity — the Euclidean transformations then only form a small subset,
- leads to a pure time-like 4-velocity $u^{\alpha} = (1, u^i)$, which can never turn into a space-like quantity during its evolution through the manifold.

Now, since the geometry is fixed, the procedure to obtain UFI within Newtonian physics is then simply defined as:

1. Write the Newtonian equations in the inertial (3+1)D Cartesian form.
2. Rewrite them into the corresponding 4D form using the geometrical structure of the Newtonian space-time manifold $\mathscr{N}$.
3. Make the transition from inertial Cartesian to arbitrary space-time coordinates by replacing the partial derivative with the covariant derivative $\partial_{\alpha} \to \nabla_{\alpha}$.

For the ensemble-averaged Navier-Stokes equations the result reads

$$\nabla_\alpha \langle u^\alpha \rangle = 0, \tag{1}$$

$$\langle u^\beta \rangle \nabla_\beta \langle u^\alpha \rangle = -h^{\alpha\beta} \nabla_\beta \langle p \rangle + \nu h^{\beta\gamma} \nabla_\beta \nabla_\gamma \langle u^\alpha \rangle - \nabla_\beta \tau^{\alpha\beta}. \tag{2}$$

By construction these equations stay form-invariant under arbitrary space-time coordinate transformations $\tilde{x}^\alpha = \tilde{x}^\alpha(x^\beta)$ in which the time coordinate, up to an additive constant, transforms as an absolute quantity $\tilde{x}^0 = x^0$. The covariant derivative is defined as $\nabla_\alpha u^\beta = \partial_\alpha u^\beta + \Gamma^\beta_{\alpha\lambda} u^\lambda$. But most importantly these equations reveal the novel information that the average and the fluctuating 4-velocities evolve differently within $\mathcal{N}$: The average 4-velocity $\langle u^\alpha \rangle = (1, \langle u^i \rangle)$ is a pure time-like vector which can never turn space-like, while the fluctuating 4-velocity $u'^\alpha = (0, u'^i)$ is a pure space-like vector which can never turn time-like. Thus the Reynolds-stress tensor $\tau^{\alpha\beta} = \langle u'^\alpha u'^\beta \rangle$ is a pure space-like 4-tensor, which has to be respected during modelling. This information, which serves as an important modelling restriction, is completely missing in the usual (3+1)D framework.

3 Proposal for an Invariant Nonlinear Eddy Viscosity Model

The aim is to close the Reynolds-stress tensor algebraically $\tau^{\alpha\beta} = \tau^{\alpha\beta}(\mathcal{V})$ with $\mathcal{V}$ as a local functional argument set of averaged variables. Its modelling restrictions are: i) contravariant tensor of rank 2, ii) pure space-like tensor, iii) symmetric tensor and iv) that it carries the dimension of velocity squared. Hence, two more restrictions, namely i) and ii), are gained than in the usual (3+1)D formulation.

If the motivation in closing the averaged Navier-Stokes equations (1)-(2) is only to make use of the sole information these equations can supply, the most basic and general ansatz for a closure up to an order of first gradients would be to use all those variables for which the system is being solved for

$$\mathcal{V} = \{ \langle P^\sigma \rangle, \langle u^\sigma \rangle ; \nabla_\lambda \langle P^\sigma \rangle, \nabla_\lambda \langle u^\sigma \rangle \}, \text{ where } \langle P^\alpha \rangle := h^{\alpha\beta} \nabla_\beta \langle p \rangle. \tag{3}$$

But for a complete description of an incompressible, isothermal turbulent flow state one needs at least a 2-equation model. Thus using the Kolmogorov phenomenology between turbulent kinetic energy and its dissipation rate as well as a subsequent Lie-group symmetry analysis [2] the list is forced to be

$$\mathcal{V} = \left\{ \frac{1}{\mathcal{K}^{1/2}} \langle u^\sigma \rangle ; \frac{\mathcal{K}^{1/2}}{\mathcal{E}} \nabla_\lambda \mathcal{K}, \frac{\mathcal{K}^{3/2}}{\mathcal{E}^2} \nabla_\lambda \mathcal{E}, \frac{\mathcal{K}^2}{\mathcal{E}^2} \nabla_\lambda \langle P^\sigma \rangle, \frac{\mathcal{K}}{\mathcal{E}} \nabla_\lambda \langle u^\sigma \rangle \right\} \tag{4}$$

in which the averaged pressure gradient of first order has to be excluded, and in which the invariant turbulent kinetic energy $\mathcal{K} = \frac{1}{2} \langle u'^\alpha u'_\alpha \rangle$ together with its invariant (pseudo-)dissipation rate $\mathcal{E} = \nu h^{\mu\lambda} \langle \nabla_\mu u'^\alpha \nabla_\lambda u'_\alpha \rangle$ have to be used for dimensional consistency in the remaining variables, including their own first gradients.

Further dimensional analysis and dominant balance arguments pave the way for the limit in a high local turbulent Reynolds number $\mathrm{Re}_T = \frac{\mathscr{K}^2}{\nu\mathscr{E}} =: \frac{\nu_T}{\nu} \gg 1$, i.e. the way for a general manifest invariant high-Re_T eddy viscosity model

$$\nabla_\alpha \langle u^\alpha \rangle = 0, \tag{5}$$

$$\langle u^\beta \rangle \nabla_\beta \langle u^\alpha \rangle = -h^{\alpha\beta} \nabla_\beta \langle p \rangle - \nabla_\beta \tau^{\alpha\beta}, \tag{6}$$

$$\langle u^\alpha \rangle \nabla_\alpha \mathscr{K} = -k^{\langle u \rangle}_{\beta\lambda} \tau^{\lambda\alpha} \nabla_\alpha \langle u^\beta \rangle - \mathscr{E} + \nabla_\lambda \mathscr{D}^\lambda_{(\mathscr{K})}, \tag{7}$$

$$\langle u^\alpha \rangle \nabla_\alpha \mathscr{E} = \mathscr{P}^\alpha_{(1)\beta} \nabla_\alpha \langle u^\beta \rangle + \mathscr{P}^{\alpha\beta}_{(2)\lambda} \nabla_\alpha \nabla_\beta \langle u^\lambda \rangle - \Psi + \nabla_\lambda \mathscr{D}^\lambda_{(\mathscr{E})}. \tag{8}$$

Next to the Reynolds-stress tensor 5 additional unclosed quantities have to be modelled in the transport equations for $\mathscr{K}$ and $\mathscr{E}$. For constructing a consistent invariant eddy viscosity model of an overall quadratic non-linearity the corresponding polynomial tensor expansions, resulting from Tensor Invariant Theory [3], as well as its invariants in $\mathscr{V}$, must be truncated at quadratic order, except for the two $\mathscr{E}$-production terms which already have to be truncated at linear order, which then results into a zero contribution for the second production term.

Now, a striking difference between the 4D and the common (3+1)D approach is the appearance of two time-like invariants $\mathscr{I}_1 = \frac{1}{\mathscr{E}} \langle u^\alpha \rangle \nabla_\alpha \mathscr{K}$, $\mathscr{I}_2 = \frac{\mathscr{K}}{\mathscr{E}^2} \langle u^\alpha \rangle \nabla_\alpha \mathscr{E}$ among the 12 quadratic-order invariants belonging to the tensor set $\mathscr{V}$. In particular, these two invariants represent the kinematic left hand sides of the $\mathscr{K}$- and $\mathscr{E}$-transport equations respectively, which eventually can account for turbulent memory effects. Also striking are the expansions for the space-like diffusion vectors which include time-like terms as proportional to $\langle u^\alpha \rangle \nabla_\alpha \langle P^\lambda \rangle$ and $\langle u^\alpha \rangle \nabla_\alpha \langle u^\lambda \rangle$. Finally the quadratic expansion in the Reynolds-stress tensor already allows for a term proportional to $(h^{\alpha\rho} h^{\beta\sigma} + h^{\beta\rho} h^{\alpha\sigma}) \nabla_\rho \nabla_\lambda \langle p \rangle \cdot \nabla_\sigma \langle u^\lambda \rangle$ being able to capture secondary flow effects, for which in current nonlinear eddy viscosity models a higher non-linearity is needed. For example in axially rotating pipe flow with rotation rate $\dot{\omega}$ the necessary off-diagonal shear component of the Reynolds-stress tensor $\tau^{r\varphi}$ for generating a non-zero mean swirl velocity $\langle u_\varphi \rangle$ is then given by $\tau^{r\varphi} \sim -\frac{\dot{\omega}}{r} \partial_r^2 \langle p \rangle + \frac{\dot{\omega}}{r^2} \partial_r \langle p \rangle$, where the swirl is maintained by the mean radial pressure gradient. In general, the mean pressure gradient shows itself as being a promising closure variable which can account for modelling non-local effects in turbulence.

References

1. Frewer, M.: More clarity on the concept of material frame-indifference in classical continuum mechanics. Acta Mechanica (2008), doi:10.1007/s00707-008-0028-4
2. Oberlack, M.: A unified approach for symmetries in plane parallel turbulent shear flows. J. Fluid Mech. 27, 299–328 (2001)
3. Spencer, A.J., Rivlin, R.S.: The theory of matrix polynomials and its applications to the mechanics of isotropic continua. Arch. Rat. Mech. Anal. 2, 309–336 (1958)

A Hybrid URANS/LES Approach Used for Simulations of Turbulent Flows

Karel Fraňa and Jörg Stiller

Abstract. A hybrid model based on the unsteady Reynolds averaged Navier-Stokes approach represented by the one-equation Spalart-Allmaras model and the Large Eddy Simulation called Detached Eddy Simulation (DES) was applied for turbulent flow simulations. This turbulent approach was implemented into the flow solver based on the Finite-Element Method with pressure stabilized and streamlines upwind Petrov-Galerkin stabilization techniques. The effectiveness and robustness of this updated solver is successfully demonstrated at benchmark calculation represented by an unsteady turbulent flow past a cylinder at Reynolds number 3900. Results such as velocity fields and the flow periodicity, Reynolds stress tensor and eddy viscosity and pressure coefficient distributions are discussed and relatively good agreement was found to direct numerical simulations and experiments.

1 Introduction

Turbulent flow regimes appear in a wide range of the technical applications and studies of such flows are a complex problem and require computational capacity and knowledge of physics and mathematics. Currently, there are various approaches adopted for turbulent flow simulations. One of them is a hybrid URANS/LES model which combines the advantages of different approaches to achieve time effective and sufficiently accurately methods for simulation of complex turbulent flows. For instance, the URANS method can not reproduce mostly the structures of wide-ranging

Karel Fraňa
Technical University in Liberec, Department of Power Engineering Equipment, Studentska 2, 461 11, Liberec, Czech Republic
e-mail: karel.frana@seznam.cz

Jörg Stiller
Technische Universität Dresden, Institute of Aerospace Engineering, 01062, Dresden, Germany
e-mail: joerg.stiller@tu-dresden.de

spatial and time scale, but as discussed in [1] it should correctly predict the growth and separation of boundary layers and accurately capture the Reynolds stress tensor. The Large Eddy Simulation resolves a significant fraction of the turbulent motions directly on the grid and then provides a powerful approach for capturing the Reynolds stresses in separated regions. To combine the most favorable aspects of URANS and LES, the hybrid technique, for instance, the DES model was proposed. The closure here is based on a simple modification to the Spalart-Allmaras model [2] and takes advantage of the good performance of RANS models in the thin shear layers for which the model was precisely calibrated. The objective of this paper is the implementation of the DES model into an existing computational Finite Element Code [6] and [7] in order to achieve a robust computational tool adopting further parallel and grid adaptation techniques. The performance of the turbulent model validation is carried out on the benchmark representing the unsteady turbulent flows past a cylinder. This commonly adopted benchmark was studied numerically [3], [5] and experimentally [4].

2 Problem Formulation

An incompressible unsteady turbulent flow with constant density and kinematic viscosity past a cylinder at Reynolds number approximately 3.900 (based on the diameter) is considered as a benchmark for validations of the hybrid URANS/LES model. This flow is governed by Navier-Stokes Equations in which the turbulent eddy viscosity ν_t provided numerically by the turbulent model appears as part of an effective viscosity in the diffusion term.

Considering the DES model, the calculation of the turbulent eddy viscosity is based on the modified eddy viscosity $\tilde{\nu}$ in the Spalart-Allmaras model and is calculated using the transport equation in the form as follows

$$\frac{D\widetilde{\nu}}{Dt} = c_{b1}\widetilde{S}\widetilde{\nu} + \frac{1}{\sigma}\left[\nabla\cdot((\nu+\widetilde{\nu})\nabla\widetilde{\nu}) + c_{b2}(\nabla\widetilde{\nu})^2\right] - c_{w1}f_w\left[\frac{\widetilde{\nu}}{d}\right]^2 \tag{1}$$

Equation 1 must be closed with the auxiliary relations f.g. in [2]. Taking into consideration the local equilibrium conditions in the turbulent flow, the production term is practically balanced by the destruction term. This so called local balance leads to the relation $\nu \approx \tilde{S}d^2$ and if the wall distance is replaced by the filter width Δ thus the previous relations become similar to the relation used in the Smagorinsky subgrid-scaled model applied in the Large Eddy Simulation. The wall distance can be replaced by a characteristic length scale $\tilde{d}$ proportional to Δ so that

$$\widetilde{d} \equiv min(d, C_{DES}\Delta), \quad \Delta \equiv max(\Delta x, \Delta y, \Delta z) \tag{2}$$

This approach is called the DES and it is a combination between the classical RANS and LES computational approach. The recommended value for the adjustable parameter is about $C_{DES} = 0.65$.

3 Results

Figure 1 depicts a snapshot of the x-component of the velocity field scaled by the inlet velocity u_{in}. The flow periodicity in the Karman vortex street represented by dimensionless frequency (Strouhal number) was detected approximately 0.219. Mean velocity profiles at positions x/D 1.54 and 2.02 confronted with DNS results [3] are illustrated in Figure 2 and relatively good agreement was found.

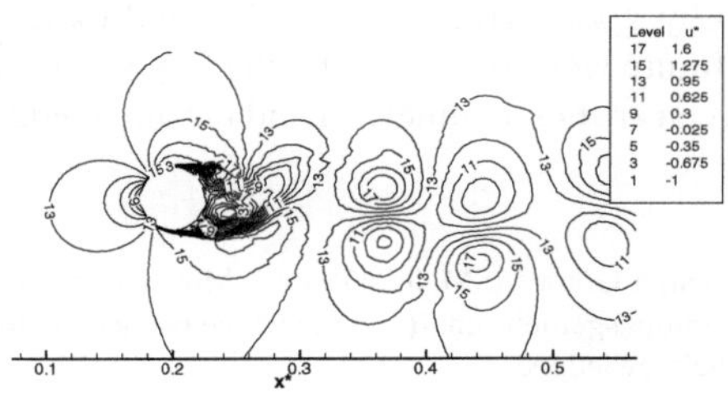

Fig. 1 Snapshot of the x-component of the velocity field

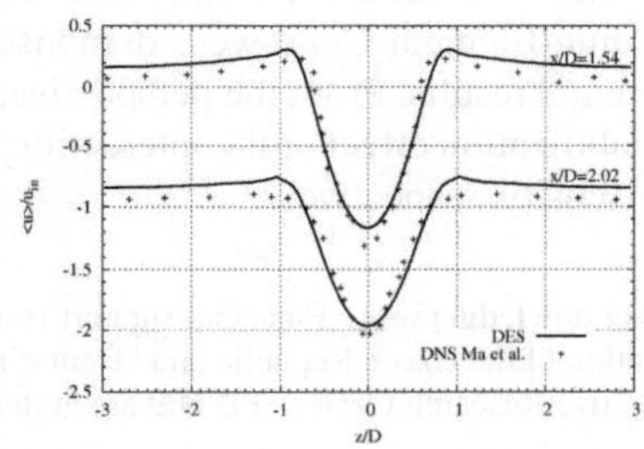

Fig. 2 Mean velocity profiles at positions x/D 1.54 and 2.02

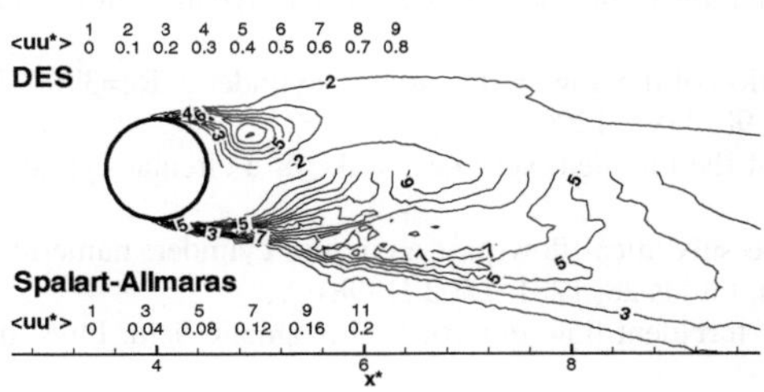

Fig. 3 The $< u'_x u'_x >$ component of the subscale stress

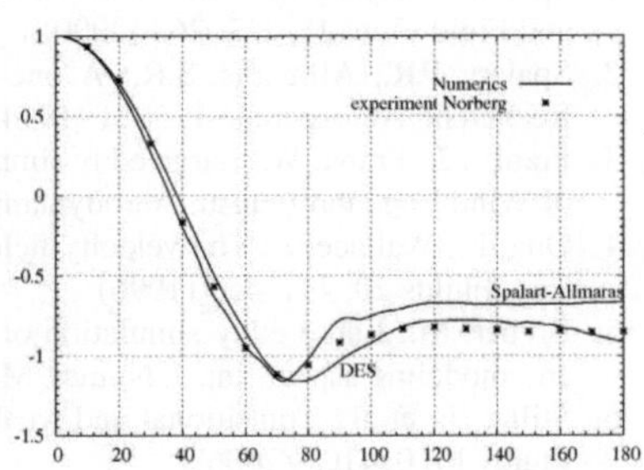

Fig. 4 The pressure coefficient c_p along the cylinder surface

The normal component $< u'_x u'_x >$ of the subscale Reynolds stress distribution scaled by the inlet velocity u^2_{in} is captured in Figure 3. Whereas the magnitude of the $< u'_x u'_x >$ ranges between 0 up to 0.2 in Spalart-Allmaras model, in DES the same variable reaches up to 0.8 and the peak of the subscale Reynolds stress appears in the wake after the cylinder. Figure 4 shows a pressure coefficient around the cylinder surface and the DES calculation was in good agreement to experiments provided by [8]. In the Spalart-Allmaras calculation, the pressure coefficient was slightly overestimated in the wake of the cylinder.

4 Conclusion

The turbulent flow past a cylinder was calculated by URANS approach represented by the Spalart-Allmaras model and the hybrid URANS/LES approach so called Detached Eddy Simulation. As main results, the frequency of the flow periodicity, the velocity field and Reynolds stress tensor and eddy viscosity distributions were presented. The robustness and correctness of the implemented turbulent models in the Finite-Element-Code were demonstrated by good agreement to DNS and experimental results. From the perspective, the future work will consider the impact of the stabilization effect or the interaction between artificial numerical and turbulent eddy viscosity, respectively.

Acknowledgement. Financial support from Research Grant MSM 4674788501 Ministry of Education of the Czech Republic and "Deutsche Forschungsgemeinschaft" in the frame of the Collaborative Research Center SFB 609 are gratefully acknowledged.

References

1. Spalart, P.R.: Strategies for Turbulence Modeling and Simulations. International Journal of Heat and Fluid Flow 21, 252–263 (2000)
2. Spalart, P.R., Allmaras, S.R.: A one-equation turbulence model for aerodynamic flows. La Recherche Aerospatiale 1, 5–21 (1994)
3. Franke, J., Frank, W.: Large eddy simulation of the flow past a circular cylinder at Re=3900. J. of Wind Eng. and Industr. Aerodynamics 90, 1191–1206 (2002)
4. Ong, L., Wallace, J.: The velocity field of the turbulent very near wake of a circular cylinder. Exp. Fluids 20, 441–453 (1996)
5. Breuer, M.: Large eddy simulation of the subcritical flow past a circular cylinder: numerical and modeling aspect. Int. J. Numer. Meth. Fluids 28, 1281–1302 (1998)
6. Stiller, J., et al.: Transitional and weakly turbulent flow in a rotating magnetic field. Phys. of Fluids 19, 074105 (2006)
7. Stiller, J., et al.: A parallel PSPG Finite Element Method for direct Simulation of Incompressible flow. In: Danelutto, M., Vanneschi, M., Laforenza, D. (eds.) Euro-Par 2004. LNCS, vol. 3149, pp. 726–733. Springer, Heidelberg (2004)
8. Norberg, C.: Effects of Reynolds number and low-intensity free stream turbulence on the flow around a circular cylinder. Department of Applied Thermoscience and Fluid Mech., Chalmers University of Technology, Gothenburg, Sweden, Publ. 87, 2 (1987)

Anisotropic Synthetic Turbulence with Sweeping Generated by Random Particle-Mesh Method

Malte Siefert and Roland Ewert

Abstract. We develop an efficient method to generate convecting synthetic turbulence in complex mean flows. In this article we present the extension to anisotropic turbulence, realistic spectra and sweeping effects, i.e. the advection of inertial range structures by the energy containing large scales. We show that due to this formulation sweeping effects can be included and the spatial-temporal field shows the similarity of Kraichnan's sweeping hypothesis.

1 Introduction

In many situations the full spatial-temporal field is needed to study the properties of a turbulent flow. Although it can in principle be calculated by the numerical solution of the Navier-Stokes equation, there are reasons for using instead a synthetic turbulent field generated by a Monte-Carlo method. This approach can be much faster and one has direct access to many of its statistical properties. Furthermore this approach can be motivated by the fact that often certain properties do not depend on the dynamical details.

Recently, Ewert [1] has proposed a random particle-mesh method in conjunction with a moving average procedure. The main idea was to account for convection effects, arbitrary mean flows and complex boundaries which are essential for aeroacoustics. This method has been applied to a variety of aeroacoustic applications such as sound produced by the interaction between turbulence and solid boundaries and edges, jet noise, combustion noise and tone scattering in turbulent flows [1].

Malte Siefert
DLR (German Aerospace Center), Braunschweig, Germany
e-mail: malte.siefert@dlr.de

Roland Ewert
DLR (German Aerospace Center), Braunschweig, Germany
e-mail: roland.ewert@dlr.de

In this article we extend the method and focus on spatial-temporal properties as well as on anisotropy of the turbulent field. Sweeping time effects, i.e. the advection of inertial range structures by the energy containing large scales are included and anisotropy of the Reynolds-stress tensor as well as of the integral length scales in different directions are taken into account.

2 Convective Moving-Average Method

In the following we use the Reynolds decomposition of the flow field $\mathbf{U} = \mathbf{u}_0 + \mathbf{u}$ with $\langle \mathbf{u} \rangle = 0$. The input of the method are the integral length scales l_i in the different directions, the Reynolds-stress tensor $\langle u_i u_j \rangle$ and the mean velocity field $\mathbf{u}_0$, which has to be provided by appropriate models, for example by Reynolds-averaged simulations (RANS). The output is the fluctuating part $\mathbf{u}$ of the Reynolds decomposition.

The basis is a hierarchy of stochastic Gaussian distributed fields $\eta_i^{(n)}(\mathbf{x})$ with vanishing mean, which are spatially and mutually uncorrelated and obey a convection equation:

$$\langle \eta_i^{(n)}(\mathbf{x}) \rangle = 0 \tag{1}$$

$$\langle \eta_i^{(m)}(\mathbf{x}) \eta_j^{(n)}(\mathbf{x}') \rangle = \delta_{mn} c_{ij}^{(n)} \delta(\mathbf{x} - \mathbf{x}') \tag{2}$$

$$\frac{D}{Dt} \eta_i^{(n)}(\mathbf{x}) = 0, \tag{3}$$

where $\frac{D}{Dt} \equiv \frac{\partial}{\partial t} + (\mathbf{U} \cdot \nabla)$ is the substantial derivative. The third equation (3) means that the white field is convected on the field $\mathbf{U}$. Each field n is convoluted with a Gaussian filter kernel[1]

$$g^{(n)}(\mathbf{x}) = \mathrm{e}^{-\frac{\pi}{2} \sum_i \left(\frac{x_i}{l_i^{(n)}} \right)^2} \tag{4}$$

and can therefore be associated with length scales $l_i^{(n)}$ in direction of the x_i-coordinates. This results in a hierarchy of fields $\phi_i^{(n)}$ with different spatial correlation lengths. The amplitudes of the fields and length scales are given by $c_{ij}^{(n)} = \beta^{n\alpha} c_{ij}^{(0)}$ and $l^{(n)} = l^{(0)}/\beta^n$, where $\beta > 1$ and α is the exponent of the energy spectrum, e.g. $\alpha = -5/3$ for the Kolmogorov energy-spectrum.

The stream function of the turbulent field $\mathbf{u} = \nabla \times \boldsymbol{\phi}$ is then given by the superposition of the fields

$$\phi_i = \sum_{n=0}^{N-1} \phi_i^{(n)} = \sum_{n=0}^{N-1} g^{(n)} * \eta_i^{(n)}. \tag{5}$$

[1] The coordinate system is chosen such that the convolution is along the principal axis of the length-scale tensor.

The matrix c^0_{ij} of Eq. (2) has to be chosen such that the correct Reynolds-stress tensor is represented:

$$\langle u_i u_l \rangle = V^{(0)} \frac{\pi}{2} \frac{1}{(l_k^{(0)})^2} \varepsilon_{ijk} \varepsilon_{lmk} \frac{1-\beta^{N(\alpha-1)}}{1-\beta^{\alpha-1}} c_{jm}^{(0)}. \tag{6}$$

$V^{(0)} \equiv l_1^{(0)} l_2^{(0)} l_3^{(0)}$ is the correlation volume. This is an equation for the unknown matrix $c_{ij}^{(0)}$. The integral length scales $l_i = \int_0^\infty \langle u_i(\mathbf{x}+r\mathbf{e}_i) u_i(\mathbf{x}) \rangle dr / \langle u_i^2(\mathbf{x}) \rangle$ in the three directions are connected to the largest scales by

$$l_i = l_i^{(0)} \frac{1-\beta^{N(\alpha-2)}}{1-\beta^{\alpha-2}}. \tag{7}$$

Note that l_i and c_{ij} can depend on $\mathbf{x}$ as long as they vary on scales larger than l_i.

The term 'particle-mesh' refers to the numerical realization of the above formulation [1]: The convected white field is realized by means of particles, which are convected by the turbulent field and carry the random values. This field is projected onto an auxiliary mesh, on which the convolution is performed.

3 Numerical Results

We generate turbulent fields with the above described method. First we look at the spatial properties of the generated synthetic field ($\beta = 2$, $\alpha = -5/3$, $N = 10$). Fig. 1 shows the turbulent kinetic energy distribution and the corresponding one-dimensional spectrum.

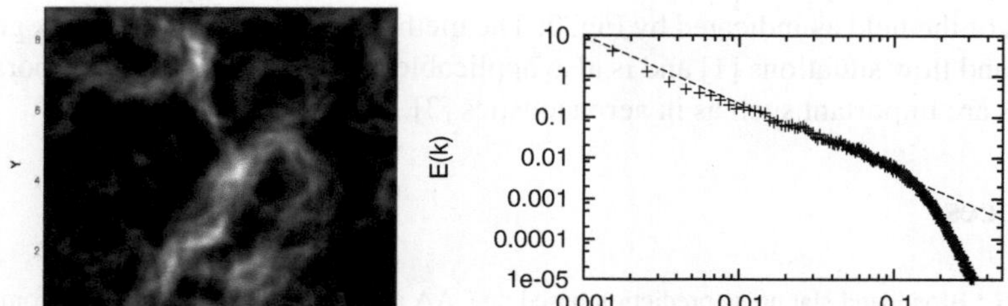

Fig. 1 Left: Instantaneous turbulent kinetic energy $\frac{1}{2}\mathbf{u}^2$ of the field. Right: Spectrum in comparison to the Kolmogorov's '$-5/3$'-law. The field is generated by a superposition of ten Gaussian correlated fields with $\alpha = -5/3$, $\beta = 2$, $N = 10$.

Next we discus the spatial-temporal properties of the field. It is known that the sweeping mechanism leads to a similar law of the spatial-temporal correlations [2]:

$$\frac{\langle u_x(\mathbf{x}+r\mathbf{e}_x,t+\tau)u_x(\mathbf{x},t)\rangle}{\langle u_x(\mathbf{x}+r\mathbf{e}_x,t)u_x(\mathbf{x},t)\rangle} \equiv C\left(\frac{u_{\mathrm{rms}}\tau}{r}\right) \tag{8}$$

where the sweeping velocity $u_{\mathrm{rms}}^2 = \langle \mathbf{u}^2 \rangle$ is the mean square of the velocity fluctuations. Fig. 2 shows the correlation functions C in dependence of the normalized time scale $v_{\mathrm{rms}}\tau/r$ for different spatial distances r. The curves collapse into one indicating that similarity (8) holds.

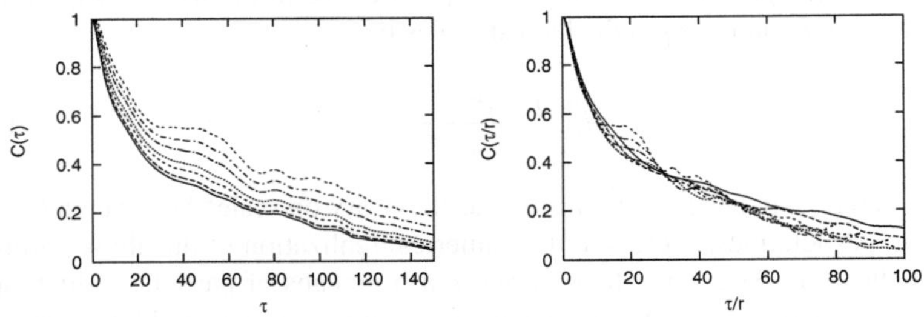

Fig. 2 Time-correlations according to Eq. (8) plotted with respect to τ (left) and to the similar variable $u_{\mathrm{rms}}\tau/r$ (right). The scale r is varied in the interval $r \in [2,5]$. Parameters are $l = 2$, $k = 0.1$. For each curve 10^6 data points are sampled with time step $\Delta t = 0.1$.

4 Conclusion

We have presented an efficient method to generate anisotropic, incompressible synthetic turbulent fields including convection by a superposition of Gaussian filtered fields. The isotropic case is recovered for $l_i \equiv l$ and diagonal $c_{ij} \equiv \delta_{ij}c$. Numerical results were given by using a random particle-mesh method. Due to the convection formulation in Eq. (3) sweeping effects could be included in the spatial-temporal dynamics of the field as indicated by Fig. 2. The method can be used in complex geometries and flow situations [1] and is also applicable for situations where temporal properties are important such as in aeroacoustics [3].

References

1. Ewert, R.: Broadband slat noise prediction based on CAA and stochastic sound sources from a fast random particle-mesh (RPM) method. Comp. Fluids 35 (2007)
2. Kraichnan, R.H.: Kolmogorov's hyotheses and eulerian turbulence theory. Phys. Fluids 7, 1723–1734 (1964)
3. Siefert, M., Ewert, R.: A stochastic source model for turbulent noise prediction including sweeping time dynamics. In: Proceedings of the Acoustic conference, Paris, France (2008)

LES and Hybrid LES/RANS Study of Flow and Heat Transfer around a Wall-Bounded Short Cylinder

D. Borello, G. Delibra, K. Hanjalić, and F. Rispoli

1 Introduction

The flow in plate-fin-and-tube heat exchangers is featured by interesting dynamics of vortical structures, which, due to close proximity of bounding walls that suppress instabilities, differs significantly from the better-known patterns around long cylinders. Typically, several distinct vortex systems can be identified both in front and behind the pin. Their signature on the pin and end-walls reflects directly in the local heat transfer. The Reynolds numbers is usually moderate and the incoming flow is non-turbulent, transiting to turbulence on or just behind the first or few subsequent pin/tube rows. Upstream from the first pin a sequence of several horseshoe vortices attached to the bounding wall is created, while the unsteady wake contains also multiple vortical systems which control the entrainment of fresh fluid and its mixing with the hot fluid that was in contact with the heated surfaces [1]. The conventional CFD using standard turbulence models, as practiced by heat exchangers industries, falls short in capturing the subtle details of the complex vortex systems. A fine-grid LES can provide accurate solutions, but for more complex configurations and higher Re numbers a hybrid RANS/LES using a coarser grid seems a more rational option, provided it can capture all important flow and vortical features.

In order to shed more light on the flow structures, their role in heat transfer and the capabilities of a simple hybrid model to return the salient flow feature, we conduct in parallel a well-resolved LES, and coarse mesh hybrid and URANS simulations of initially non-turbulent flow over a single heated short cylinder bounded by infinite walls.

2 Models and Numerics

The flow considered mimics the cold experiments of [1]. A cylinder with a diameter $d = 50$ mm is placed in the center of a rectangular duct. The solution domain extends $20d$ in the streamwise (X) and $14d$ (Y) in the spanwise direction, with the height h (in Z direction) of $0.4d$ (Fig.1.a). The Reynolds number (based on d and the inlet bulk velocity) is equal to 6150.

Dipartimento di Meccanica ed Aeronautica, 'Sapienza' University of Rome, Roma, Italy
e-mail: borello@dma.ing.uniroma1.it

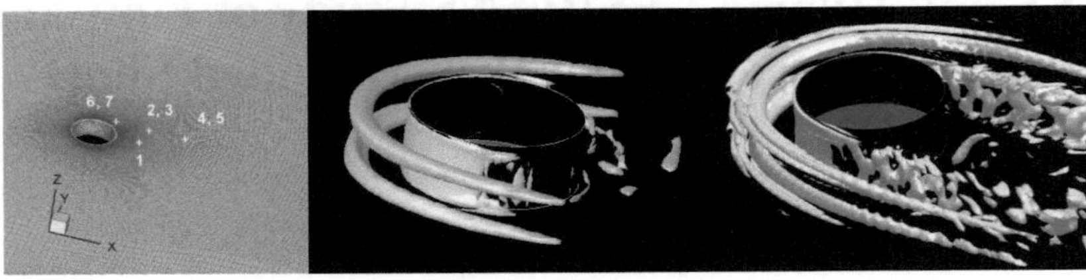

Fig. 1 Computational grid with details of monitoring points (left); pressure laplacian for hybrid (mid) and LES (right)

The simulations were performed with the in-house parallelized finite-volume computational package T-FlowS, using unstructured grid clustered in the wall regions, with 600k cells for LES and 120k cells for URANS and hybrid simulations. The LES was closed with the dynamic sgs model, whereas an elliptic-relaxation eddy-viscosity model (k-ε-ζ-f), was applied for the URANS and as the wall-model in the hybrid RANS/LES, matched in the latter with the dynamic LES in the outer flow regions, [2]. The convection terms were discretized with a second-order CDS in LES, and the SMART scheme for URANS and hybrid methods. The pressure-velocity coupling was treated with the SIMPLE algorithm, whereas the time marching was performed with a three-step fully implicit scheme, with the typical nondimensional time step of 0.012. No-slip conditions were imposed on all walls, convective flow conditions at the outlet and constant variables profiles at the inlet. As the turbulence intensity was unknown, a value of 3% was assumed, this being the only uncertainty in simulating the experiment. For the URANS and hybrid LES/RANS a ratio ν_t/ν=100 was presumed for evaluating the inflow ε. A thermal LES simulation is carried out assuming heating of the cylinder with a heat flux of 14,8 W/m2 (corresponding to the nondimensional value of 0.00243). The inlet temperature was 298 K.

3 Discussion

The case represents a challenge to turbulent models because of the low Reynolds number imposed. In fact in LES simulation the y^+ value in the mid of channel ($Y = 7, Z = 0.2$) half diameter before the stagnation point ($X = 9$), is about 30. This turbulence level accessible only to well-tuned low-RE-number RANS models, but it is suitable for assessing the potential of the proposed hybrid LES/RANS model [2]. Visualization of $\nabla^2 p$=20 shows complex turbulent structures both in hybrid and LES simulations (see Fig.1.b,c). Although the hybrid model does not reproduce the spectrum of structures in the wake and the double horseshoes pattern as does LES, it seems capable of capturing the major pattern in the recirculation region. The analysis of the flow upstream the stagnation line at 5% of cylinder height shows the presence of one horseshoe vortex near the endwall (see Fig.2.a). All simulations return similar behavior as in the experiment, as seen from the instantaneous velocity plots in Fig.2.b,c. On the other hand, instantaneous plots of the measured streamlines in mid XZ section show the presence of more complex horseshoes vortices system

Fig. 2 Details of the velocity streamlines in a plane parallel to the endwall (at 5% of channel height); experiments (left), hybrid (mid) and LES (right)

Fig. 3 Details of the velocity streamlines in a plane normal to the endwalls (at 50% of channel span); experiments (left), hybrid (mid) and LES (right)

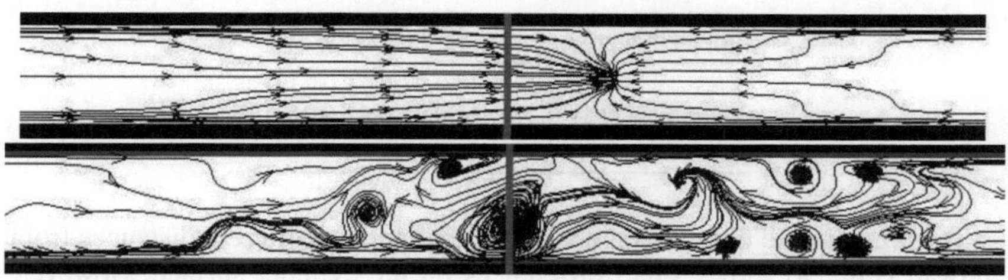

Fig. 4 Instanteous streamlines in a section placed at X=12; up - hybrid; down - LES

(Fig.3.a). LES simulation returns a similar pattern (see Fig.3.c), while hybrid tends to suppress the counter rotating vortices appearing between the horseshoe vortices location (Fig.3.b). In Fig.3.b,c the red line show the location of the measurement plane of Fig.2.

In Fig.4 an analysis of the wake region is presented for hybrid and LES. We refer to a cross section normal to steamwise velocity placed $2d$ downstream from the cylinder center ($X = 12$). The vertical lines indicate the mid-channel position ($Y = 7$). In URANS simulation (not shown here) the center of the wake remains always near the mid of the duct, while in hybrid simulations the wake shows an unsteady flapping between left and right of the midline (Fig.4.a). This indicates the vortex shedding under the effect of the LES forcing. LES results (Fig.4.b) shows a more complex and unsteady vortical structure due to the longer recirculation; in the right part it is possible to recognize the imprint of the two horseshoe vortices structures (see Fig.1.c) near the two endwalls. In Tab.1 we compare Strouhal numbers returned by URANS, hybrid and LES. The monitoring points are indicated in Fig.1.a and are related to positions placed near the mid plane and the endwalls.

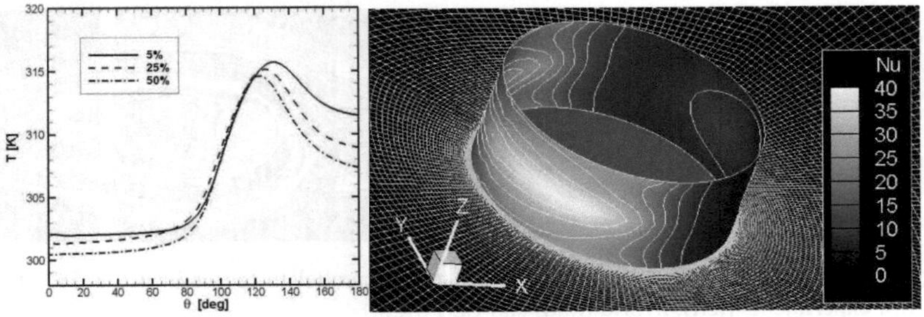

Fig. 5 Temperature around the cylinder at three different distance from the lower endwall (left) and distribution of the Nusselt number around the cylinder (right)

Hybrid model returns the Strouhal number very close to that of LES, while URANS gives a somewhat lower values at all the locations.

Table 1 Strouhal number in monitoring points

Simulation	1	2	3	4	5	6	7
LES	0.22	0.23	0.21	0.23	0.24	0.23	0.23
Hybrid	0.21	0.21	0.21	0.22	0.22	0.21	0.21
URANS	0.18	0.18	0.18	0.18	0.18	0.18	0.18

The temperature distribution around the cylinder at different distances from the endwall (Fig.5.a) shows small differences on the cylinder surface. However, the circumferential distribution is very non-uniform with peak temperature at around 110 deg corresponding to flow separation. Both of these features are reflected in the Nu distribution around the cylinder (Fig.5.b).

4 Conclusions

The analysis demonstrated that hybrid model is capable of capturing a major part of the turbulence spectrum despite missing some small structures, returning credible prediction of the flow upstream from the cylinder, wake flapping and Strouhal number in close according with LES. Simulation of the thermal field showed the importance of the reproducing the dominant vortical structures for obtaining credible predictions of the thermal field and Nu number.

References

1. Sahin, B., Ozturk, N.A., Gurlek, C.: Horseshoe vortex studies in the passage of a model plate fin and tube heat exchanger. Int. J. Heat and Fluid Flow 29, 340–351 (2008)
2. Delibra, G., Borello, D., Hanjalić, K., Rispoli, F.: U-RANS of flow in pinned passages relevant to gas-turbine blade cooling. ETMM7, selected for publication IJHFF (2008)

Stochastically Forced Laminar Plane Couette Flow: Non-normality and Hydrodynamic Fluctuations

George Khujadze, Martin Oberlack, and George Chagelishvili

Abstract. The background of three dimensional (3D) hydrodynamic/vortical fluctuations in a stochastically forced, laminar and incompressible plane Couette flow is simulated numerically. It was found that the fluctuating background in the flow has the following characteristics: The hydrodynamic fluctuations show the non-exponential, transient growth; an anisotropy of the fluctuating velocity field increases with the shear rate; existence of the streamwise structural regularities (coherent structures) with the characteristic length-scale of the order of a channel width; appearance of the nonzero velocity cross-correlations; Symmetry breaking of the spanwise reflection of the dynamical processes due to the stochastic forcing.

1 Introduction

The resolution of the paradox of a subcritical transition in shear flows (plane Couette or pipe Poisuille flows, for example) is a long-standing problem in fluid mechanics. It is well-known that the plane Couette flow is linearly stable at all Reynolds numbers but in practice always becomes turbulent at sufficiently large Reynolds number (see for example the review [1]). On the other hand we also know that the stochastic forcing is inherent for the environmental and engineering flow systems. During the past two decades it has become evident that thermally excited fluctuations in fluids in the non-equilibrium steady states are always spatially long ranged even in the absence of convection or turbulence. By recent studies it has been shown that the relatively small stochastic forcing of the laminar Couette flow in the linear regime

George Khujadze · Martin Oberlack
Chair of Fluid Dynamics, Technische Universität Darmstadt, Hochschulstr. 1, Darmstadt, Germany
e-mail: khujadze@fdy.tu-darmstadt.de, oberlack@fdy.tu-darmstadt.de

George Chagelishvili
Abastumani Astrophysical Observatory, Kazbegi Ave. 2A, Tbilisi, Georgia
e-mail: g.chagelishvili@astro-ge.org

leads to the high level of the fluctuation energy production (see [2] and references therein).

2 DNS of the Stochastically Forced Plane Couette Flow

Consider an incompressible plane Couette flow ($\mathbf{U}_0 = (Ax_2, 0, 0)$) in 3D with shear parameter A, channel half-width L and Reynolds number Re based on the mean centerline velocity and the half-width of the channel ($Re \equiv AL^2/\nu$). In the laminar case this flow is slightly non-equilibrium and the fluctuations can be neglected beyond the linear order. Consequently, the linearized equations for the stochastically forced small fluctuations can be written in the Cartesian coordinates as

$$\frac{\partial u_i(\mathbf{r},t)}{\partial x_i} = 0, \tag{1}$$

$$\left(\frac{\partial}{\partial t} + Ax_2\frac{\partial}{\partial x_1}\right) u_i(\mathbf{r},t) + Au_2(\mathbf{r},t)\delta_{i1} = -\frac{1}{\rho_0}\frac{\partial p(\mathbf{r},t)}{\partial x_i} + \nu\Delta u_i(\mathbf{r},t) + \frac{\partial s_{ij}(\mathbf{r},t)}{\partial x_j}, \tag{2}$$

$$\text{No-slip boundaries:} \quad u_i(x_1, \pm L/2, x_3, t) = 0.$$

We denote the point (x_1, x_2, x_3) in a Cartesian frame by $\mathbf{r}$. ρ_0 is the uniform fluid density. $u_i(\mathbf{r},t)$ and $p(\mathbf{r},t)$ are the components of the velocity fluctuations ($i = 1,2,3$) and the pressure fluctuations respectively, Δ is the Laplacian and $s_{ij}(\mathbf{r},t)$ corresponds to the spontaneous strain tensor. The last term containing $s_{ij}(\mathbf{r},t)$ defines the stochastic forcing of the system. Statistical properties of the spontaneous strain tensor are modeled in accordance with the Fluctuation-Dissipation theory [3]:

$$\langle s_{ij}(\mathbf{r},t)s_{kl}(\mathbf{r}',t')\rangle = \frac{2T\nu}{\rho_0}\left[\delta_{ik}\delta_{jl} + \delta_{il}\delta_{kj} - \frac{2}{3}\delta_{ij}\delta_{kl}\right]\delta(\mathbf{r}-\mathbf{r}')\delta(t-t'). \tag{3}$$

For the numerical study of the flow we use the pseudo-spectral code [4]. Because of the spectral method we have to implement the stochastic forcing in the wave-number space. Hence, using Fourier transform

$$s_{ij}(\mathbf{r},t) = \int S_{ij}(\mathbf{k},t) exp^{\imath\mathbf{kr}} d\mathbf{k}, \tag{4}$$

one can obtain the correlator function for spontaneous strain tensor in the wave-number space:

$$\langle S_{ij}(\mathbf{k},t)S_{kl}(\mathbf{k}',t')\rangle = \frac{2T\nu}{\rho_0}\left[\delta_{ik}\delta_{jl} + \delta_{il}\delta_{kj} - \frac{2}{3}\delta_{ij}\delta_{kl}\right]\delta(\mathbf{k}-\mathbf{k}')\delta(t-t'). \tag{5}$$

To implement in the code the stochastic forcing satisfying the statistical characteristics of $S_{ij}(\mathbf{k},t)$ (eq. 4) we introduce the following numerical realization:

$$S_{ij}(\mathbf{k},t) \equiv \varepsilon \left\{ \begin{array}{ccc} \sqrt{\frac{4}{3}}cos[2\pi\phi_1(\mathbf{k},t)] & cos[2\pi\phi_2(\mathbf{k},t)] & cos[2\pi\phi_3(\mathbf{k},t)] \\ cos[2\pi\phi_2(\mathbf{k},t)] & \sqrt{\frac{4}{3}}cos[2\pi\phi_1(\mathbf{k},t)+\frac{2\pi}{3}] & cos[2\pi\phi_4(\mathbf{k},t)] \\ cos[2\pi\phi_3(\mathbf{k},t)] & cos[2\pi\phi_4(\mathbf{k},t)] & \sqrt{\frac{4}{3}}cos[2\pi\phi_1(\mathbf{k},t)+\frac{4\pi}{3}] \end{array} \right\}, \quad (6)$$

where $\varepsilon \equiv \sqrt{8T\nu/\rho_0}$ is a measure of the stochastic forcing, $\phi_1(\mathbf{k},t), \phi_2(\mathbf{k},t)$, $\phi_3(\mathbf{k},t)$ and $\phi_4(\mathbf{k},t)$ are random numbers in the range of $[0,1]$.

Linear Navier-Stokes equations with stochastic forcing were numerically simulated with the box $(L_x = 6\pi) \times (L_y = 2) \times (L_z = 2\pi)$ and the grid $256 \times 217 \times 128$ for the following flow parameters $[A] = 1,3$, $[L] = 1$, $[\nu] = \sqrt{1/300}$, i.e. for the Reynolds number $Re = 300$. The initial flow-field perturbations were set to zero. Consequently, in the system flow-field perturbations were caused solely due to the stochastic forcing. The performed linear simulations show that variance reaches a finite statistically stationary level at $[t] > 150$, which greatly exceeds the variance level resulting from the balance between energy accumulated from stochastic forcing and energy dissipated by the normal modes. The reason of this is a transient extraction of the background shear energy by eddy fluctuations that is a consequence of the non-normality of the linear dynamics of channel flows. This result is in agreement with the studies of a stochastically forced shear flows [2].

3 Results

The results are presented in Figures 1 and 2. One-point velocity correlations $\langle u_i u_j \rangle$ and mean velocity profile are presented in Figure 1. The figure shows the existence of nonzero cross-correlations of velocity components. The last plot (f) represents the linear mean velocity profile of the *laminar* Couette flow.

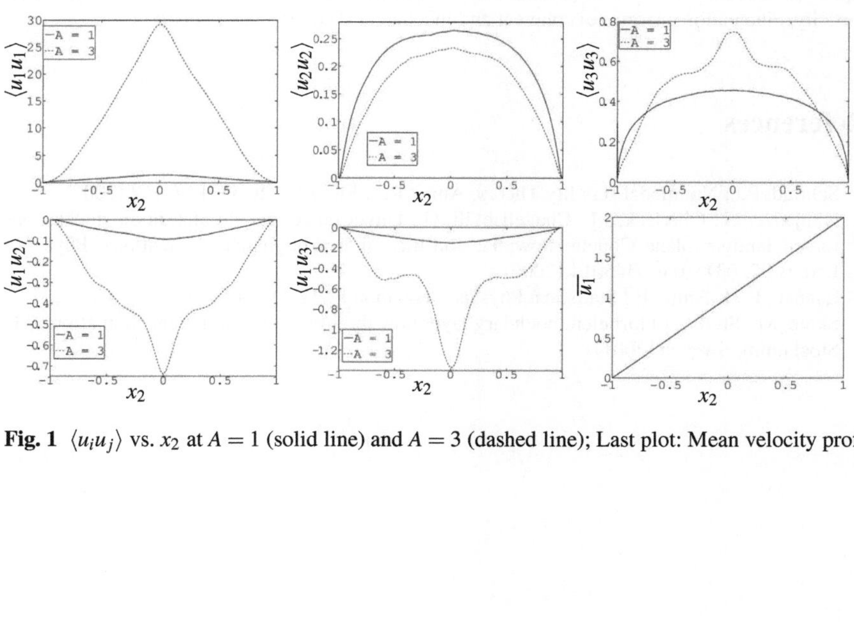

Fig. 1 $\langle u_i u_j \rangle$ vs. x_2 at $A = 1$ (solid line) and $A = 3$ (dashed line); Last plot: Mean velocity profile.

Predominance of the streamwise *non-constant* structural regularities in the fluctuating background with the characteristic length-scale of the order of the channel width are displayed in Figure 2. It displays the velocity components, wall-normal vorticity and energy of fluctuations in $3D$ domain. One of the main result of the presented work is the different configurations of the different components of the fluctuation velocity field (Figure 2, plots a,b and c) due to the non-normality of the linear dynamics of the shear flows. Different component of velocity perturbation extract the energy from the mean flow not by the classical/exponential but by algebraic laws different for each component. This circumstance leads to the different characteristic configurations and scales of the hydrodynamic fluctuation background. The plot (d) represents the coherent part of the wall-normal vorticity. The last two plots (e,f) in this figure show the coherent and incoherent (white noise) parts of the fluctuation energy. All these plots show that the background of the hydrodynamic fluctuations in the laminar plane Couette flow is anisotropic and has well pronounced peculiarities due to the non-normality of the dynamical processes. The main result of the work is that the high level of vortical perturbations can be produced in the system by the intrinsic stochastic forcing (hydrodynamic fluctuations) of the flow.

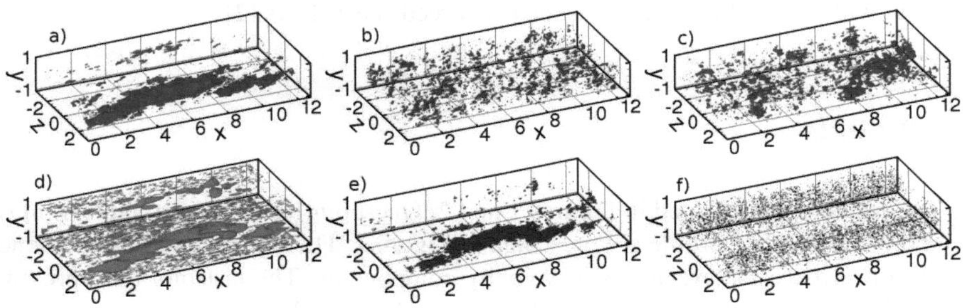

Fig. 2 (a) Streamwise, (b) wall-normal and (c) spanwise veloctity fluctuations, (d) Wall-normal vorticity. Fluctuation energy: (e) coherent and incoherent (f) parts (white noise).

References

1. Schmid, P.J.: Nonmodal stability Theory. Annu. Rev. Fluid Mech. 39, 129–162 (2007)
2. Khujadze, G., Oberlack, M., Chagelishvili, G.: Direct numerical simulation on stochastically forced laminar plane Couette flow: Peculiarities of hydrodynamic fluctuations. Phys. Rev. Letters 97, 0345011–0345014 (2006)
3. Landau, L., Lifshitz, E.: Statistical Physics. Pergamon Press, Oxford (1980)
4. Skote, M.: Studies of turbulent boundary layer flow through direct numerical simulation, KTH, Stockholm, Sweden (2001)

Reynolds Stress Model Based on the RDT Equations and Turbulence Dynamics in the Aerodynamic Nozzle

V. L. Zimont and V.A. Sabelnikov

1 The Paradigm of the Turbulence Model

The well known challenge of turbulence modeling is the inclusion of the effects of the large-scale structure in the one-point model equations. We proposed an approach [1,2] to the Reynolds stresses modeling that is based on an unclosed equation in terms of the spectral tensorr $\Phi_{i,j}(\mathbf{k})$ deduced in [3]:

$$\frac{d\Phi_{ij}}{dt} - E_{\alpha\beta}\frac{\partial k_\alpha \Phi_{ij}}{\partial k_\beta} + E_{i\alpha}\Phi_{\alpha j} + E_{j\alpha}\Phi_{i\alpha} - E_{j\alpha}\overline{u_i u_\alpha}$$
$$-2E_{\alpha\beta}\frac{k_\alpha}{k^2}(k_i\Phi_{\beta j} + k_j\Phi_{i\beta}) + \cdots = 0, \qquad (1)$$

where $E_{ij}(t) = \partial U_i/\partial x_j$ and dots denote missed nonlinear and dissipation terms approximated in our model. Using well known expressions for the Reynolds stresses $\overline{u_i u_j}$ and integral scale L in terms of $\Phi_{i,j}(\mathbf{k})$ we can write the model invariant equations for the case of homogeneous turbulence as follows:

$$\frac{d\overline{u_i u_j}}{dt} = -E_{i\alpha}\overline{u_\alpha u_j} - E_{j\alpha}\overline{u_i u_\alpha} + 2E_{\alpha\beta}\int (k_\alpha/k^2)(k_i\Phi_{\beta j} + k_j\Phi_{i\beta})d^3\mathbf{k}$$
$$-Au^3/L - B(u/L)(\overline{u_i u_j} - u^2\delta_{ij}) \qquad \text{(a)},$$
$$\frac{dLu^2}{dt} = E_{\alpha\beta}(\pi/4)[\int k^{-1}(\partial(k_\alpha\Phi_{\gamma\gamma})/\partial k_\beta - \Phi_{\beta\alpha} - \Phi_{\alpha\beta})d^3\mathbf{k} - 0.5Au^3] \quad \text{(b)}. (2)$$

In Eqs.(2a,b) we use Kolmogorov [4] and Rotta [5] approximations for the dissipation and nonlinear exchange terms usual in Reynolds stress models, but we keep the

---

V. L. Zimont
CRS4 Research Center, Polaris, 09010 Pula(CA), Italy
e-mail: zimont@crs4.it

V.A. Sabelnikov
ONERA, 91761 Palaiseau, France
e-mail: vladimir.sabelnikov@onera.fr

two-points linear terms, which are controlled essentially by the large-scale structure of turbulence, A and B are empirical constants ~ 1. The key point of our approach is a physical hypothesis we use for large-scale structure modeling in Eqs.(2). It is based on solutions of the linear rapid deformation theory (RDT), i.e. solutions of Eqs.(1) without pointed terms following to the Townsend analysis of experimental data [6,7]. He found that experimental dimensionless (using actual turbulence energy and integral scale) correlation functions $B_{ij}(\mathbf{r})$ are very close to RDT solutions using effective distortion of actual nonuniform flows. It means that the large-scale structure in both cases is similar thought the RDT solutions themselves are not valid at real distortion rates. This hypothesis results simple (but not invariant, i.e. depending on a used coordinate system) expressions for Φ_{ij} in terms of Φ^*_{ij} (index * refers to the RDT solutions):

$$\Phi_{ij}(\mathbf{k}L)/(\overline{u_i u_j}) = \Phi^*_{ij}(\mathbf{k}L^*)/(\overline{u_i u_j})^* . \qquad (3)$$

We use below a non-invariant approximation for modeling turbulence in nozzles. The invariant form is presented in [1]. Very important is the fact that all integrals in Eqs.(2), which are controlled by the large-scale turbulence, do not depend on assumed isotropic initial spectrum $\Phi^*_{ij0} = (E^*_0(k)/4\pi k^2)\Delta_{ij}$ of the accompanied RDT problem. In the case of a flow with nonuniform turbulence l.h.s. of all equations is $D/Dt = \partial/\partial t + u_\alpha \partial/x_\alpha$ and in Eqs.(2) we added gradient diffusion terms. An idea to use RDT for turbulence modeling was proposed independently in [9] and developed in several papers, see a review paper [10]. The main difference between the structure-based [11,12] and our models is that we postulate the identity of the large scale structures of actual turbulence and solution of the corresponding uniform RDT problem. We strongly believe that this physical assumption cardinally simplify the problem and permit to avoid arising serious mathematical problems.

2 Turbulence in Nozzles

Here we apply our approach to a practical problem of dynamics of initially isotropic turbulence moving with the flow through a nozzle, which becomes strongly anisotropic due to flow distortion. We assume the one-dimensional scheme of a compressible nozzle flow where the density $\rho(x)$ and speed $U_1(x)$ are described by the one-dimensional equations, while the lateral components are as follows:

$$U_2(\mathbf{x}) = f(x_1)x_2\,,\; U_3(\mathbf{x}) = f(x_1)x_3\,,\; \text{where} \quad f(x_1) = (-0.5/\rho)d(\rho U_1)/dx_1\,. \qquad (4)$$

The turbulence dynamics is controlled by joint influence of distortion, nonlinear interaction and dissipation and in many cases of supersonic flows turbulence can be treated as incompressible. As $\overline{u_i^2} = \int \Phi ii(\mathbf{k})d^3\mathbf{k}$ and $L_i = \int_0^\infty B_{ii}(r_i)dr_i = \pi \int F_{ii}(\mathbf{k})\delta(k_i)d^3\mathbf{k}$ the equations in terms of $\overline{u_i^2}$ and i-component L_i of the integral scale $L = (L_\alpha u_\alpha^2)/(3u^2)$, ($u^2 = u_\alpha u_\alpha/3$), using Eq.(1) are as follows:

$$U_i \frac{d\overline{u_i^2}}{dx_i} = U_1 \overline{u_1^2} \frac{d[ln(\rho U_1/\rho_0 U_{10})]}{dx_1} - 2\nu \int k^2 \Phi_{ii}(\mathbf{k}) d^3\mathbf{k} + \int \Pi_{ii}(\mathbf{k}) d^3\mathbf{k} -$$
$$U_1 \frac{d[ln(\rho U_1^3/\rho_0 U_{10}^3)]}{dx_1} \left(u_i^2 - \int \frac{k_1 k_i}{k^2} (\Phi_{1i}(\mathbf{k}) + \Phi_{i1}(\mathbf{k})) d^3\mathbf{k} \right), \quad (5)$$

$$U_1 d(L_i \overline{u_i^2})/dx_1 = -L_i \overline{u_i^2} \partial U_i / \partial x_i + \pi \int \Gamma_{ii}(\mathbf{k}) \delta(k_i) d^3\mathbf{k} - 2\nu \int k^2 \Phi_{ii}(\mathbf{k}) d^3\mathbf{k}. \quad (6)$$

The nonlinear terms Γ_{ij} and Π_{ij} from Eq.(1) are analyzed in [13], index "0" refers to the channel entrance. Our final equations are presented in terms of $\overline{u_1^2}$ and u^2 as the latter equation does not contain the spectral tensor. Using approximations of nonlinear and dissipative terms we have instead of Eqs.(5 and (6) the following equations:

$$U_1 d\overline{u_i^2}/dx_1 = U_1 \overline{u_i^2} \cdot dln\mu/dx_1 - A(u\overline{u_i^2}/L) - B(u/L)(u_1^2 - u^2), \quad (7)$$

$$U_1 d\overline{u^2}/dx_1 = U_1 u^2 d[ln(\rho U_1/\rho_0 U_{10})]/dx_1 - (1/3) U_1 \overline{u_1^2} d[ln(\rho U_{10}^3)]dx_1 - Au^3/L, \quad (8)$$

$$U_1 d(L_i \overline{u_1^2}/dx_1) = -L_i \overline{u_1^2} \partial U_i / \partial x_i - B(u/L)(L_i \overline{u_i^2} L u^2) - 0.5 A \overline{u_i^2} u, \quad (9)$$

$$\mu = \overline{u_1^2}^* / \overline{u_1^2 0} = (3/4e_1^2)[1/(\alpha - 1) + (2 - \alpha)/(1 - \alpha)^{3/2} Arth(\sqrt{1 - \alpha})] \quad (10)$$

where $e_i = U_1/U_{10}, \alpha = (\rho_0 U_{10}^3)/(\rho U_1^3)$. Eq.(10) is the analytical solution of the linear theory for our case $\alpha < 1$ [14]. We notice that Eqs.(7-10) are written in the specific "natural" coordinate system and they have no the invariant form. But it simplifies the problem and seems reasonable in this case.

We compare our theoretical predictions (I) with the Uberoi [15] measurements of r.m.s turbulent fluctuations in incompressible flows; (II) with our measurements of the dispersion for an estimation of turbulent diffusion coefficient in accelerating subsonic and supersonic flows.

(I) Figs. 1 shows behavior of the turbulence, which is generated by the grid along the channel that consists first on a part with the constant area (dissipation), later on the nozzle with the length A (formation of strong anisotropy) and finally on the channel of lower constant area (tendency to isotropy and dissipation). In Figs.(1, 2) $\varepsilon_1 = (\overline{u_1^2})^{1/2}/U_{10}$ and $\varepsilon_2 = (\overline{u_2^2})^{1/2}/U_{10}$ are plotted. The mesh size M is used to have dimensionless coordinate. Solid lines show our calculations ($A = 1$ and $B = 0.6$), dashed line is the linear theory [13], marks show the experimental data from [15].

(II) Prediction of the lateral diffusion coefficient $D_2 \approx (\overline{u_2^2})^{1/2} L_2$ is significant as it controls mixing. The linear theory and our model predict the dimensionless lateral turbulent diffusion coefficient $D(x) = D_2(x)/D_{20}$. In the undisturbed flow $d\sigma/dx_1 = 2D_2/U_1$ (σ^2 is the dispersion), in nozzles it is more convenient to use a generalized dispersion $\Sigma^2(x_1) = (F(x_{10})/F(x_1))\sigma^2$, which obeys the formula $d\Sigma^2/dx_1 = 2(F(x_{10})/F(x_1)) D_2(x_1)/U_1(x_1)$. Here F is the channel area, x_{10} is the coordinate of the nozzle entrance. Figs. (3) and (4) demonstrate $D(\Delta x)$ (upper curves) and $\Sigma^2(\Delta x)$ (lower curves). Linear results do not depend on the initial intensity of turbulence ε_0. The model results correspond to $\varepsilon_0 = 0.05, 0.10, 0.15$.

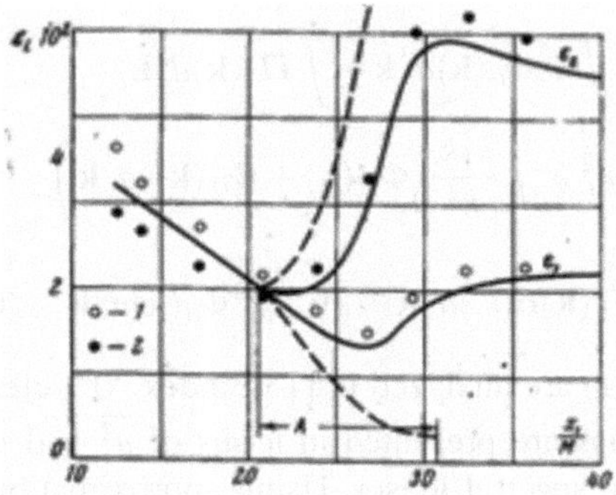

Fig. 1 Contraction degree 16

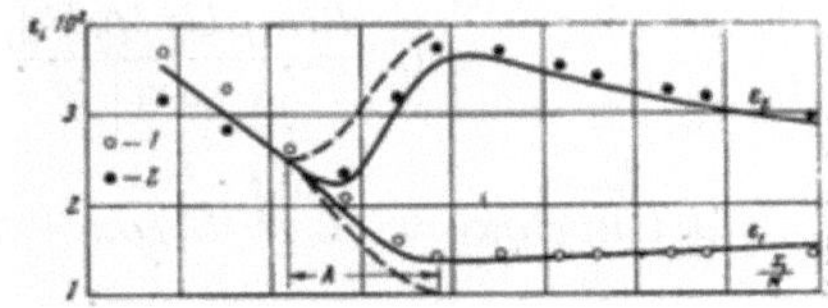

Fig. 2 Contraction degree 4

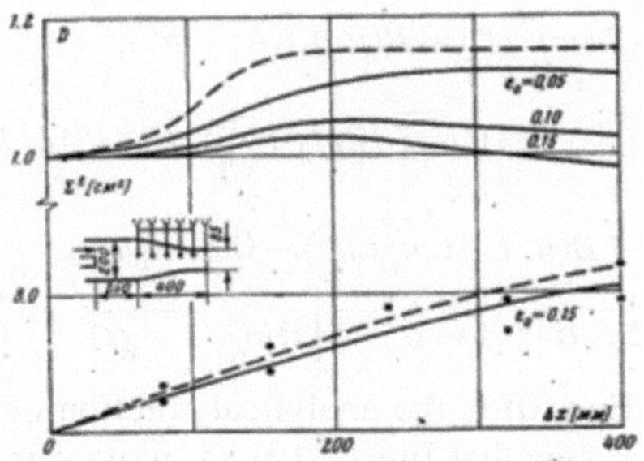

Fig. 3 The subsonic nozzle

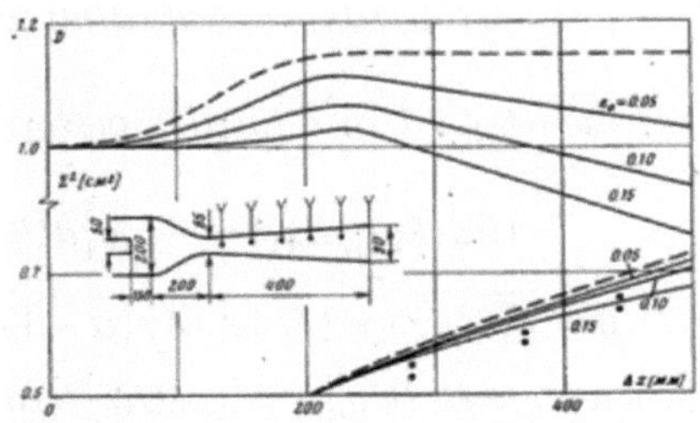

Fig. 4 The supersonic nozzle $M_{exit} = 1.4$

References

1. Zimont, V.L., Sabelnikov, V.A.: Sov. Phys. Dokl. 20(5), 324–326 (1975)
2. Zimont, V.L., Sabelnikov, V.A.: Fluid Mechanics - Soviet Research 7(2), 39–47 (1978)
3. Batchelor, G.K., Proudman, J.: Quart. J. Mech. Appl. Math. 7, 83–103 (1954)
4. Kolmogorov, A.N.: Doklady AN. SSSR 30, 299–303 (1941)
5. Rotta, J.C.: Zeitschrift f. ph. 131(1), 51–77 (1951)
6. Townsend, A.A.: The Structure of Turbulence Shear Flow. Cambridge University Press, Cambridge (1956)
7. Townsend, A.A.: J. of Fluid Mech. 41, 13–46 (1970)
8. Keffer, J.F., Kawall, J.G., Hunt, J.C.R., Maxey, M.R.: J. Fluid Mech. 86, 465–490 (1978)
9. Reynolds, W.C., Kassinos, S.C.: Proc. R. Soc. Lond. A 451(1941), 87–104 (1995)
10. Cambon, C., Scott, J.: Annual Review of Fluid Mechanics 31, 1–53 (1999)
11. Kassinos, S.C., Langer, C.A., Haire, S.L., Reynolds, W.C.: Int. J. Heat Fluid Flow 21, 599 (2000)
12. Poroseva, S.V., Kassinos, S.C., Langer, C.A., Reynolds, W.C.: Phys. Fluids 14(4), 1523–1532 (2002)
13. Batchelor, G.K.: The Theory of Homogeneous Turbulence. Cambridge University Press, Cambridge (1953)
14. Ribner, H.S., Tucker, M.: NACA Report 1113 (1953)
15. Uberoi, M.S.: J. Aero. Sci. 23, 754–764 (1956)

Heat Transfer Modulation by Microparticles in Turbulent Channel Flow

Alfredo Soldati, Francesco Zonta, and Cristian Marchioli

Abstract. In this paper, we analyze the heat transfer modulation produced by the addition of small particles to the base fluid in particle-laden channel flow. We performed direct numerical simulations without gravity at shear Reynolds number $Re_\tau = 150$ and molecular Prandtl number $Pr = 3$ (liquid-solid flow), using the Eulerian-Lagrangian point-particle approach and considering full (momentum and energy) coupling between the fluid and the particles. For thermally-developing flow, we notice significant changes (up to 10 %) to the fluid heat transfer which can be attributed to particle distribution not yet in equilibrium with turbulence. This transient state promotes the *extra* transfer mechanisms that modify strongly the overall heat fluxes. Results for fully-developed flow show that these changes tend to become negligible once the hydrodynamic and thermal equilibrium is achieved.

1 Introduction

Heat transfer enhancement is a fascinating subject with extremely interesting possibilities for application. One option to increase heat transfer is to devise a heat transfer media constituted by a base fluid in which suitably-chosen heat transfer agents, precisely micro and nano particles, are injected. In this way, the fluid can be a standard fluid characterized by simplicity of use and well-known properties, like water, and the heat transfer agents can be heavy-metal, high-heat-capacity, dispersed particles. Current literature trends show the potentials of such heat transfer media,

---

Alfredo Soldati
Centro Interdipartimentale di Fluidodinamica e Idraulica and Dipartimento di Energetica e Macchine, Universita‘ degli Studi di Udine, Via delle Scienze 208, 33100 Udine, Italy
e-mail: alfredo.soldati@uniud.it

Francesco Zonta
Dipartimento di Energetica e Macchine, Universita‘ degli Studi di Udine, Via delle Scienze 208, 33100 Udine, Italy
e-mail: francesco.zonta@uniud.it

Cristian Marchioli
Dipartimento di Energetica e Macchine, Universita‘ degli Studi di Udine, Via delle Scienze 208, 33100 Udine, Italy
e-mail: marchioli@uniud.it

yet the mechanisms which govern the turbulent heat transfer among the fluid and the particles are far from being fully understood due to the complicacy and the cost of experiments. Further complicating effects are represented by the particle inertia and the particle thermal inertia, which are additional parameters. Recently, we have started a research project whose strategic goals are to investigate the heat transfer mechanisms in micro and nanofluids and to devise a suitable numerical methodology to analyse their behavior. The first preliminary results were presented in a recent paper [1] in which we exploited Direct Numerical Simulation (DNS) and Lagrangian Particle Tracking (LPT) to study the influence of dispersed micrometer size particles on the turbulent heat transfer mechanisms in liquid-solid channel flow. To the best of our knowledge, that was the first attempt to account for these mechanisms all together treating particles as active heat transfer agents which interact both with the temperature field and the velocity field of the base fluid. In this paper we complement those results by performing a quantitative analysis of the heat fluxes occurring in the flow when proper energy and momentum coupling terms are incorporated in the governing equations of both phases.

2 DNS/LPT Methodology

In this study a DNS of fully-developed channel flow with heat transfer is performed. Assuming that the fluid is incompressible and Newtonian, the governing fluid momentum and temperature transport equations (in dimensionless form) read as:

$$\frac{\partial u_i}{\partial t} + u_j \frac{\partial u_i}{\partial x_j} = \frac{1}{Re^*} \frac{\partial^2 u_i}{\partial x_j{}^2} - \frac{\partial p}{\partial x_i} + \delta_{1,i} + f_{2w}, \quad \frac{\partial T}{\partial t} + u_j \frac{\partial T}{\partial x_j} = \frac{1}{Re^* Pr} \frac{\partial^2 T}{\partial x_j^2} + q_{2w},$$

where u_i is the ith component of the velocity vector, p is the fluctuating kinematic pressure, $\delta_{1,i}$ is the mean pressure gradient that drives the flow, T is the temperature, Re^* is the shear (or friction) Reynolds number and Pr is the Prandtl number. The momentum-coupling term f_{2w} and the energy-coupling term q_{2w} are defined in terms of momentum and energy flux per unit mass. The fluid equations are solved using a pseudo-spectral Fourier-Chebyshev method in which time integration is performed using an explicit two-stage Euler/Adams-Bashforth scheme for convective terms and an implicit Crank-Nicolson method for the viscous terms. The time step used is $dt^+ = 0.045$ in wall units. The reference geometry consists of two infinite flat parallel walls; the origin of the coordinate system is located at the center of the channel and the $x-$, $y-$ and $z-$ axes point in the streamwise, spanwise and wall-normal directions respectively. The calculations are performed on a computational domain of size $4\pi h \times 2\pi h \times 2h$ in x, y and z, respectively. Periodic boundary conditions are imposed on both velocity and temperature in the homogeneous x and y directions; at the wall, no-slip condition is enforced for the momentum equation whereas constant temperature condition is adopted for the energy equation.

Large samples of $\mathcal{O}$ $(10^5 - 10^6)$ heavy particles with diameter $d_p = 4$ μm or $d_p = 8$ μm and with density $\rho_p = 19.3 \cdot 10^3$ kg m^{-3} (gold in water) are injected into the flow at concentration high enough to have significant two-way coupling effects in both the momentum and energy equations but negligible particle-particle interactions. The particle dynamics is described by the following set of ordinary differential equations for velocity and temperature:

$$\frac{d\mathbf{u_p}}{dt} = -\frac{\mathbf{u}_p - \mathbf{u}}{\tau_p}\left(1 + 0.15Re_p^{0.687}\right), \qquad \frac{dT_p}{dt} = \frac{Nu}{2} \cdot \frac{(T_f - T_p)}{\tau_T},$$

where $\mathbf{u}_p = d\mathbf{x}_p/dt$ is the particle velocity (and $\mathbf{x_p}$ is the particle position), $\mathbf{u}$ is the fluid velocity at particle position, τ_p is the particle response time, T_p is the temperature of the particle, Nu is the Nusselt number, T_f is the temperature of the fluid at particle position, and τ_T is the particle thermal response time. In this study, we considered uniform temperature inside the particle since the particle Biot number is $\mathscr{O}(10^{-6})$. The dimensionless values of τ_p and τ_T, namely the particle Stokes numbers, are $St = 1.56$ and $St_T = 0.5$ for the smaller particles, and $St = 6.24$ and $St_T = 2$ for the larger particles. Particle equations are solved using a 4th-order Runge-Kutta scheme for time integration and 6th-order Lagrangian polynomials to interpolate the fluid velocity and fluid temperature at particle position. Particles are treated as pointwise, rigid spheres (point-particle approach) that have random initial distribution. Particle initial velocity and temperature are set equal to those of the fluid at the particle position. Periodic boundary conditions are imposed on particles moving outside the computational domain in the homogeneous directions, whereas perfectly-elastic collisions are assumed when the particle hit the wall.

3 Results and Discussion

The wall-normal heat flux of the fluid is defined as $q = \rho c_p \overline{w'T'} - k \cdot \partial T/\partial z$, where the two terms on the right-hand side represent the turbulent heat flux and the laminar (or conductive) heat flux, respectively. In the expression above, w' and T' are the fluid velocity and temperature fluctuations, and their product is time averaged. In Fig. 1 we analyze the behavior of the unsteady dimensionless heat flux in the unladen flow to that observed in the laden flow for increasing values of the particle mass fraction, Φ_m. Profiles are normalized by the value of $q/\rho c_p$ at the wall in the unladen flow case and were calculated by averaging both in space, over the homogeneous directions, and in time, over a time span of 450 wall time units starting at time $t^+ = 3800$ after particle injection in the flow. For the 4 μm particles (red profiles), the addition of particles does not seem to produce significant changes in the heat flux at the wall. Away from the wall ($20 < z^+ < 150$) an increase of Φ_m appears to decrease the heat flux. Taking the unladen flow case as reference, in the $\Phi_m = 3.28 \cdot 10^{-3}$ case the heat flux increases of about 3 % in the core region of the flow, whereas in the $\Phi_m = 2.6 \cdot 10^{-2}$ case the flux is even lower. A similar trend is observed for the 8 μm particles (blue profiles), even if the situation seems to improve in terms of particle-induced heat enhancement: the heat flux in the outer flow region increases up to 10% compared to the unladen flow with $\Phi_m = 2.6 \cdot 10^{-2}$ and by slightly less than 3% with $\Phi_m = 2.1 \cdot 10^{-1}$. In the wall region, heat enhancement is always less than 3%. This behavior can be explained considering that higher mass loadings produce modifications of the flow Reynolds stresses such that the magnitude of the temperature variance is reduced and the turbulent heat flux of both fluid and particles is dampened. Let us now examine how the transient results obtained for thermally-developing flow change once the steady state is approached, namely when the wall-normal profiles tend towards a constant and uniform value. In our simulations, a statistically steady state is approached after about 8500 wall time units upon particle release. In Fig. 2 we show the modifications to the steady-state heat flux produced by changes in the particle diameter: there is no clearcut evidence of a persistent effect of particles on heat transfer enhancement. The same is observed when particle concentration is changed (not shown). In our opinion, this result can be attributed to the connection existing between the heat exchange process and the particle wall accumulation process. At this point of particle dispersion, the outer flow region has been depleted of particles since a major proportion has already settled in the viscous sublayer. The number of active heat transfer agents left in the center of the channel is too small to produce significant heat exchange between the two phases; conversely particles in the near-wall region remain trapped for very long times, reach a condition of hydrodynamic and thermal equilibrium with the surrounding fluid and eventually act as an additional thermal resistance both between the walls and the fluid and between the fluid and the particles: the main consequence of this behavior is that we may expect changes in the heat flux as long as inertia is decoupling particle motion from fluid motion. As soon as particles reach equilibrium through mutual interaction with the surrounding fluid environment the heat exchange between the two phases tends to relax to minor differences compared with the unladen flow situation.

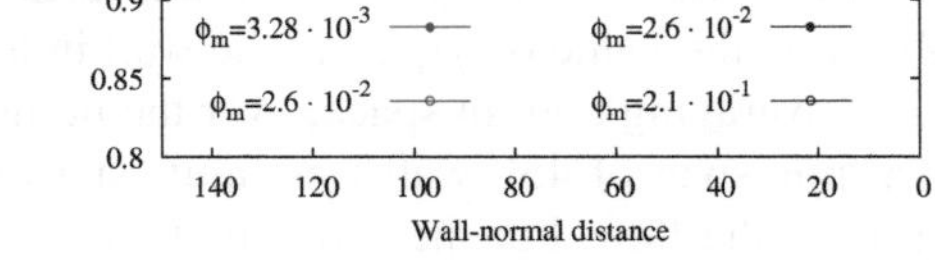

Fig. 1 Particle size and concentration effects on total heat flux for thermally-developing flow.

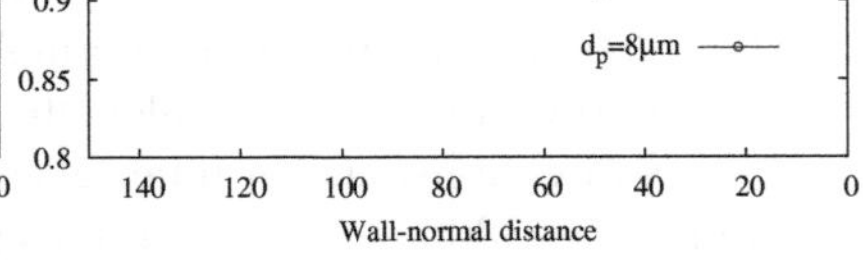

Fig. 2 Effect of particle size on total heat flux for thermally-steady flow ($\Phi_m = 2.6 \cdot 10^{-2}$).

References

1. Zonta, F., et al.: Direct numerical simulation of turbulent heat transfer modulation in micro-dispersed channel flow. Acta Mechanica 195, 305–326 (2008)

Particle Diffusion in Stably Stratified Flows

Geert Brethouwer and Erik Lindborg

Abstract. Numerical simulations are used to study the vertical dispersion of fluid particles in homogeneous turbulent flows with a stable stratification. The results of direct numerical simulations are in good agreement with the relation for the long time fluid particle dispersion, $\langle \delta z^2 \rangle = 2\varepsilon_P t/N^2$, derived by [6], though with a small dependence on the buoyancy Reynolds number. Here, $\langle \delta z^2 \rangle$ is the mean square vertical particle displacement, ε_P is the dissipation of potential energy, t is time and N is the Brunt-Väisälä frequency. A simulation with hyperviscosity is performed to verify the relation $\langle \delta z^2 \rangle = (1+\pi C_{PL})2\varepsilon_P t/N^2$ for shorter times, also derived by [6]. The agreement is reasonable and we find that $C_{PL} \sim 3$. The onset of a plateau in $\langle \delta z^2 \rangle$ is observed in the simulations at $t \sim E_P/\varepsilon_P$ which scales as $4E_P/N^2$, where E_P is the potential energy.

1 Introduction

The flows in the ocean, atmosphere, estuaries and lakes are often stably stratified which has a large impact on mixing and dispersion. Several researchers have examined the vertical dispersion of fluid particles in stratified flows in order to obtain a better understanding of mixing in geophysical flows. For example, [7], [3] and [8] used respectively kinematic simulations, rapid distortion theory and a Langevin model to study fluid particle dispersion in stratified flows. Direct numerical simulations (DNS) of decaying stratified turbulent flows were used by [4] and [9] to study vertical dispersion. The only DNS of fluid particle dispersion in a stationary stratified turbulent flow was carried out by [1]. In general, it is found that the mean square of the vertical fluid particle displacement, $\langle \delta z^2 \rangle$, reaches or shows the onset

---

Geert Brethouwer
Linné Flow Centre, KTH Mechanics, SE-100 44 Stockholm, Sweden
e-mail: geert@mech.kth.se

Erik Lindborg
Linné Flow Centre, KTH Mechanics, SE-100 44 Stockholm, Sweden
e-mail: erikl@mech.kth.se

of plateau where either $\langle \delta z^2 \rangle \sim E_K/N^2$ or $\langle \delta z^2 \rangle \sim \langle w^2 \rangle /N^2$. Here, E_K is the turbulent kinetic energy, w is the vertical velocity fluctuation and N is the Brunt-Väisälä frequency. However, not much is known about the long time dispersion. Van Aartrijk [1] observed in some of their simulations that $\langle \delta z^2 \rangle \sim t$ at long times t and argued that this phenomenon is due to the changing density of the fluid particles. This long time behaviour was also predicted by [8].

Recently, [6] derived analytical relations for $\langle \delta z^2 \rangle$ in statistically stationary and stratified homogeneous turbulent flows governed by the Boussinesq equations. By integrating the governing equations along a fluid particle trajectory and neglecting one term using scaling arguments, we derived

$$\langle \delta z^2 \rangle = \frac{2}{N^2} \left[\varepsilon_P t \left(1 - \mathscr{O}(\mathscr{R}^{-1/2}) \right) + 2E_P \right] \tag{1}$$

for $t \gg E_P/\varepsilon_P$. Here, ε_P is the dissipation of potential energy, E_P is the potential energy, $\mathscr{R} = \varepsilon_K/\nu N^2$ is the buoyancy Reynolds number, ε_K is the turbulent kinetic energy dissipation and ν is the viscosity. Adiabatic dispersion, represented by the last term in (1), gives a finite contribution to long time dispersion. The first term on the right-hand-side, which represents diabatic dispersion due to density changes of the fluid particles, leads to $\langle \delta z^2 \rangle \sim t$ for $t \to \infty$. In geophysical flows generally $\mathscr{R} \gg 1$ and consequently the $\mathscr{O}(\mathscr{R}^{-1/2})$-term in (1) can be neglected. However, in laboratory experiments or numerical simulations this term can give a significant contribution since $\mathscr{R}$ is then not always very large.

Strongly stratified turbulence with a high Reynolds number has an anisotropic inertial range at scales larger than the Ozmidov length scale [2][5]. Assuming the existence of such an inertial range, we derived

$$\langle \delta z^2 \rangle = \frac{2}{N^2} \varepsilon_P t \left[1 + \pi C_{PL} - \mathscr{O}(\mathscr{R}^{-1/2}) \right], \tag{2}$$

for $N^{-1} \ll t \ll E_P/\varepsilon_P$. Using documented observations, [6] estimated that the constant $C_{PL} \approx 3$. The adiabatic dispersion, $\langle \delta z^2 \rangle = 2\pi C_{PL} \varepsilon_P t/N^2$, gives then the dominant contribution to dispersion in this period. The aim of our study is to verify (1) and (2) by numerical simulations.

2 Numerical Simulations and Results

We have carried out DNS of homogeneous turbulence with different stratifications. The Boussinesq equations are solved with a standard pseudospectral method. The large velocity scales are forced to obtain a statistically stationary flow. We use 256 modes in the horizontal direction and 96 up to 256 modes in the vertical direction in the DNS discussed here. More details about the code and the forcing can be found in [2] and [5]. Table 1 lists the Reynolds and Froude number and $\mathscr{R}$ of the DNS, and shows that we have simulated flows with a strong as well as weak stratification. In

Table 1 Parameters of the DNS. The Reynolds and Froude number are defined as $Re = E_K^2/\nu\varepsilon_K$ and $F_h = \varepsilon_K/NE_K$ respectively. Here, $\mathscr{R} = ReF_h^2 = \varepsilon_K/\nu N^2$.

run	Re (10^3)	F_h	$\mathscr{R}$
A	2.1	0.02	0.9
B	2.5	0.06	9.3
C	2.8	0.11	37
D	2.4	1.6	6200

the simulations we track a large number of fluid particles. An interpolation scheme is used to obtain the particle velocity at its position.

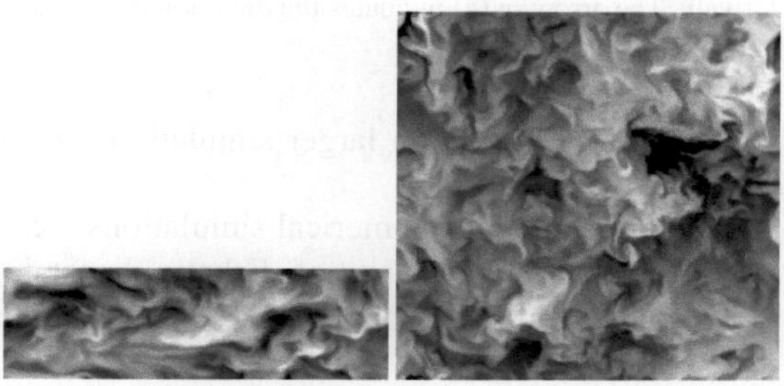

Fig. 1 Snapshot of the buoyancy field in a vertical plane in run B (left) and run D (right).

Fig. 1 show snapshots of the buoyancy field in a vertical plane in run B and D which have a strong and weak stratification respectively. In the strongly stratified case the buoyancy field is highly anisotropic and layers develop while in the weakly stratified case it is much more isotropic.

Fig. 2(a) shows the time development of $\langle \delta z^2 \rangle$ in the DNS. The relations (1) and (2) are also plotted in the figure. Initially, the fluid particle dispersion is ballistic with $\langle \delta z^2 \rangle \sim t^2$. Thereafter, the dispersion is mainly adiabatic. We see the onset of a plateau in run A, B and C where $\langle \delta z^2 \rangle \approx 4E_P/N^2$ at $t \sim E_P/\varepsilon_P$, in agreement with (1). This plateau is the result of the constraint put on the adiabatic dispersion by the stable stratification. However, there is no period where $\langle \delta z^2 \rangle$ follows the dispersion relation (2). Nevertheless, we can observe that the computed $\langle \delta z^2 \rangle$ approaches (2) when the Froude number decreases. It is probably necessary to perform a simulation with a much higher Reynolds number and stronger stratification in order to observe this behaviour, but this is unfortunately not within the reach of DNS. At longer times, $\langle \delta z^2 \rangle$ starts growing faster again in run A, B and C because diabatic dispersion becomes the dominant contribution. We see that in all simulations $\langle \delta z^2 \rangle \to 2\varepsilon_P t/N^2$ at $t \gg E_P/\varepsilon_P$ and that $\langle \delta z^2 \rangle$ comes closer to this limit when $\mathscr{R}$ increases. These latter results are consistent with (1).

We have also carried out a simulation of stratified turbulence with hyperviscosity because in such a simulation the inertial stratified range is more clear [5]. Fig. 2(b) shows the result of this simulation. We see that $\langle \delta z^2 \rangle$ approaches (1) at long times. Moreover, at shorter times $\langle \delta z^2 \rangle$ approaches (2), but there is still no clear range

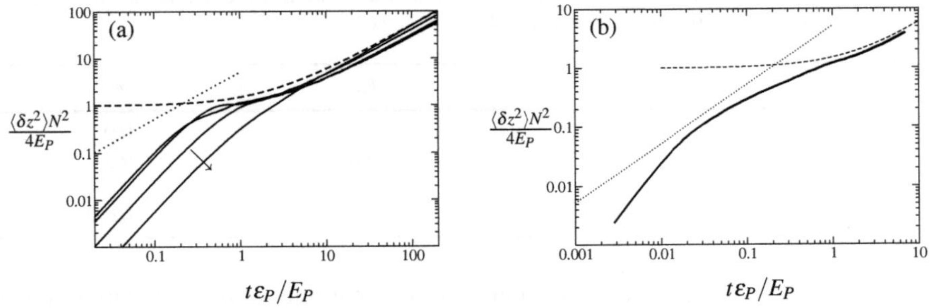

Fig. 2 Time development of $\langle\delta z^2\rangle N^2/4E_P$ *vs.* $t\varepsilon_P/E_P$ in the DNS (a) and hyperviscosity simulation (b). The dotted, dashed and solid lines show relations (2), (1) and simulation results respectively. The arrow in (a) indicates the direction of increasing $\mathscr{R}$.

where $\langle\delta z^2\rangle \sim t$. Even larger simulations are probably needed to reveal such an inertial behaviour.

To conclude, our numerical simulations are in good agreement with (1), which points out the importance of diabatic dispersion on long time scales. At shorter times, the simulation with hyperviscosity is in reasonable agreement with (2).

Acknowledgement. G. B. received financial support from the Swedish Research Council. Computational resources were provided by the Swedish National Infrastructure for Computing.

References

1. van Aartrijk, M., Clercx, H.J.H., Winters, K.B.: Single-particle, particle-pair, and multiparticle dispersion of fluid particles in forced stably stratified turbulence. Phys. Fluids 20, 25104 (2008)
2. Brethouwer, G., Billant, P., Lindborg, E., Chomaz, J.-M.: Scaling analysis and simulation of strongly stratified turbulent flows. J. Fluid Mech. 585, 343–368 (2007)
3. Kaneda, Y., Ishida, T.: Suppression of vertical diffusion in strongly stratified turbulence. J. Fluid Mech. 402, 311–327 (2000)
4. Kimura, Y., Herring, J.R.: Diffusion in stably stratified turbulence. J. Fluid Mech. 328, 253–269 (1996)
5. Lindborg, E., Brethouwer, G.: Stratified turbulence forced in rotational and divergent modes. J. Fluid Mech. 586, 83–108 (2007)
6. Lindborg, E., Brethouwer, G.: Vertical dispersion by stratified turbulence. J. Fluid Mech. (2008) (in press)
7. Nicolleau, F., Vassilicos, J.C.: Turbulent diffusion in stably stratified non-decaying turbulence. J. Fluid Mech. 410, 123–146 (2000)
8. Pearson, H.J., Puttock, J.S., Hunt, J.C.R.: A statistical model of fluid-element motions and vertical diffusion in a homogeneous stratified turbulent flow. J. Fluid Mech. 129, 219–249 (1983)
9. Venayagamoorthy, S.K., Stretch, D.D.: Lagrangian mixing in decaying stably stratified turbulence. J. Fluid Mech. 564, 197–226 (2006)

Anisotropic Clustering of Inertial Particles in Shear Turbulence

P. Gualtieri, F. Picano, and C.M. Casciola

Abstract. Dynamics of inertial particles has been thoroughly analyzed for statistically homogeneous and isotropic flows where clustering occurs below the Kolmogorov scale. Since anisotropy is strongly depleted through the inertial range, the advecting field anisotropy may be expected in-influential for the small scale features of particle configurations. By addressing direct numerical simulations (DNS) of a statistically steady particle-laden homogeneous shear flow, we find instead that the small scales of the particle distribution are strongly affected by the geometry of velocity fluctuations at large scales even in the range of scales where isotropization of velocity statistics occurs.

1 Motivation and Results

Transport of inertial particles is involved in several fields of science, e.g. droplets growth in clouds or technological applications such as injection systems of internal combustion engines or for sediments accumulation in pipelines.

Inertial particles differ from perfectly Lagrangian tracers due to inertia which prevents them from following the flow trajectories. The main effect consists of "preferential accumulation" that leads to small-scale clustering for locally homogeneous and isotropic flows [1] and to turbophoresis in wall-bounded flows i.e. preferential spatial segregation at the boundary [3].

Inhomogeneity is essential to have spatial segregation, though anisotropy is probably the effective engine of the process. The two features are strongly entangled in wall bounded flows. A special flow exists however–the statistically steady homogeneous shear flow–which retains most of the anisotropic dynamics of wall bounded flows still preserving, spatial homogeneity. This flow shares with the wall-layer

---

P. Gualtieri · F. Picano · C.M. Casciola
Dipartimento di Meccanica e Aeronautica Università di Roma "La Sapienza", Via Eudossiana 18, 00184 Roma, Italy
e-mail: p.gualtieri@caspur.it

streamwise vortices and turbulent kinetic energy production mechanisms. Velocity fluctuations are strongly anisotropic at the large scales driven by production while, for smaller separations, the classical energy transfer mechanisms become effective in inducing re-isotropization.

The main contribution of the present paper is the quantitative evidence that particle distributions do not necessarily reduce their anisotropy at small scales. Rather it may even grow below the Kolmogorov length where the velocity field, is smooth and almost isotropic. By considering the spherical distribution function see [4], its angular decomposition is used to evaluate the relative importance of its different components showing that anisotropy is a leading order effect which may persist down to vanishing separations. A visual impression of the instantaneous spatial distribution of particles is provided in figure 1, where slices of the domain in selected coordinate planes are displayed for three different Stokes numbers. Particle clustering is apparent for $St_\eta = 1$ (mid panel). The particle distribution exhibits many voids, correlated with high enstrophy regions and intertwined thin "stretched" regions at the border

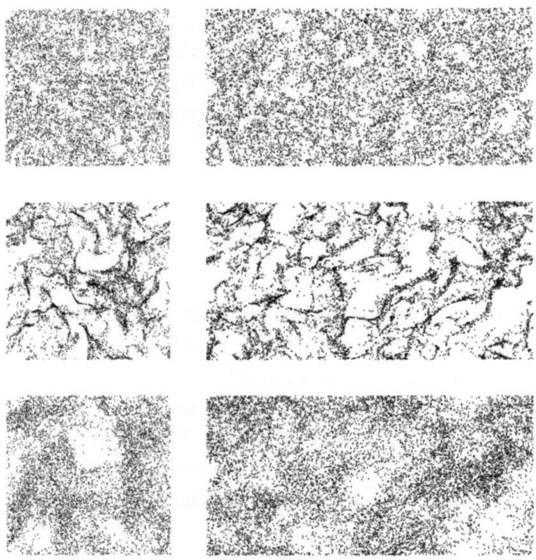

Fig. 1 Snapshots of particle positions for increasing Stokes number based on Kolmogorov timescale, from top to bottom: $St_\eta = 0.1, 1, 10$. Left column thin slice in the $y-z$ plane; right column slice in the $x-y$ plane. The slice thickness is of the order of a few Kolmogorov scales. The mean flow is in the x direction $U = Sy$ with S the mean shear. DNS data with a resolution of $384 \times 384 \times 192$ collocation points in a $4\pi \times 2\pi \times 2\pi$ periodic box. The Kolmogorov scale is $\eta = 0.02$ corresponding to $K_{max}\eta = 3.1$. $L_s/\eta \simeq 35$ where $L_s = \sqrt{\varepsilon/S^3}$ is the shear scale being ε the mean energy dissipation rate. Taylor-Reynolds number $\mathrm{Re}_\lambda = \sqrt{5/(\nu\varepsilon)}\langle u_\alpha u_\alpha\rangle \simeq 100$ and shear strength $S^* = S\langle u_\alpha u_\alpha\rangle/\varepsilon = (L_0/L_s)^{2/3} \simeq 7$ where L_0 is the integral scale. Five populations of $N_p = 3 \cdot 10^5$ particles each are evolved with $St_\eta = 0.1, 0.5, 1.0, 5.0, 10.0$, see [4].

of the voids where a large particle concentration is achieved. Recently particle clustering have been described in terms of other mechanisms such as the sweep-stick mechanism described in [7] which takes into account more complex topological information of the fluid acceleration field. Qualitatively, peak clustering occurs near this Stokes number to decrease both for larger and smaller values, see top and bottom panels of figure 1. Despite of the above similarities with isotropic cases, the homogeneous shear flow manifest specific features associated with the large scale anisotropy. The shear-induced orientation is apparent from the bottom-left/top-right alignment of the sheet-like arrangement of particles in the shear plane $x-y$, see the right panels of figure 1. Purpose of the present paper is to provide an evaluation of this anisotropy effects on clustering, along the same line proved successful for the isotropic case.

The main statistical tool is the radial distribution function (RDF) of particle pairs $g(r)$ which is a function of radial distance r, see e.g. [2] where the RDF is dealt with for isotropic flows. The concept is extended to anisotropic cases by considering the number of pairs $d\mu_r = \nu_r(r,\hat{\mathbf{r}})d\Omega$ contained in a spherical cone of radius r, with axis along the direction $\hat{\mathbf{r}}$ and solid angle $d\Omega$. By this definition the number of pairs in the ball $\mathscr{B}_r$ is $N_r = \int_\Omega \nu_r d\Omega$, hence $dN_r/dr = \int_\Omega d\nu_r/dr\, d\Omega$. We define the Angular Distribution Function (ADF) as

$$g(r,\hat{\mathbf{r}}) = \frac{1}{r^2}\frac{d\nu_r}{dr}\frac{1}{n_0}, \tag{1}$$

which retains information on the angular dependence of the distribution. The RDF is the spherical average of the ADF $g(r) = 1/(4\pi)\int_\Omega g(r,\hat{\mathbf{r}})d\Omega$ and is shown in figure 2, where a scaling behavior is apparent in the range $r/\eta \in [.1:1]$. From the figure, particles with $St_\eta \sim 1$ exhibit maximum accumulation, i.e. the RDF diverges at a faster rate as r is decreased.

The RDF quantitatively confirms the overall impression gained from the visualizations of figure 1. The strong anisotropy apparent in those plots, however, needs a description in terms of the more complete ADF which allows for a systematic evaluation of anisotropy. For given separation r, its angular dependence can be resolved in terms of spherical harmonics, $g(r,\hat{\mathbf{r}}) = \sum_{j=0}^{\infty}\sum_{m=-j}^{j} g_{jm}(r)Y_{jm}(\hat{\mathbf{r}})$. In this notation, the classical RDF $g(r)$ is the projection of the ADF on the isotropic sector $j=0$, namely $g_{00}(r) \equiv g(r) = \int_\Omega g(r,\hat{\mathbf{r}})Y_{00}(\hat{\mathbf{r}})d\Omega$.

The ADF provides a quantitative account of the anisotropy induced by the fluid velocity field on the disperse phase and can be effectively used to parameterize the level of anisotropy through the scales in terms of the Stokes number.

The plot of figure 2 gives the normalized amplitude of the most energetic anisotropic mode in absolute value–$|g_{2-2}|/g_{00}$–for our set of Stokes numbers $St_\eta = 10, 5, 1, 0.5, 0.1$ ranging from heavy to light particles. Focusing on the heaviest particles, $St_\eta = 10, 5$, the relative amplitude of the strongest anisotropic mode first increases towards the small scales to reach a maximum at $r \sim \ell_c$. Below this scale the anisotropy level decreases, until the very small scales become essentially isotropic.

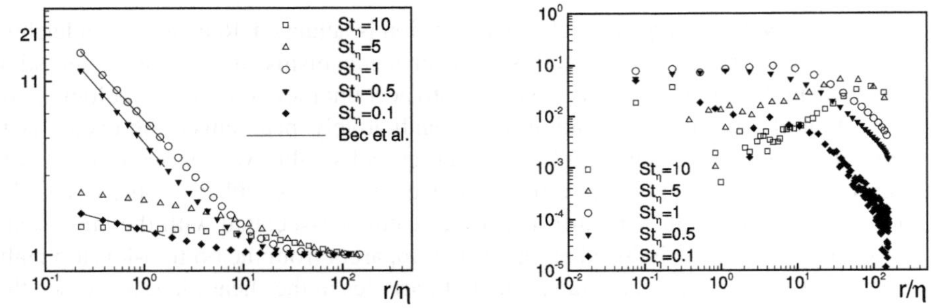

Fig. 2 Left: radial distribution function *vs* separation, for different Stokes number. Right: ratio between the most energetic anisotropic sector $(2,-2)$ normalized by isotropic sector as a function of separation for different Stokes number.

Particles with smaller Stokes numbers behave in a different way. The anisotropy substantially increases to reach a saturation at small scales as Kolmogorov scale is approached. It keeps an almost constant value below η. In other words, the clustering process maintains its anisotropic features even below the dissipative scale for sufficiently small Stokes number particles. Remind that the overall clustering process described by g_{00} is here characterized by a singular exponent α. The saturation observed on the ratio g_{2-2}/g_{00} implies that the dominating anisotropic contribution inherits the same behavior, $g_{2-2} \propto r^{-\alpha}$. Hence, large scale anisotropy of the advecting field originates the anisotropy of the singular clustering process of light particles. Looking at the data for our lightest population, we cannot exclude that the singularity exponent of the most significant anisotropic sector may even be larger than those inferred from the RDF. Conversely, heavy particles appear to preferentially concentrate on finite size patches, where they are more or less evenly and isotropically distributed. The behavior of light inertial particles is completely different from that of velocity fluctuations which manifest two distinct isotropy recovery rates, a smaller one observed in the production range above the shear scale and a larger rate in the inertial transfer below, [6]. As a matter of fact, isotropy is always recovered at dissipative scales, provided the scale separation L_s/η is large enough, i.e. at sufficiently large Reynolds number. This trend towards a recovery of isotropy is not observed in strong small scale-clustering, which in the range from L_s to η shows instead a substantial increase of directionality.

References

1. Bec, J., et al.: Phys. Rev. Lett. 98 (2007)
2. Sundaram, S., Collins, L.R.: J. Fluid Mech. 335 (1997)
3. Rouson, D.W.I., Eaton, J.K.: J. Fluid Mech. 428 (2001)
4. Gualtieri, P., et al.: J. Fluid Mech. (in press, 2009)
5. Shotorban, B., Balachandar, S.: Phys. Fluids 18 (2006)
6. Casciola, C.M., et al.: Phys. Fluids 19 (2007)
7. Goto, S., Vassilicos, J.C.: Phys. Rev. Lett. 100 (2008)

Spatial Evolution of Inertial Particles in a Turbulent Pipe Flow

G. Sardina, F. Picano, and C.M. Casciola

Abstract. The dynamics of small inertial particles transported by a turbulent flow is crucial in many engineering applications. For instance internal combustion engines or rockets involve the interaction between small droplets, chemical kinetics and turbulence. Small, diluted particles, much heavier than the carrier fluid, are essentially forced only by the viscous drag i.e. the Stokes drag (gravity, feedback on fluid and collisions are neglected). The difference between particle velocity V and fluid U produces various anomalous phenomena such as small-scale clustering or preferential accumulation at the wall even for incompressible flows. To stress the interaction between wall bounded flows and particle dynamics we have performed a direct numerical simulation of a fully-developed particle-laden pipe flow. Seven different populations of particles are injected at a fixed location on the axis of the pipe and their evolution is analyzed for a streamwise extension of $200R$ (with R the pipe radius) to asses the onset of turbophoresis.

1 Instantaneous Visualization

Friction Reynolds number of the simulation is about 200, while the injection rate is fixed at 900 particles per $\Delta t = R/U$ for each population. The whole domain is depicted in the top panel of figure 1. Particles injected at the axis are dispersed by turbulent motions until they reach the wall where accumulate because of turbophoretic effects. Note the trend towards a homogeneous distribution at large distance from the injection point.

The bottom panel represents the particles in the early stages from the injection point, the colours show the "smoke lines" traced by the heavy particles with different response time $\tau_p = 18 d_p^2 \rho_p / (\rho_f \nu)$ (d_p, ρ_p are the particle diameter and density

G. Sardina · F. Picano · C.M. Casciola
Dip. di Meccanica e Aeronautica University "Sapienza", Roma
e-mail: gaetano.sardina@uniroma1.it, francesco.picano@uniroma1.it, carlomassimo.casciola@uniroma1.it

respectively; ρ_f and ν are the fluid density and viscosity). Lines are corrugated by the fluid velocity fluctuations resulting in a different particle dispersion. This effect is due to the value of the Stokes number St^+ that is the ratio between τ_p and the characteristic time of near wall region ν/U^{*2}.

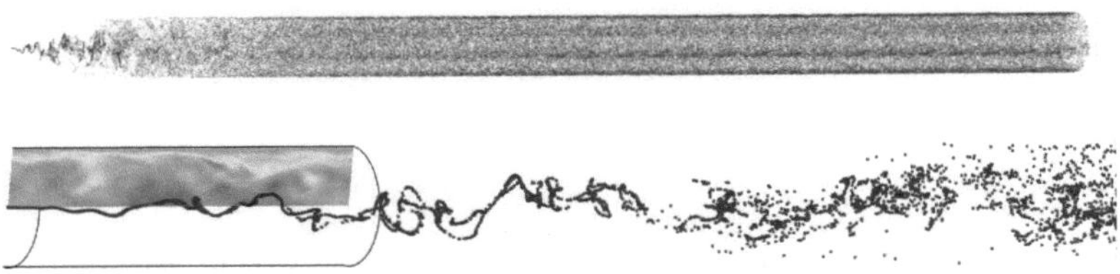

Fig. 1 Snapshots of particle positions ($St^+ = 0.1$, 10, 100 green, blue, and red spheres). Top panel, view of the whole computational domain plotted in arbitrary aspect ratio; bottom panel, enlargment of the near field region with colour encoded istantaneous axial fluid velocity isocontours.

More generally it is possible to distinguish a developing region and a far field. In figure 2 are plotted two cross-flow slices in these two different zones. In the near region ($25R$, left panel) rod-like structures and void regions are present. All particles (except the lightest ones) show a preferential accumulation near the wall. In particular the heaviest ($St^+ = 100$) are distributed around the wall more intermittently compared to the blue ones ($St^+ = 10$).

On the contrary, in the developed region particles fill the whole domain, although preferential accumulation is present near the wall expecially for the heavier particles $St^+ = 10 \div 100$. While turbophoretic effects are negligible for lighter particles.

2 Particles Concentration and Shannon Entropy

The mechanisms of particle dispersion and accumulation is a function of the Stokes number, the panels in figure 3 display the mean concentration for two Stokes numbers $St^+ = 0.1, 10$ as a function of the axial distance at four characteristic wall normal distance i.e. viscous, buffer, log layer and bulk region. In left panel ($St^+ = 0.1$) a uniform distribution for the whole section is reached at $z/R \cong 50$, below this value, the transient state can be observed with the gradual replenishment of all zones from the axis to wall. A different behavior characterizes particles with $St^+ = 10$ which present the maximum wall accumulation. In this case, by observing the developing region, we note that the trend of particles towards the wall is not gradual, but at almost $z/R \cong 15$ the concentration assumes the minimum value in the section in the buffer layer. Hence in this zone the coupling between fluid motions and particles dynamics tend to expell the particles away from this region. This phenomenon is related to the strong interactions between particles and the coherent structures characterizing the buffer region [3].

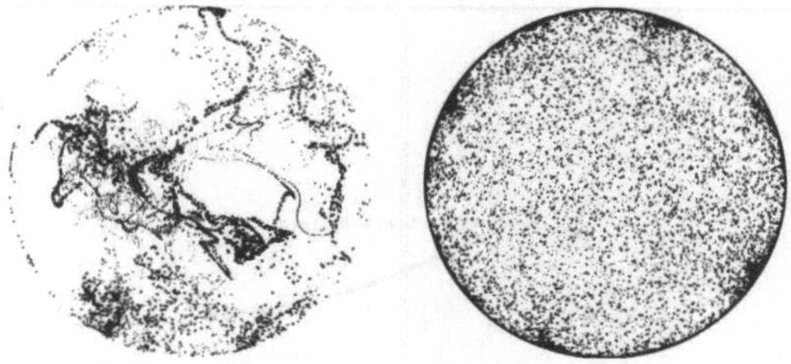

Fig. 2 Instantaneous particle positions in a thin cross flow slice of thickness $\sim R/10$ (same color coding as in fig. 1, symbol size increasing with St^+). Left, snapshot in the near field (distance from the releasing station $25R$); right, far field (distance $200R$).

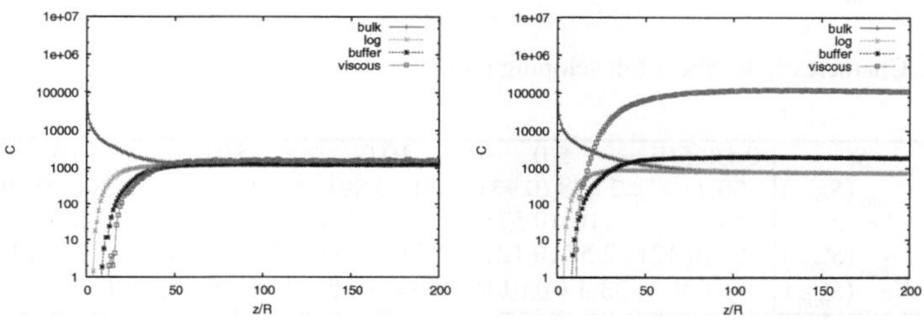

Fig. 3 Mean particle concentration vs axial distance z/R in the four regions of the wall. Left plot $St^+ = 0.1$, Right plot $St^+ = 10$.

In order to characterize with a single global index the particle dispersion along the developing region and the preferential accumulation in the far field, we make use of the entropy associated to mean particle concentration. The particle entropy is defined by:

$$S(z) = -\sum_i \frac{N_i(z)}{N(z)} \ln \frac{N_i(z)}{N(z)}, \tag{1}$$

where N_i is the mean particle number staying in a shell with volume $\Delta z\pi(r_{out}^2 - r_{inn}^2)_i$, where i denotes the cell centered at r_i. All the shells assume the same volume. N indicates the total number of particles at a given axial slice.

In figure 4 the entropy, normalized the corresponding value of the uniform distribution, is reported as a function of the axial distance. The distance where entropy is maximum $L_{S_{max}}$ denotes the location where particles tend to a homogeneous distribution in the section, while the end of developing region can be distinguished by L_{S_∞} where the entropy becomes constant. The maximum rate of dispersion and of preferential accumulation can be represented by the maximum (S'_{max}) and minimum (S'_{min}) of the axial derivative of entropy respectively whose locations are described in terms of two characteristic lengths $L_{S'_{max}}$ and $L_{S'_{min}}$, see table 1. From figure 4, all the particles start from the null value because they are injected in a very small region on the axis. Then the lightest particles ($St^+ \leq 1$) reach a uniform and homogeneous spatial distribution at $z \simeq 50R$. Intermediate particles ($St^+ = 5$, $St^+ = 10$) achieve a maximum of entropy smaller than that of the homogeneous distribution, see table 1.

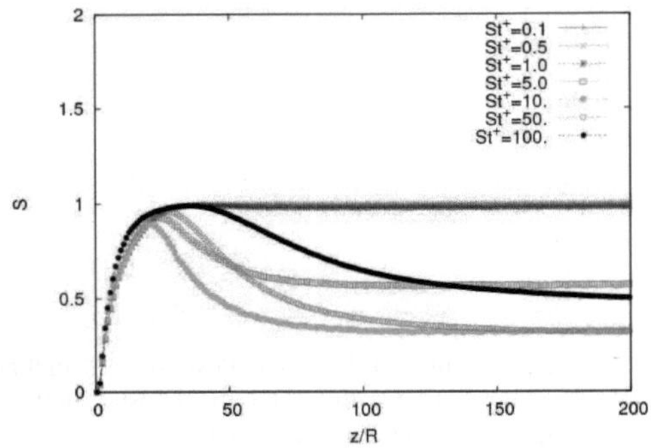

Fig. 4 Entropy of the mean particle concentration vs z/R for each particle population.

Table 1 Charactestic lenghts of developing region.

St^+	0.1/0.5/1.	5.0	10.0	50.0	100.0
$L_{S_{max}}(S_{max})$	50. (1.)	23.18 (0.93)	20. (0.89)	27. (0.96)	34. (0.99)
$L_{S_\infty}(S_\infty)$	50. (1.)	110. (0.57)	120. (0.32)	170. (0.31)	> 200. (*)
$L_{S'_{max}}(S'_{max})$	2.57 (0.12)	2.57 (0.12)	2.57 (0.13)	2.52 (0.14)	2.54 (0.15)
$L_{S'_{min}}(S'_{min})$	- (-)	33.4 (-0.014)	29.6 (-0.026)	44.7 (-0.015)	61.8 (-0.0074)

Downstream, for $z/R \geq 110/120$, the entropy decreases up to two different constant values, (0.57 and 0.32,respectively), denoting a non-uniform particle concentration across the section. As far as the heaviest particles ($St^+ = 50$ and $St^+ = 100$) are concerned, we note that the peak value increases with the Stokes number assuming the unity for the heaviest ones. Then the entropy decreases, although more slowly compared to the medium size particles. In fact particles with $St^+ = 50$ reach a developed state further than $170R$, while those with $St^+ = 100$ do not reach the asymptotic value of entropy within the simulation domain. So apart for lighter particles, preferential accumulation is evidenced by the not unitary value of entropy in the far field.

These results show that particles with different St^+ can achieve similar level of preferential accumulation, although at very different distances from the injection point. This phenomenon highlights the importance of DNS spatially evolving particle-laden wall flows.

References

1. Portela, L.M., et al.: Numerical study of the near-wall behaviour of particles in turbulent pipe flow. Powder Technology 125, 149–157 (2002)
2. Marchioli, C., et al.: Direct numerical simulation of particle wall transfer and deposition in upward turbulent pipe flow. Journal of Multiphase Flow 129, 1017–1038 (2003)
3. Rouson, D.W.I., Eaton, J.K.: On the preferential concentration of solid particles in turbulent channel flow. Journal of Fluid Mechanics 428, 149–169 (2001)

Direct Numerical Simulation of Particle Interaction with Coherent Structures in a Turbulent Channel Flow

C.D. Dritselis and N.S. Vlachos

Abstract. The interaction of small, solid particles with turbulent coherent structures near the walls of a vertical channel is numerically investigated using direct numerical simulation (DNS) together with Lagrangian particle tracking (LPT) and a point-force approximation for the feedback effect of the particles on the fluid. Results from conditional sampling show that the diameter and the streamwise extent of the mean vortices are increased due to the momentum exchange between the two phases, depending on the particle inertia, gravitational settling and interparticle collisions.

1 Introduction

Turbulent flows with dispersed particles widely occur in the environment and in engineering applications, exhibiting several complex interactions, many of which still remain poorly understood. Several experimental [1, 2, 3, 4, 5, 6] and numerical studies [7, 8, 9, 10, 11, 12, 13] have shown that at sufficient mass loadings, particles can modify the turbulent kinetic energy of the fluid. Modification of fluid turbulence due to the presence of particles is the primary topic of the present paper. The interest is focused on the effect of heavy, small diameter particles on the near-wall quasi-streamwise vortices of a fully developed turbulent channel flow. A conditional sampling scheme [14] is used to educe the dominant coherent structures, and to address the effect of particle inertia, gravity and interparticle collisions.

A combination of DNS/LPT simulations with a point-force approximation were performed for a low Reynolds number value of 180, based on the wall friction velocity $u_{\tau 0}$ of the particle-free flow and the half distance of the walls h. Large ensembles

C.D. Dritselis
University of Thessaly, Athens Avenue 38334, Volos, Greece
e-mail: dritseli@mie.uth.gr

N.S. Vlachos
University of Thessaly, Athens Avenue 38334, Volos, Greece
e-mail: vlachos@mie.uth.gr

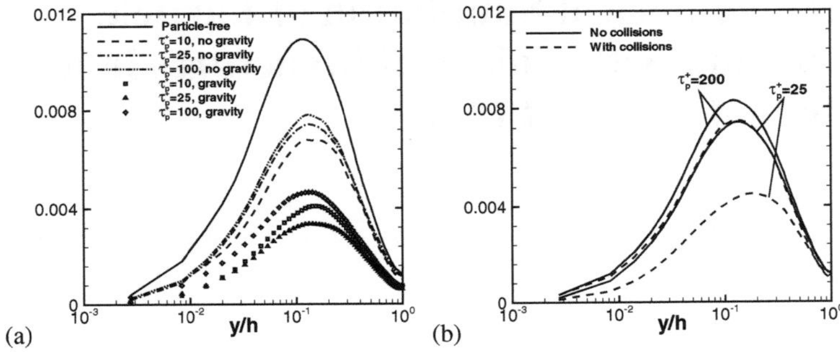

Fig. 1 Distributions of $\lambda_{2,rms}$ for several cases with noncolliding (a) and colliding particles (b).

of particles with dimensionless response times $\tau_p^+ = (Sd_p^2/18\nu)u_{\tau 0}^2/\nu = 10, 25, 100$ and 200 wall units were used at an average mass fraction of $\phi_m = 0.2$, where d_p is the particle diameter, S, is the density ratio between the two phases (= 7333 for copper particles in air), and ν is the fluid kinematic viscosity. The particle motion was determined by the fluid drag and gravity forces. Elastic, binary interparticle collisions were also considered by using geometrical criteria. The undisturbed fluid velocity in the particle equations of motion was approximated by the velocity that results from the numerical solution of the Navier–Stokes equations. Further details about the fluid flow solver and the Lagrangian particle tracking scheme (e.g. interparticle collisions and point-force models) can be found in [13] and references therein.

The λ_2-criterion [15] was used to detect vortical regions with negative streamwise vorticity in several instantaneous flow fields, where λ_2 is the second greatest eigenvalue of the $S_{ik}S_{kj} + \Omega_{ik}\Omega_{kj}$ tensor, and S_{ij}, Ω_{ij} are the strain and rotation rate tensors, respectively. Such regions are considered as coherent structures, if the axis (i.e. the locus of local minima of λ_2 in connected regions) has a streamwise extent of at least 150 wall units in $10 \leq y^+ \leq 40$ and each axis point is within a cone of 60° relative to its previous one. The flow field around every structure is obtained by ensemble averaging of the fluid velocity, vorticity, pressure, particle force and concentration, after proper alignment of the identified structure at the middle point in the streamwise direction. Any instantaneous event having a cross-correlation coefficient with the ensemble average structure smaller than 0.3 was discarded.

2 Results

Figure 1 shows the time-averaged distributions of $\lambda_{2,rms}$ for various particle-laden cases. All distributions of $\lambda_{2,rms}$ are similar to each other, but a reduction is observed for the particle-laden cases, which is more apparent for particles with smaller inertia. Simultaneously, the local maximum of $\lambda_{2,rms}$ is moved further away from the wall. Similar observations can be made for the effect of gravitational settling and interparticle collisions on $\lambda_{2,rms}$. The rms of λ_2 is actually a statistical indicator of

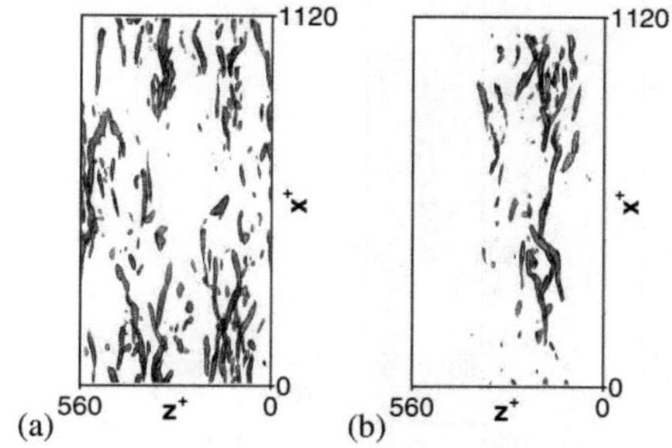

Fig. 2 Top view of instantaneous structures ($y^+ \leq 60$) by plotting a characteristic λ_2 isosurface for the particle-free case (a) and for a representative particle-laden case (b).

the flow regions populated by vortical structures. The reduction in $\lambda_{2,rms}$ indicates that the number of structures located in the buffer layer is decreased with a corresponding reduction in the particle response time.

A confirmation of this hypothesis is shown in Fig. 2, where the population of instantaneous structures in representative flow fields is significantly reduced by the presence of the particles, while an elongation of these events in the streamwise direction is also observed. The fact that the maxima of $\lambda_{2,rms}$ occur at larger y positions indicates that the vortical events are shifted at larger distances from the wall due to the momentum exchange between the two phases. The dominant structure orientation is little affected by the particles, and these events are found to be nearly aligned in the streamwise direction for all cases. This was also confirmed by the normalized correlation coefficients of $-\lambda_2$ with each component of vorticity ω_i (not shown here), which revealed that the cross-correlation between $-\lambda_2$ and ω_x was higher than the others in the buffer layer, independent of the particle inertia, gravity or interparticle collisions.

Figure 3 shows the top view of the ensemble average coherent structures for several cases by plotting a characteristic λ_2 isosurface (equal to 10% of its minimum value in each flow case). For the particle-free case, the diameter and length of the structure are approximately 22 and 200 wall units, respectively. Its axis has an incidence of about 9° in the x-y plane and is tilted approximately 4° in the x-z plane. For the particle-laden cases, the mean coherent structures have larger diameters and streamwise extents as compared to that of the particle-free flow. The inclination and tilting angles are little affected by the momentum exchange between the two phases, in agreement with previous observations of the instantaneous structures in Fig. 2.

3 Conclusions

Results from conditional sampling of DNS data of particle-laden turbulent channel flows with momentum exchange between the two phases show that gravitational

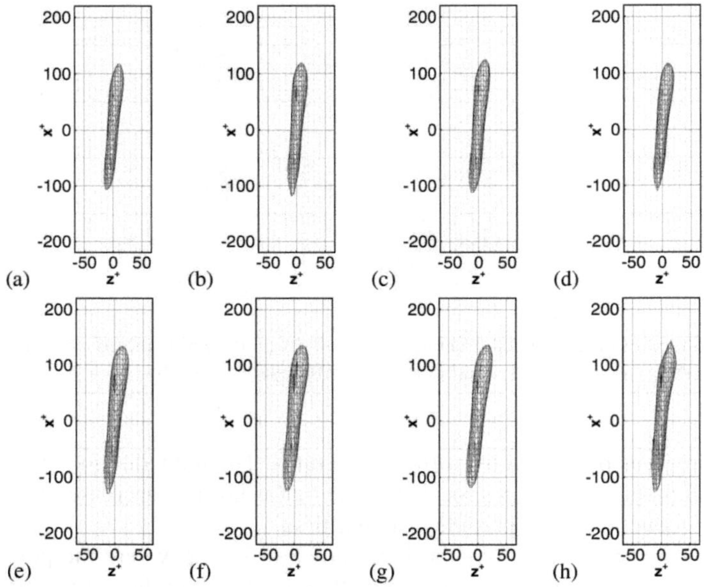

Fig. 3 Top view of the mean coherent structures: (a) particle-free case, [(b), (e)] $\tau_p^+ = 10$, [(c), (f)] $\tau_p^+ = 25$, [(d), (g)] $\tau_p^+ = 100$ and (h) colliding particles with $\tau_p^+ = 25$, (b) to (d) without gravitational settling and (e) to (g) with gravitational settling.

settling and interparticle collisions significantly affect the mean quasi-streamwise vortex. These effects increase slip velocity between the two phases, and consequently the particle feedback force by which momentum exchange takes place, leading to modifications in fluid turbulence, reflected in the features of the structures.

References

1. Tsuji, Y., Morikawa, Y.: J. Fluid Mech. 120, 385–405 (1982)
2. Tsuji, Y., Morikawa, Y., Shiomi, H.: J. Fluid Mech. 139, 417–434 (1984)
3. Sun, T.-Y., Faeth, G.M.: Int. J. Multiphase Flow 12, 99–114 (1986)
4. Parthasarathy, R.N., Faeth, G.M.: J. Fluid Mech. 220, 485–514 (1990)
5. Rogers, C.B., Eaton, J.K.: Phys. Fluids A 3, 928–937 (1991)
6. Kulick, J.D., Fessler, J.R., Eaton, J.K.: J. Fluid Mech. 277, 109–134 (1994)
7. Squires, K.D., Eaton, J.K.: Phys. Fluids A 2, 1191–1203 (1990)
8. Elghobashi, S., Truesdell, G.H.: Phys. Fluids A 5, 1790–1801 (1993)
9. Boivin, M., Simonin, O., Squires, K.D.: J. Fluid Mech. 375, 235–263 (1998)
10. Sundaram, S., Collins, L.R.: J. Fluid Mech. 379, 105–143 (1999)
11. Pan, Y., Banerjee, S.: Phys. Fluids 8, 2733–2755 (1996)
12. Li, Y., McLaughlin, J.B., Kontomaris, K., Portela, L.: Phys. Fluids 13, 2957–2967 (2001)
13. Dritselis, C.D., Vlachos, N.S.: Phys. Fluids 20, 055103 (2008)
14. Jeong, J., Hussain, F., Scoppa, W., Kim, J.: J. Fluid Mech. 332, 185–214 (1997)
15. Jeong, J., Hussain, F.: J. Fluid Mech. 285, 69–84 (1995)

Asymmetries in Turbulent Rayleigh-Bénard Convection

Ronald du Puits, Christian Resagk, and André Thess

Abstract. We report high-resolution local temperature measurements in the upper and the lower boundary layer of turbulent Rayleigh–Bénard (RB) convection. The measurements were undertaken in air ($Pr = 0.71$) at constant aspect ratio $\Gamma = 1.13$ and variable Rayleigh number $5 \times 10^5 < Ra < 10^{12}$. The primary purpose of the work is to preserve a comprehensive data set of the temperature field against which various phenomenological theories and numerical simulations can be tested. In our talk we show the mean temperature profiles $\vartheta_h(z)$ at the heating and $\vartheta_c(z)$ at the cooling plate. We demonstrate, that the corresponding profiles do not collapse even at low Rayleigh numbers and that the measured bulk temperature ϑ_b strongly deviates from the mean between the heating and the cooling plate $\vartheta_{b,t}$. Normalizing all temperatures to the difference between the plates, the deviation remains constant for all $Ra > 2 \times 10^{11}$.

1 Introduction

Thermal convection is an ubiquitous type of flows in nature. This class of highly turbulent flows is characterized by very strong velocity and temperature fluctuations compared with their mean. For a systematic study of these flows the Rayleigh-Bénard (RB) system – a closed box with a heated bottom plate and a cooled top plate as well as adiabatic sidewalls – represents a well-defined model, which has been investigated in great detail in the past [2, 4]. For a long time this system was considered as a symmetric problem as long as the applied temperature difference between both horizontal plates is small enough to satisfy the Boussinesq approximation. However, recent experimental, numerical and theoretical works (see e.g. [1, 5]) showed that this assumption is not invariably justified. The authors of those

Ronald du Puits, Christian Resagk and André Thess
Ilmenau University of Technology, POB 100 565, D-98684 Ilmenau, Germany
e-mail: ronald.dupuits@tu-ilmenau.de, christian.resagk@tu-ilmenau.de, andre.thess@tu-ilmenau.de

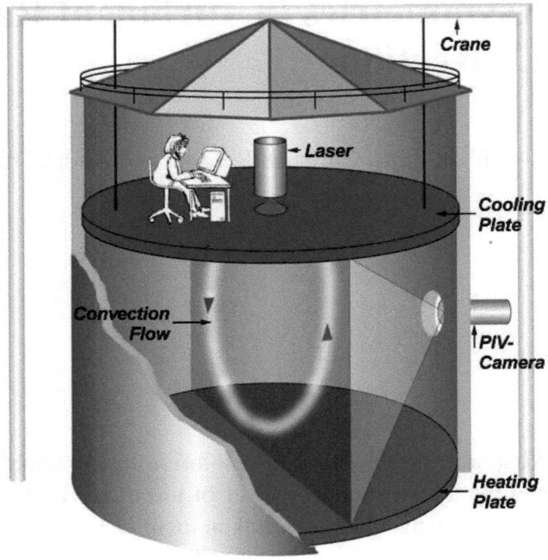

Fig. 1 Experimental facility: Sketch of the large scale RB facility ("Barrel of Ilmenau")

works found that the temperature measured in the bulk of the RB cell ϑ_b deviates from the predicted average $\vartheta_{b,t} = 1/2(\vartheta_{hp} + \vartheta_{cp})$ of the temperatures of the heating and the cooling plate, ϑ_{hp} respectively ϑ_{cp}. Insights into a potential asymmetry of the temperature field close to both horizontal boundaries are still missing but those differences might be crucial for the scaling of the global heat transport through the cell.

2 Experimental Results

We present temperature measurements at a large-scale RB experiment simultaneously undertaken at the top and the bottom plate using ultra-small microthermistors with a size of 125 μm. The convection apparatus - known as the "Barrel of Ilmenau" - is a cylindrical box with an inner diameter of $D = 7.15$ m. It is filled with air ($Pr = 0.71$). An electrical heating plate at the bottom and a free hanging cooling plate at the top trigger the convective flow. Top and bottom plate are maintained at a very homogeneous temperature circulating water inside. The distance H between both plates can be adjusted between 0.05 m $< H <$ 6.30 m. In order to guarantee the adiabatic boundary condition at the sidewall the experiment is shielded with an electrical compensation heating system. This facility permits the investigation of convective airflows in a parameter range from $Ra = 10^5$ ($\Gamma = D/H = 143$), $\Delta\vartheta = 4$ K to $Ra = 10^{12}$ ($\Gamma = 1.13, \Delta\vartheta = 60$ K). A sketch of the experimental facility is shown in figure 1. A more detailed description can be found in [3].

In a first series of experiments we fixed the aspect ratio at the lowest possible value of $\Gamma = 1.13$ and we changed the Rayleigh number between $5 \times 10^5 < Ra < 10^{12}$ by the variation of the temperature difference applied to the heating and the cooling plate from $\Delta\vartheta = 2.4$ K to $\Delta\vartheta = 60$ K. We measured the profiles of the mean temperature $\bar{\vartheta}(z)$ simultaneously at both plates along the central axis of the convection cell. The measurements were undertaken using ultra-small microthermistor of a size of $s = 125\ \mu$m. Each of them was mounted on a high precision positioning system and could be moved in steps of $\Delta z = 10\ \mu$m. Time series of the temperature, each of them 1.5 hours long, were recorded at 35 different z-positions up to a maximum distance $z = 150$ mm from the surface of the horizontal plates. From those time series the mean values were computed and their composition yielded to the profiles.

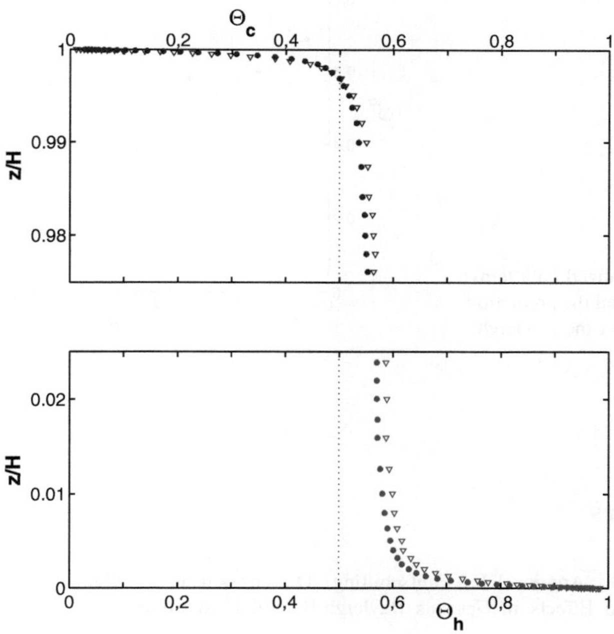

Fig. 2 Profiles of the normalized mean temperature $\Theta_c(z)$ at the cooling plate (top) and $\Theta_h(z)$ at the heating plate (bottom) for two different Rayleigh numbers $Ra = 3.2 \times 10^{11}$ (open triangles) and $Ra = 8.3 \times 10^{11}$ (full circles).

Figure 2 shows the corresponding profiles of the mean defect temperature $\Theta_h(z) = (\vartheta_h(z) - \vartheta_{cp})/\Delta\vartheta$ at the centre of the heating plate and $\Theta_c(z) = (\vartheta_c(z) - \vartheta_{cp})/\Delta\vartheta$ at the centre of the cooling plate for two different Rayleigh numbers $Ra = 3.2 \times 10^{11}$ and $Ra = 8.3 \times 10^{11}$ within the Boussinesq approximation and beyond. For both cases the normalized bulk temperature $\Theta_b = (\vartheta_b - \vartheta_{cp})/\Delta\vartheta$ significantly exceeds the prediction $\Theta_{b,t} = 0.5$. Plotting the temperatures $\Theta_h(z = 150\,\text{mm})$ and $\Theta_c(z = 150\,\text{mm})$ over Rayleigh, the deviation $\Delta\Theta = \Theta_b - \Theta_{b,t}$ from the prediction tends to an asymptotic value of $\Delta\Theta = 0.55$ with increasing Rayleigh number but there isn't any significant transition beyond the limit of the Boussinesq approximation. The growth of $\Delta\Theta = 0.55$ for low Rayleigh numbers, where the total heat flux inside the cell is very small, is possibly caused by an additional heat input from the compensation heating system at the sidewall. However, a final explanation requires a further analysis of our data.

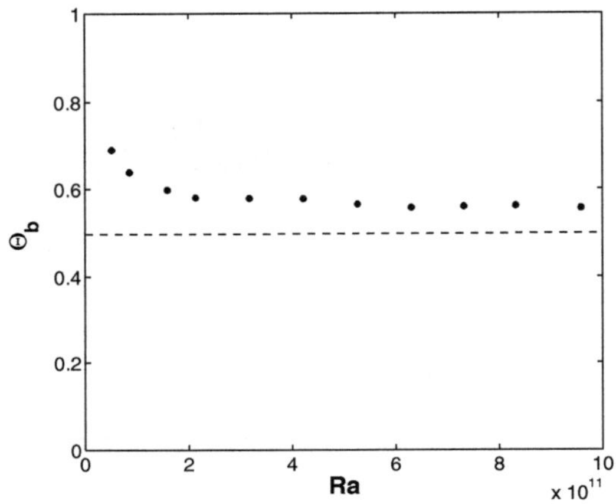

Fig. 3 Normalized bulk temperature Θ_b and the prediction $\Theta_{b,t} = 0.5$ over the Rayleigh number.

References

1. Ahlers, G., Araujo, F.F., Funfschilling, D., Grossmann, S., Lohse, D.: Non-Oberbeck-Boussinesq Effects in Gaseous Rayleigh-Bénard Convection. Phys. Rev. Lett. 98, 054501 (2007)
2. Castaing, B., Gunaratne, G., Heslot, F., Kadanoff, L., Libchaber, A., Thomae, S., Wu, X.-Z., Zaleski, S., Zanetti, G.: Scaling of hard thermal turbulence in Rayleigh-Bénard convection. J. Fluid Mech. 204, 1–30 (1989)
3. Du Puits, R., Resagk, C., Tilgner, A., Busse, F.H., Thess, A.: Structure of thermal boundary layers in turbulent Rayleigh-Bénard convection. J. Fluid Mech. 572, 231 (2007)
4. Siggia, E.D.: High Rayleigh Number Convection. Annu. Rev. Fluid Mech. 26, 137–168 (1994)
5. Sun, C., Cheung, Y.-H., Xia, K.-Q.: Experimental study of the viscous boundary layer properties in turbulent RB convection. J. Fluid Mech. 605, 79–113 (2008)

LES of Riblet Controlled Temporal Transition of Channel Flow

S. Klumpp, M. Meinke, and W. Schröder

Abstract. Numerical investigations of the impact of riblets on natural transition in a temporal evolving channel flow have been performed. A set of riblet geometries are used for two different Reynolds numbers. The results show that transition is accelerated by all riblets compared with a smooth surface. This acceleration is caused by an amplification of the two-dimensional Tollmien-Schlichting (TS) waves by riblets, where the amplification magnitude dependeds on the specific geometry of the riblets.

1 Introduction

Surface structures, so-called riblets consisting of tiny grooves aligned with the main flow direction are well known for reducing friction drag in turbulent flow [1]. The influence of riblets on laminar-turbulent transition is, however, not fully understood yet. A few experimental investigations, e.g. [2, 5], indicate that there is a delaying effect on natural transition in zero-pressure gradient boundary layers if the wall is covered by riblets. On the other hand, Ladd *et al.* [4] found no effect of riblets on transition. A reason for this disagreement might be a different spacing in wall units of the used riblets in these investigations. While the dimensionless distance was $s^+ < 10$ in [4] a higher value of $s^+ \approx 20$ was used in [2, 5]. Natural transition occurs at a low degree of freestream turbulence associated with the development of two-dimensional Tollmien-Schlichting (TS) waves. In the beginning these waves are linearly amplified, followed by a three-dimensional stage in which Λ- and hairpin vortices occur and finally, the turbulent breakdown is observed. Grek *et al.* [2] concluded from their hot-wire anemometry measurements that the two-dimensional waves are amplified by the riblets, whereas the three-dimensional structures are

S. Klumpp · M. Meinke · W. Schröder
Institute of Aerodynamics, RWTH Aachen University, Germany
e-mail: office@aia.rwth-aachen.de

damped. However, no detailed data of the flow close to the wall in the transitional regime is available. A delayed transition by riblets would provide an additional drag reduction capability of riblets.

To gain more insight in the effect of riblets on transition two sets of large-eddy simulations (LES) of temporally developing channel flow have been performed, where the channel walls are either clean or covered with riblets of different geometries. These geometries are shown in Figure 1 with their corresponding dimensionless tip spacings s^+. Geometries 1 and 2 are similar but scaled by a factor of 1.93. The geometries 1 to 3 have been chosen such that they can be manufactured by rolling processes in dimensions appropriate for technical applications. A geometry similar to case 3 was used in an experimental investigation by Litvinenko et al. [5] proving its transition delay capability. The Reynolds number (Re) based on the centerline velocity U_{cl} and the channel half-height h is $Re = 5000$ for the first set of simulations, for which all riblet surfaces have been investigated, and $Re = 15560$ for the second set, where besides the clean surface only the riblet geometries 2 and 3 have been investigated with respect to the Reynolds number. The computational domain used at $Re = 5000$ extends $5.61h \times 2h \times 2.99h$ in the streamwise, wall normal, and spanwise direction and is resolved using 2.7×10^6 grid points, whereas a domain of $6.98h \times 2h \times 2.17h$ resolved by 11.9×10^6 grid points is used at $Re = 15560$. To initialize the simulation at $Re = 5000$ a Poiseuille flow with a superposed two-dimensional TS wave and two three-dimensional oblique waves following the setup of Schlatter [6] is chosen. The streamwise wavelength of the two-dimensional TS wave is $\lambda_{x,2d} = 5.6h$, that of the three-dimensional waves is $\lambda_{x,3d} = 5.61h$ in the streamwise and $\lambda_{z,3d} = 2.99h$ in the spanwise direction. At $Re = 15560$ the wavelengths $\lambda_{x,2d} = 6.98h$, $\lambda_{x,3d} = 6.98h$, and $\lambda_{z,3d} = 2.17h$ correspond to the most unstable disturbances. The profiles of the amplitudes of the superposed waves are obtained using a standard Chebyshev collocation method involving the solution of the Orr-Sommerfeld and Squire equations. At this state a formation of Λ-vortices will occur immediately after the flow has been initialized and a K-type transition, as shown in Figure 2, will take place.

2 Results

To validate the numerical method a simulation with clean walls at $Re = 5000$ is performed. Figure 3 shows the evolution of the Reynolds number Re_τ based on the friction velocity compared with the DNS data of Schlatter [6]. Good agreement of the point of transition onset indicates a proper simulation of the transition mechanism. Expanding the computational domain to three times the original length in the streamwise direction does not change the point of transition onset, which indicates the characteristic length scales to be resolved sufficiently also in the smaller computational domain.

Figure 4 shows the evolution of Re_τ versus the non-dimensional time t for several wall configurations at $Re = 5000$. All riblet surfaces lead to an acceleration of transition. For a closer investigation, the evolution of the maximum velocities occurring in a cross flow section normal to the streamwise direction are taken into account. Figure 5 shows the development of the maximum wall normal velocity $|v_{max}|$, which represents the evolution of the two-dimensional TS waves. All riblet geometries amplify these waves compared with the clean surface, which agrees with the findings of Grek *et al.* [2]. The magnitude of the amplification varies for the different geometries. Overall, the stronger the TS waves are amplified, the earlier the transition occurs. The maximum spanwise velocity $|w_{max}|$, which indicates the evolution of the three-dimensional disturbances, is shown in Figure 6. Until $t \approx 40$, $|w_{max}|$ is nearly identical for all surfaces. Beyond this point the gradients of $|w_{max}|$ are ordered by the time transition occurs. A faster growing three-dimensional disturbance corresponds to an earlier transition. A damping of three-dimensional structures by riblets, as described by Grek *et al.* [2] and Litvinenko *et al.* [5], is not found. Due to the amplification of the two-dimensional TS waves, which dominates the transition at this setup, there is no clear evidence of the impact of riblets on the three-dimensional disturbances. Note, this result cannot be generalized, i.e., when external flows are considered [3].

Some new findings on the impact of riblets on transition have been made. The results of case 1 and 2 indicate that similar but scaled riblets have the same effect on natural transition at least in the investigated range of s^+. In general, the effect of riblets on transition has been found to be dependent on the riblet geometry, where riblet geometries with sharp edge amplify TS waves stronger than riblets with rounded tips. Due to lack of space the results of $Re = 15560$ are not presented in this paper. However, they are similar to the $Re = 5000$ findings, which indicates the effect of riblets on natural transition to be Reynolds number independent in the investigated range.

Acknowledgement. The authors gratefully acknowledge the financial support of the joint project "RibletSkin" by the Volkswagen Foundation, Hannover, Germany.

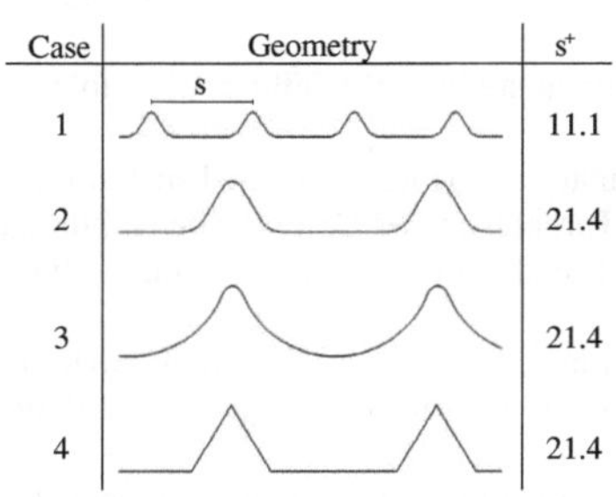

Fig. 1 Riblet geometries and their non-dimensional tip spacings

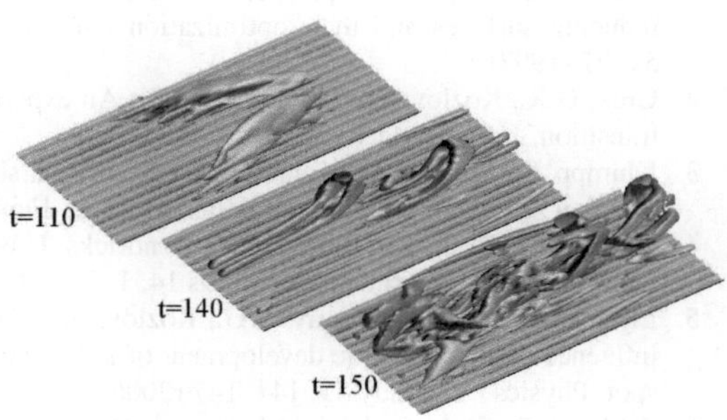

Fig. 2 Illustratrion of a Λ- and hairpin vortices and a turbulent spot at three different times during the simulation of case 2 at $Re = 5000$ visualized by λ_2-contours

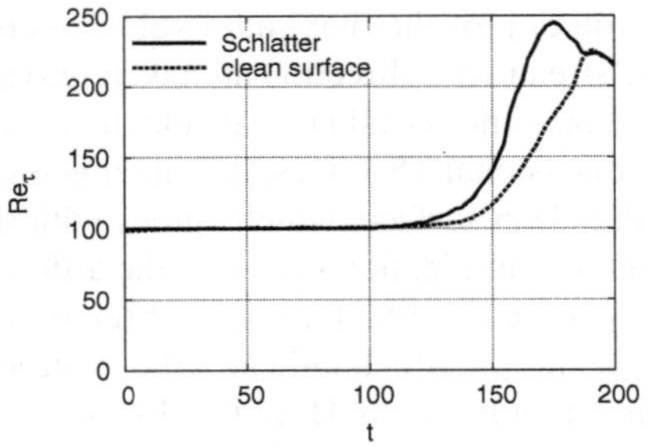

Fig. 3 Re_τ vs. t for the clean surface case compared with the DNS data of Schlatter [6]

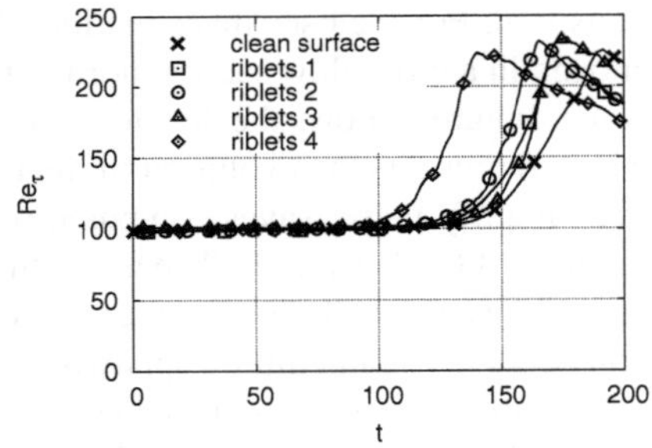

Fig. 4 Re_τ vs. t for several surface configurations at $Re = 5000$

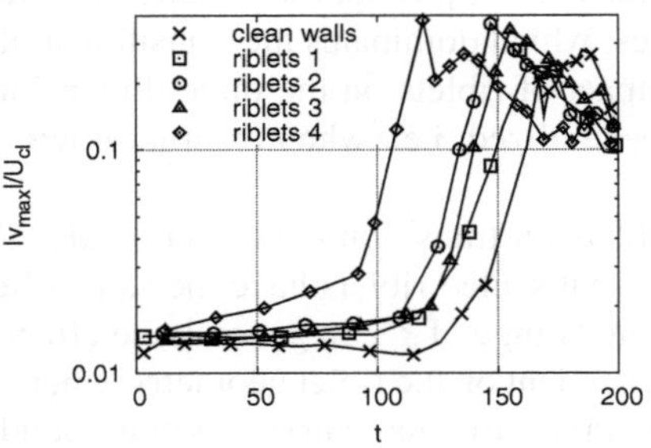

Fig. 5 Maximum wall normal velocity $|v_{max}|$ as a function of t for several surface configurations at $Re = 5000$

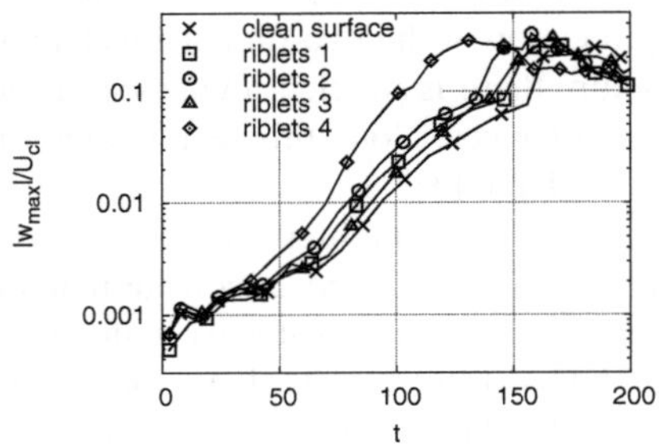

Fig. 6 Maximum spanwise velocity $|w_{max}|$ as a function of t for several surface configurations at $Re = 5000$

References

1. Bechert, D.W., Bruse, M., Hage, W., van der Hoeven, J.G.T., Hoppe, G.: Experiments on drag-reducing surfaces and their optimization with an adjustable geometry. J. Fluid Mech. 338, 59–87 (1997)
2. Grek, G.R., Kozlov, V.V., Titarenko, S.V.: An experimental study of the influence of riblets on transition. J. Fluid Mech. 315, 31–149 (1996)
3. Klumpp, S., Meinke, M., Schröder, S.: Numerical simulation of riblet controlled spatial transition in a zero-pressure gradient boundary layer. Flow, Turbulence and Combustion (submitted)
4. Ladd, D.M., Rohr, J.J., Reidy, L.W., Hendricks, E.W.: The effect of riblets on laminar to turbulent transition. Experiments in Fluids 14, 1–2 (1993)
5. Litvinenko, Y.A., Chernoray, V.G., Kozlov, V.V., Loefdahl, L., Grek, G.R., Chun, H.H.: The influence of riblets on the development of a Λ structure and its transformation into a turbulent spot. Physics - Doklady 51, 144–147 (2006)
6. Schlatter, P., Stolz, S., Kleiser, L.: Computational simulation of transitional and turbulent shear flow. In: Progress in Turbulence, Proceedings iTi Conference on Turbulence (2005)

Evolution of a Boundary Layer from Laminar Stagnation-Point Flow towards Turbulent Separation

Bernhard Scheichl and Alfred Kluwick

Abstract. We present the most relevant recent findings that allow for a rational time-averaged description of laminar–turbulent transition of an incompressible nominally two-dimensional and steady boundary layer along the impermeable surface of a rigid blunt body. Rigorous application of matched asymptotic expansions for sufficiently high values of both the Reynolds number and a turbulence-level gauge parameter shows that the presence of a leading-edge stagnation point is associated with the generation of a turbulent shear layer that exhibits an asymptotically small stream-wise velocity deficit. Remarkably, however, the turbulence intensity never reaches its theoretically possible maximum that conforms to fully developed turbulent flow.

1 Motivation and Problem Formulation

Although desirable for engineering applications, a reliable prediction of the flow past a more-or-less bluff body for realistically high values of the globally formed Reynolds number, *Re*, is presently still unavailable, notwithstanding the undeniable ongoing progress of computational methods like Direct Numerical and Large-Eddy Simulation. That is, a self-consistent comprehensive description of the time-mean process of gross separation that is essentially derived from first principles in the limit $Re \to \infty$ yet provides a challenging problem of theoretical hydrodynamics. As an interesting step forwards, here we focus on the asymptotic structure of the attached boundary layer (BL) evolving from the stagnation point, *S*, of the external flow.

Let the natural coordinates x, y along and perpendicular to the (perfectly smooth) body surface with origin in *S*, the corresponding Reynolds-averaged velocity components u, v, and the Reynolds stresses be non-dimensional with respect to a typical half-diameter L of the obstacle, the speed U of the incident parallel free-stream

Bernhard Scheichl · Alfred Kluwick
Institute of Fluid Mechanics and Heat Transfer, Resselgasse 3/E322, A-1040 Vienna, Austria
e-mail: bernhard.scheichl@tuwien.ac.at, alfred.kluwick@tuwien.ac.at

flow, and the uniform fluid density. Denotes ν the constant kinematic viscosity of the fluid, we accordingly define $Re := UL/\nu$ and assume that $Re \gg 1$. Then the external flow, $[u,v] = [u_0, v_0]$, has to be sought in the class of potential flows with separating streamlines, see [3]. It imposes the surface speed $u_e = u_0|_{y=0}$ at the outer edge of the BL upstream of separation. The level of turbulence intensity in the BL is measured by the so-called turbulence-level gauge factor T. To be more precise, cf. [1], [2], the inclusion of the Reynolds shear stress, σ, in the classical (Prandtl-type) BL equations, suggests to express the order of magnitude of σ as αu^2, where $\alpha := TRe^{-1/2}$. Furthermore, we disregard free-stream turbulence and make the following assumptions for the developed turbulent BL, merely based on physical reasoning:

1. each flow layer is characterised by a single typical turbulent velocity scale u_t, such that all components of the Reynolds stress tensor are of $O(\alpha u_t^2)$ there (hypothesis of locally isotropic turbulence);
2. in the fully turbulent main portion of the BL, where σ is much larger than the viscous shear stress, the local streamwise velocity defect $u_e - u$ is of $O(u_t)$;
3. in the viscous near-wall region, characteristic of wall-bounded turbulent flows and where σ and the viscous shear stress are of comparable magnitude, $u = O(u_t)$.

2 Asymptotic Analysis

In the limit $Re \to \infty$ the region of laminar–turbulent transition shrinks to the point S, cf. [4]. In other words, this limit is associated with the occurrence of a developed turbulent BL for $x = O(1)$, which is expected to exhibit the highest turbulence intensity possible. As a consequence and physically reasonable strategy, we, therefore, study solutions of the BL equations in the limit $T \to \infty$. From the viewpoint of time- or Reynolds-averaging, a sound understanding of the process of rapid transition taking place in a small region close to S (where $|x| \ll 1$) then proves crucial for the detection of the correct asymptotic structure of the BL originating there. Noting that $u_e \sim bx + O(x^2)$, $b > 0$, as $x \to 0_+$, the aforementioned magnitude of the Reynolds shear stress σ is seen to be comparable to that of its viscous counterpart in the region of transition. One then infers that the latter has a streamwise extent of $O(1/T)$. In turn, for $X := b^{1/2}Tx = O(1)$ the leading-order BL problem is cast into the form

$$h_\zeta^2 - \rho(X/\rho)' h h_{\zeta\zeta} + X(h_\zeta h_{\zeta X} - h_X h_{\zeta\zeta}) = 1 + X\rho s_\zeta + \rho^2 h_{\zeta\zeta\zeta}\,, \tag{1a}$$

$$\zeta = 0: \; h = h_\zeta = 0, \; s = O(\zeta^3), \quad \zeta \to \infty: \; h_\zeta \to 1, \; s \to 0, \tag{1b}$$

$$X \to 0_+: \; h \sim H(\eta)\,[+O(x^2)], \; s = O(X^2), \; \rho \to 1, \tag{1c}$$

$$H'^2 - HH'' = 1 + H''', \; H(0) = H'(0) = H'(\infty) - 1 = 0. \tag{1d}$$

Herein, $[h,s](X,\zeta)$, $\rho(X)$, and $\zeta := \rho(bRe)^{1/2}y$ are quantities of $O(1)$, where h denotes a stream function such that $u \sim h_\zeta$, $\sigma \sim bX^2T^{-1}Re^{-1/2}s$, and $\rho(X) > 0$ is chosen to properly account for the development of the pronounced outer "turbulent" BL edge as $X \to \infty$. Also, the correct near-wall behaviour of σ is required by (1b).

Most important, h matches the Hiemenz flow function $H(\zeta)$ as $X \to 0_+$, which is well-known to describe purely laminar flow within a distance of $O(Re^{-1/2})$ from S.

The rigorous justification of the BL approximation adopted in the leading-order problem (1) governing the transitional flow responds to the question, why its streamwise extent of $O(1/T)$ is asymptotically larger than that normal to the surface, varying with $Re^{-1/2}$. First, one may argue intuitively as follows: in case of the remaining possibility $T = O(Re^{1/2})$ stretching of both x and y with $Re^{1/2}$ would retain the full set of the Reynolds-averaged Navier–Stokes equations. These then formally exhibit a Reynolds number equal to 1, which, in turn, does not point to a proper mechanism that triggers fully developed turbulence within the region under consideration. Hence, one infers that a BL-type approximation should be valid even for arbitrarily small values of x, which is receptive to the (presently) known scenarios of laminar–turbulent transition. Consequently, the Hiemenz flow approximation is valid in a subregion of the region of transition where $x = O(1/T)$, having an extent of $Re^{-1/2}$.

In any of these cases, however, for $X \gg 1$ but $X \ll T$ as $T \to \infty$ the relationships (1a,b) describe the flow with sufficient accuracy. They serve as the starting point for studying the turbulent BL that takes place for $x = O(1)$. There in each layer $\sigma \sim \alpha u_\tau^2 \sigma_0 + \cdots$, $\sigma_0 = O(1)$. Accordingly, $[h,s] \sim [h_0, s_0](\zeta) + \cdots$ for $\zeta = O(1)$, $X \to \infty$, which reveals two possibilities, see [2]: First, assuming $s_0 \not\equiv 0$ is consistent with a large velocity defect, i.e. $u_\tau = O(1)$ for $\zeta = O(1), x = O(1)$. Hence, matching the viscous wall region with the adjacent layer by exploiting the three assumptions stated in Sect. 1 yields $u_\tau = O(1/\ln T)$, so that $\sigma_0 \to 0$ as $\zeta \to 0$. However, this implies a mismatch for $x \to 0$ as one finds that $s_0(0) > 0$. Thus, we are left with the second case, $s_0 \equiv 0$, which predicts a small-defect BL. Here the analysis for $x \to 0$ gives $\sigma_0 \to 1$ as $\zeta \to 0$. This results in a direct match with the wall layer and a two-tiered BL, characterised by the typical logarithmic velocity law in the overlap and a single velocity scale $u_\tau \sim u_e \kappa/(2 \ln T)$ with the von Kármán constant $\kappa \doteq 0.384$. By noting that $h_1' \sim -\kappa^{-1} \ln \zeta + B$, $B > 0$, $\zeta \to 0$ and $s_1(0) = 1$, one obtains for $X \to \infty$

$$[\zeta - h, s] \sim [\gamma h_1(\zeta), \gamma^2 s_1(\zeta)], \quad \gamma = \kappa/(2 \ln X), \quad \rho \sim 1/(\Delta X \gamma), \quad \Delta = O(1). \quad (2)$$

3 Numerical Results

In order to solve the boundary–initial-value problem (1) numerically, it is supplemented with a turbulence closure that correctly accounts for the generation of the small-defect BL as outlined above by suitably modelling both the scaling function $\rho(X)$ and the Reynolds stress function $s(X,\zeta)$. Appropriate choices are given by $\rho(X) := 1/[1 + \Delta X \hat{\gamma}(X)]$, $\hat{\gamma}(X) := 1/[(2/\kappa)\ln(X e^{\kappa/2} + 1) + 1]$, and an algebraic closure of the common form $s = [\ell(X,\zeta) h_{\zeta\zeta}]^2$, where the mixing length ℓ is modelled in a slightly modified manner as usually adopted in case of a fully developed turbulent BL. For specific details of the closure and the numerical procedure see [2].

We first solve the boundary-value problem determining $[h_1, s_1](\zeta)$, the constant Δ, and, in turn, $\rho(X)$. These data initiate downstream integration of (1). The results

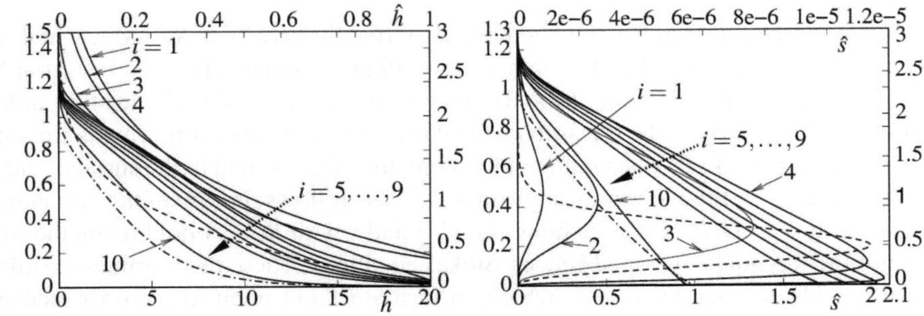

Fig. 1 Values of $\hat{h}$, $\hat{s}$ (abscissae) versus values of ζ (ordinates); top abscissae/right ordinates: $X=0$ (Hiemenz solution, *dashed*); bottom abscissae/left ordinates: $X=X^*\exp[(i/10)\ln(X_{10}/X^*)]$, $X^*:=100$, $X_{10}=4\times10^6$, $i=1,2,\ldots,10$, and $X=\infty$ (*dashed-dotted*)

for the rescaled velocity defect $\hat{h}:=(1-h_\zeta)/\hat{\gamma}$ and Reynolds stress $\hat{s}:=s/\hat{\gamma}^2$ are displayed in Fig. 1: they clearly indicate the evolution of the extremely thin wall layer, having a thickness of $O(\rho^2)$ relative to that of the outer tier, and the associated logarithmic portion of $\hat{h}$ as $[\hat{h},\hat{s}](X,\zeta)\to[h_1',s_1](\zeta)$, $X\to\infty$, according to (2).

4 Conclusions and Further Outlook

The analysis of Sect. 2 suggests that the turbulent BL sufficiently far from S is of two-tiered small-defect type, but the turbulence intensity parameter T is asymptotically smaller than $Re^{1/2}$; perfectly developed turbulence is indicated by $\alpha=O(1)$.

The triple-deck structure the small-defect BL is subjected to close to separation, cf. [1], provides a stringent rationale for the turbulent BL evolving on a bluff body being a slightly "underdeveloped" one in the limit $Re\to\infty$: in a fully developed turbulent BL the viscous wall layer is exponentially thin compared to its main region. Then the displacement effect on the irrotational flow in the upper deck exerted by the slow motion in the lower deck adjacent to the surface is too weak to induce a sufficiently strong pressure feedback there that allows for a singularity-free description of the separation process. This inconsistency is only avoided if the ratio of the widths of the inner and outer layer varies merely algebraically with Re, implying $\alpha\to0$. The detailed flow structure near separation is a topic of the current research.

References

1. Neish, A., Smith, F.T.: J. Fluid Mech. 241, 443–467 (1992)
2. Scheichl, B., Kluwick, A., Alletto, M.: Acta Mech. (2008), doi:10.1007/s00707-008-0078-7
3. Sychev, V.V., Ruban, A.I., Sychev, V.V., Korolev, G.L.: Asymptotic Theory of Separated Flows. Cambridge University Press, Cambridge (1998)
4. Zdravkovich, M.M.: Flow Around Circular Cylinders, vol. 1. Oxford University Press, Oxford (1997)

The Response of Wall Turbulence to Streamwise-Traveling Waves of Spanwise Velocity

Maurizio Quadrio

Abstract. Waves of spanwise velocity traveling along the walls of a plane turbulent channel flow are studied by Direct Numerical Simulations. We consider waves of spanwise velocity which are oscillating in time and modulated in space along the streamwise direction. Waves which slowly travel forward produce a large reduction of drag. Faster waves, with a phase speed corresponding to the convection velocity of the turbulent fluctuations at the wall, lock with the convecting near-wall turbulent structures and yield drag increase. Backward-traveling waves, on the other hand, invariably produce a drag-reducing effect. To explain the physical mechanisms involved, the generalized Stokes layer induced by the traveling waves is studied in the laminar regime.

1 Background

Observing how turbulence in the near-wall region of a boundary-layer-type flow responds to an external forcing presents a fundamental and applied interest. Here we study the effects of spanwise velocity waves traveling in the streamwise direction along the wall of a turbulent channel flow.

We consider waves of spanwise velocity that are oscillating in time and modulated in space along the streamwise direction: the phase speed makes the waves travel (forward or backward) in the same direction of the mean flow. The wall waves are described by

$$w_w(x,t) = A \sin\left(\kappa_x x - \omega t\right), \tag{1}$$

where w_w is the spanwise component of the velocity vector at the wall, x is the streamwise coordinate and t is time, A is the amplitude of the forcing, $\kappa_x = 2\pi/\lambda_x$ is the forcing wavenumber in the x direction and $\omega = 2\pi/T$ its frequency.

Maurizio Quadrio
Dip. Ing. Aerospaziale del Politecnico di Milano, Milano, Italy
e-mail: `maurizio.quadrio@polimi.it`

Such waves generalize two previously considered wall-based forcings that are known for their drag-reducing effect. One is the oscillating-wall technique:

$$w_w(t) = A \sin(\omega t), \tag{2}$$

and the other one is the standing-wave technique:

$$w_w(x) = A \sin(\kappa_x x). \tag{3}$$

The forcing (2) is a traveling wave with infinite phase speed, has been introduced by [1], and described for example in [4]. The recently studied [5] standing-wave forcing (3) is a wave with zero phase speed, that behaves similarly to (2) when x is converted to time through the convection velocity [2] of the near-wall turbulence.

In this work we probe via Direct Numerical Simulations the response of a turbulent channel flow to such a traveling-wave forcing, with the primary goal of understanding how it affects friction drag and how it behaves from an energetic viewpoint. A laminar analysis is then used to assess the properties of the generalized Stokes layer induced by the traveling waves.

2 Numerical Details

Direct Numerical Simulations (DNS) are used to examine the response of a turbulent channel flow to the wall traveling waves. The code and the architecture of the computing system, where it can run with maximum efficiency, have been described elsewhere [3]. The Reynolds number is set at $Re_\tau = hu_\tau/\nu = 200$, where u_τ is the friction velocity, and h is half the channel width. The computational domain has a streamwise length of $L_x = 6\pi h$ and a spanwise width of $L_z = 3\pi h$. 320×320 Fourier modes are used in the streamwise and spanwise directions, and 160 points discretize the wall-normal direction. The total integration time is $1000\, h/U_P$.

3 Drag Reduction and Energy Savings

Figure 1 describes the effects of the traveling waves in terms of the percentage change in friction drag. The forcing amplitude is fixed at $A^+ = 12$. Only half of the whole $\omega - \kappa_x$ plane is plotted, since (1) yields results that must be symmetric upon exchanging (ω, κ_x) with $(-\omega, -\kappa_x)$. In this plane, the phase speed c is the inverse of the slope of straight lines passing through the origin.

Each data point in figure 1 is the result of one DNS and expresses the percentage reduction of friction drag. Contours are computed after linear interpolation of the irregularly scattered points on a regular mesh.

On the two axes, the flow response was already known by previous studies for the oscillating wall and the stationary wall waves. Off the axes, local maxima and local minima form two distinct and elongated narrow regions. In the red region the flow presents a strong reduction of drag. The maximum measured drag reduction is about 48%. In the blue region, the drag increases significantly, up to 23%. The maximum change corresponds to a straight line crossing the origin, and is therefore associated with a constant value of the phase speed, $c^+ = 11$.

The efficiency of the traveling waves in affecting the friction drag is evaluated by computing the amount of external energy spent to enforce the control action. The power that an ideal device necessarily dissipates against the viscous resistance of the fluid is an upper bound for the global energetic performance. The contour map of this power (not shown) strikingly resembles the distribution of drag reduction just seen in figure 1. The direct consequence is that the traveling waves perform exceptionally well in terms of net energy saving: a maximum of 18% is measured at the present forcing amplitude, and wide ranges of κ_x and ω yield a positive net.

4 The Laminar Generalized Stokes Layer

For the oscillating wall (2), the analytical solution of the *laminar* Stokes layer has been succesfully used [4] to understand *turbulent* drag reduction. A

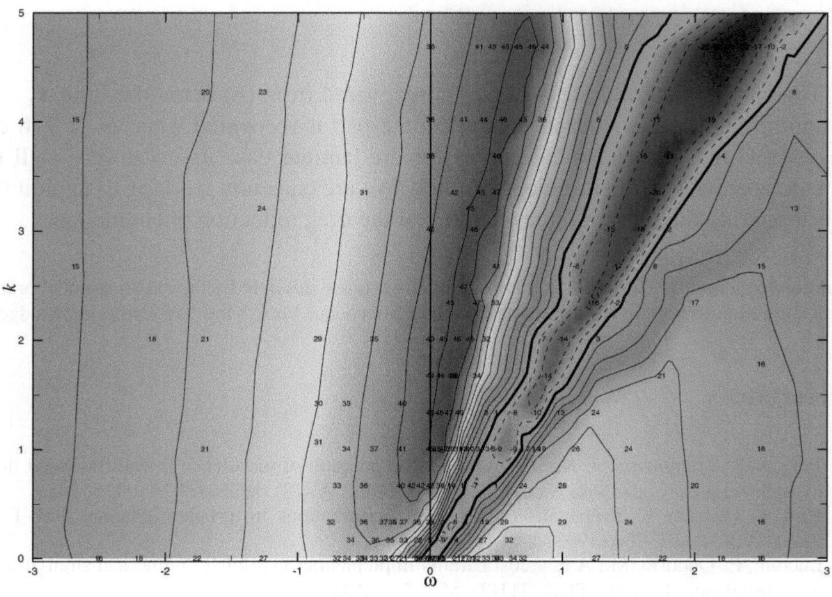

Fig. 1 Percentage change in friction coefficient as a function of wavenumber and frequency of the traveling wave (1). The numbers indicate actual DNS data, whereas color contours are the result of linear interpolation. Contours by 5% increments, thick line is zero and negative values are dashed.

space-modulated transversal boundary layer develops [5] over the standing waves (3). In the space-time case, a generalized Stokes layer is present, modulated both in space and time: under a thin-layer assumption, it takes the following analytical expression:

$$w(x,y,t) = A\Re\left\{ Ce^{2\pi i(x-ct)/\lambda_x} \mathrm{Ai}\left[e^{\pi i/6} \left(\frac{2\pi u_{y,0}}{\lambda_x \nu} \right)^{1/3} \left(y - \frac{c}{u_{y,0}} \right) \right] \right\}, \qquad (4)$$

where Ai is the Airy function of the first kind, $u_{y,0}$ is the y derivative of the u profile at the wall, $\Re$ indicates the real part and C is a complex constant. Typical profiles for the temporal, spatial and spatio-temporal Stokes layer are shown in figure 2.

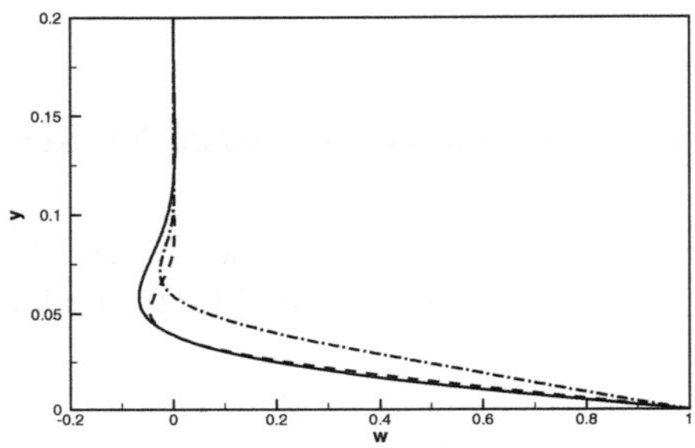

Fig. 2 Typical spanwise velocity profiles after the analytical solution (4). Continuous line: temporal case. Dashed line: spatial case. Dash-dotted line: general space-time case.

The classic temporal Stokes layer is recovered from (4) when the limit $\kappa_x \to 0$ is appropriately taken, and its spatial counterpart is recovered when $\omega \to 0$. It can be verified that the analytical curves for the laminar case describe very well the (mean) spanwise in the turbulent case too. We are currently working to exploit this for improving our current understanding of the drag reduction phenomenon.

Acknowledgement. This work is a long-term effort made possible by the standing collaboration with P.Luchini. Important contributions from Dr P.Ricco and Mr C.Viotti are also acknowledged.

References

1. Jung, W.J., Mangiavacchi, N., Akhavan, R.: Suppression of turbulence in wall-bounded flows by high-frequency spanwise oscillations. Phys. Fluids A 4(8), 1605–1607 (1992)
2. Kim, J., Hussain, F.: Propagation velocity of perturbations in turbulent channel flow. Phys. Fluids A 5(3), 695–706 (1993)
3. Luchini, P., Quadrio, M.: A low-cost parallel implementation of direct numerical simulation of wall turbulence. J. Comp. Phys. 211(2), 551–571 (2006)
4. Quadrio, M., Ricco, P.: Critical assessment of turbulent drag reduction through spanwise wall oscillation. J. Fluid Mech. 521, 251–271 (2004)
5. Viotti, C., Quadrio, M., Luchini, P.: Streamwise oscillation of spanwise velocity at the wall of a channel for turbulent drag reduction. Phys. Fluids (2008) (submitted)

Dynamics of Viscoelastic Wall Turbulence in Different Ranges of Scales

E. De Angelis, C.M. Casciola, and R. Piva

Abstract. We address the dynamics of viscoelastic wall turbulence by means of a generalization of a scale-by-scale approach extended to both an inhomogeneous and viscoelastic case. Analysing the results obtained by a series of Direct Numerical Simulations of a dilute polymer solution in a plane channel, we focus our attention on the alteration of the inertial transfer across the scales and the polymer scale-dependent term in the budget for the second order velocity structure function. We confirm that both these observables lead to a scenario were the main alteration of turbulence structure in a channel flow occurs in the buffer layer.

1 Introduction

A small amount of long chain polymers dissolved in an otherwise Newtonian flow is known to reduce dramatically drag in wall bounded flows. This corresponds to an alteration in the mean velocity profile, where the slope the log-law of the wall passes from 2.5 to 11.7 [1].

The parameter which couples the dynamics of the polymer chains and the turbulence is the ratio of two times scales, namely the Deborah (or Weissemberg) number $De_* = \tau_p/\tau_*$. Here $\tau_* = \nu/u_*^2$ is the friction time scale and τ_p is the principal relaxation time of the chain, the estimated time needed to recover equilibrium after the external strain is removed. In most of the experiments this ratio is of the order of one for dilute systems at the onset of drag reduction. Hence the internal dynamics of the

---

E. De Angelis
II Facoltà di Ingegneria, Università di Bologna
e-mail: `e.deangelis@unibo.it`

C. M. Casciola
Dipartimento di Meccanica e Aernautica, Sapienza Università di Roma
e-mail: `carlomassimo.casciola@uniroma1.it`

R. Piva
Dipartimento di Meccanica e Aernautica, Sapienza Università
e-mail: `renzo.piva@uniroma1.it`

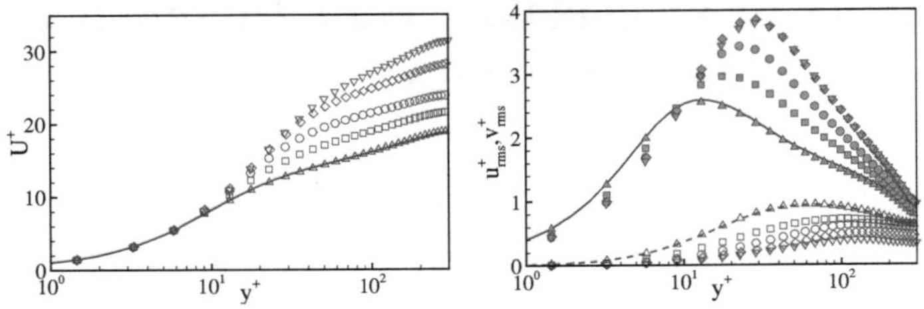

Fig. 1 Mean velocity profiles in viscous variables. In the different cases the friction Reynolds is the same, $Re_* = 300$. The solid line gives the Newtonian data. The polymeric data are: $De_* = 18$ (triangles), $De_* = 36$ (squares), $De_* = 72$ (circles), $De_* = 90$. (diamonds). Fluctuation intensities in viscous variables, streamwise $u^+_{\mathrm{rms}} = \sqrt{(u'/u_*)^2}$ (heavier symbols) and wall-normal $v^+_{\mathrm{rms}} = \sqrt{(v'/u_*)^2}$ (lighter symbols). The values of De_* are coded as in the left panel.

chain is unlikely to get substantially coupled to the turbulence. This opens the way for an hyper-simplified description of the polymer, as a system with a single internal degree of freedom, the resulting model is called Oldroid-B. In the present contribution the results obtained by the direct numerical simulations of the equations for a dilute polymer solution in a flow between planar walls are discussed. Figure 1 shows several mean profiles and fluctuation intensities, see [2] for a description of the algorithm. The various cases, at the same friction Reynolds number $Re_* = u_* h/\nu$, where h is the channel half-width, differ for the Deborah number, De_*.

2 Small Scale Dynamics and Energy Containing Scales

One of the effects of dilute polymers in turbulence is to alter the energy distribution among the scales and the dissipation mechanisms in the system, hence we firstly address the role of the polymers in the turbulent kinetic energy balance,

$$-\langle u'v'\rangle \frac{dU}{dy} - \frac{d\langle q'^2 v'\rangle}{2dy} - \frac{d\langle p'v'\rangle}{dy} + \frac{1}{Re_*}\frac{d^2\langle q'^2\rangle}{2dy^2} - \varepsilon_{\mathrm{N}} - \pi_{\mathrm{p}} + \frac{d\langle \mathbf{T}'_{\mathrm{p}} \cdot \mathbf{u}'\rangle}{dy} = 0, \quad (1)$$

where q'^2 is the fluctuation intensity, $\pi_{\mathrm{p}} = \langle \mathbf{T}'_{\mathrm{p}} : \nabla \mathbf{u}'\rangle$, represents the energy drain in favour of the miscrostructure and $\langle \mathbf{T}'_{\mathrm{p}} \cdot \mathbf{u}'\rangle$ contributes to the spatial flux in the wall normal direction, $\phi_s = \langle q'^2 v'\rangle/2 + \langle p'v'\rangle - \langle \mathbf{T}'_{\mathrm{p}} \cdot \mathbf{u}'\rangle - d\langle q'^2\rangle/(2Re_* dy)$. The energy available at a certain location, due to local production and to the spatial flux, is partly dissipated by ordinary viscosity, ε_{N}, and partly moved to the polymers, π_{P}.

For a planar channel or pipe flow, a simultaneous view of small and large scale dynamics can be achieved by a suitable generalization of the scale-energy budget [4], here extended to polymeric flows,

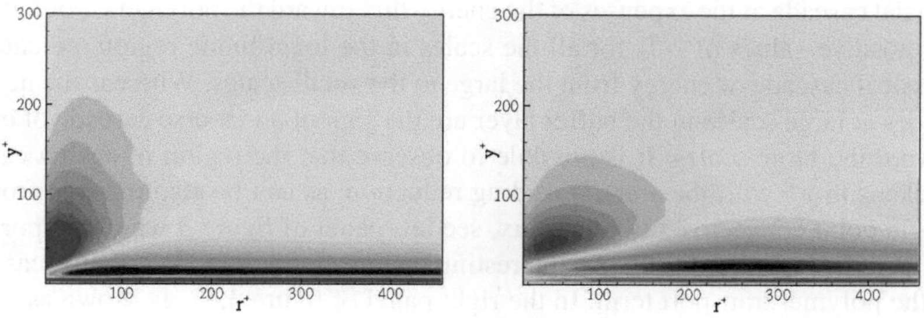

Fig. 2 Nonlinear transport term $-T_r$ plotted as a function of the scale and of the distance from the wall. Left: Newtonian case, right: $De_* = 54$. Red is positive and blue is negative.

$$\frac{\nabla_r \cdot \langle \delta q'^2 \delta \mathbf{u}' \rangle}{4} + \left[\frac{\langle \delta u' \delta v' \rangle}{2} \frac{dU}{dy} + \frac{\partial \langle v'^* \delta q'^2 \rangle}{4 \partial y} + \frac{\partial \langle \delta p' \delta v' \rangle}{2 \rho \partial y} \right] =$$

$$\left[\frac{\nu}{2} \left(\nabla_r^2 + \frac{\partial^2}{8 \partial y^2} \right) \langle \delta q'^2 \rangle - \varepsilon_{\mathrm{N}}^* \right] + \left[\left(\nabla_r \cdot \langle \mathbf{T}'^*_{\mathrm{p}} \cdot \delta \mathbf{u}' \rangle + \frac{\partial \langle \hat{\mathbf{y}} \cdot \delta \mathbf{T}'_{\mathrm{p}} \cdot \delta \mathbf{u}' \rangle}{4 \partial y} \right) - \pi_{\mathrm{p}}^* \right], \quad (2)$$

where the asterisk is defined as the mean value the extremes of the increment vector, $\hat{\mathbf{y}}$ is the wall-normal unit vector, u', v' and p' are streamwise, wall-normal velocity fluctuations and pressure fluctuation, respectively. This simplified form holds when the two points lay at the same wall-normal distance. The different terms are integrated in r-space over square domains of length $2r$ on planes parallel to the walls and normalized by the area. After integration certain terms are grouped together, see [4]. Here we focus our attention on the inertial transfer term T_r issuing from the first term on the l.h.s. and to the overall polymer transport G_e, from the round bracket in the last term on the r.h.s. In figure 2 T_r is shown as a function both of the scale r^+ and the distance from the wall $y+$ both for the Newtonian flow and the case $De_* = 54$. It is possible to observe from a general point of view the depletion of the intensities between the Newtonian and the viscoelastic case, sign of a weakening of the

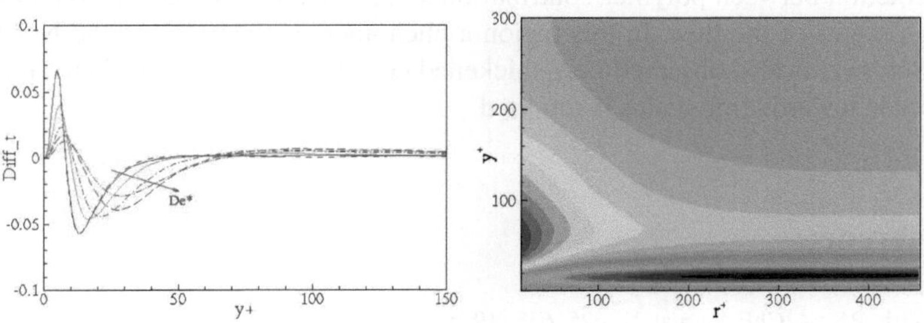

Fig. 3 Left panel: Convective transport term in the turbulent kinetic energy balance at increasing values of the Deborah number. Right: Polymer transport term G_e plotted as a function of the scale and of the distance from the wall. Red is positive and blue is negative.

inertial cascade at the expenses of the energy flux toward the polymers. For both plot the positive values of $-T_r$ for all the scales in the logarithmic region indicates the classical cascade of energy from the large to the small scales. Whereas the negative values at large scales in the buffer layer are the sign of an *inverse* cascade of energy toward the large scales. It is possible to observe that the region of such a cascade thickens in y^+ with the amount of drag reduction, as can be also inferred from the single point convective turbulent flux, see left panel of figure 3 which proportional to the large scale limit of T_r. An interesting feature emerges in the polymer case, due to the polymer transport term. In the right panel of figure 3, G_e is shown as a function both of the scale r^+ and the distance from the wall $y+$ for the case $De_* = 36$. Let us notice that the maximum in the polymer transport term is located in small scales of the buffer layer, this is the part of the channel where the scenario change with respect to the Newtonian, as previously observes. Here infact the two main contribution in the scale-by-scale budget are the production and polymer transport, namely the larger scales are dominated by production effects, the smaller scales are controlled by the polymers. The inertial term is completely subdominant, a clear indication of the effect of polymers on the structure of turbulence in the large scale range of the system, presumably at the origin of the strong alteraction of the energy containing scales observed in drag reducing flows.

3 Comments and Remarks

As reported in [3] in homogeneous isotropic turbulence, the energy flux is intercepted by the polymers at small scales and accumulated in the microstructure as elastic energy and dissipated by the relative friction between polymers and solvent.

Clearly, near a solid wall the physics is more complex. The classical mean velocity profile is substantially modified, the logarithmic slope increases to a limiting value which corresponds to the maximum drag reduction asymptote. A generalized form of scale-energy budget has been proposed for a deeper evaluation of the alteration of turbulence. As the amount of drag reduction increases, a region where the interaction between polymer contribution and production prevails as to modify the large scales of the flow. In this region a phenomenology similar to the Newtonian buffer layer is also observed i.e. a thickened portion of the channel where an inverse cascade toward large scales is reported.

References

1. Virk, P.S.: AIChE Journal 21, 625–655 (1975)
2. De Angelis, E., Casciola, C.M., Piva, R.: Computers & Fluids 31, 495–507 (2002)
3. De Angelis E., Casciola C. M., Benzi R. and Piva R.: J. Fluid Mech. 531, 1–10 (2005)
4. Marati, N., Casciola, C.M., Piva, R.: J. Fluid Mech. 521, 191–215 (2004)

Hairpin Structures in a Turbulent Boundary Layer under Stalled-Airfoil-Type Flow Conditions

Y. Maciel and M.H. Shafiei Mayam

Abstract. Hairpin structures in the outer region of a turbulent boundary layer subjected to a strong adverse pressure gradient have been studied using PIV. The external flow conditions are similar to those found on the suction side of airfoils in trailing-edge post-stall conditions. Even if the flow is very different from zero-pressure-gradient turbulent boundary layers, the gross features of the hairpin vortices and hairpin packets remain similar, even as separation is approached. The hairpin vortices are however slightly more inclined with respect to the wall, and their streamwise separation is smaller when scaled with the boundary layer thickness. The upward growth of the hairpin packets in the streamwise direction is also more important. The variations of these properties are consistent with the variations of the mean strain rates, in particular rates of streamwise contraction and wall-normal extension.

1 Introduction

In canonical turbulent wall flows, several recent studies support the existence of hairpin structures in the log and wake regions [1]. In contrast, not much is known about these structures in adverse-pressure-gradient turbulent boundary layers (APG TBL). The objective of the present study is to gain a better understanding of coherent structures in the outer region of APG TBL. The turbulent boundary layer studied evolves under external flow conditions similar to those found on the suction side of airfoils in trailing-edge post-stall conditions (figure 1). It has therefore suffered from an abrupt transition from very strong favourable pressure gradient to very strong APG, leading to a large separation zone. The statistical properties of this flow and details of the experimental set-up can be found in ref. [2].

Y. Maciel · M.H. Shafiei Mayam
Dept of Mech Eng, Laval University, Quebec City, Canada
e-mail: Yvan.Maciel@gmc.ulaval.ca, hossein.shafiei.1@ulaval.ca

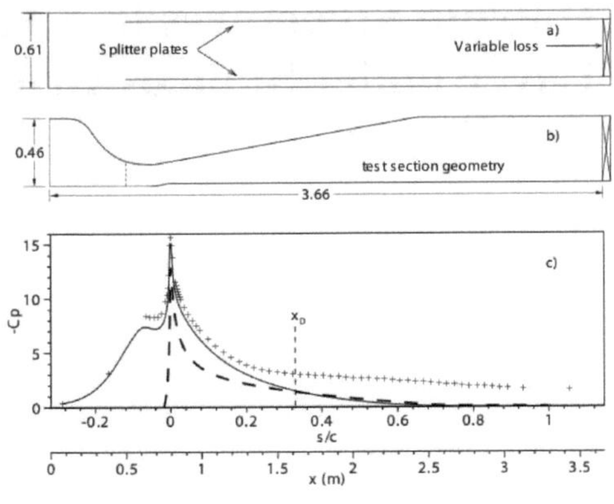

Fig. 1 a,b Respectively top view and side view of the modified test section of the boundary-layer wind tunnel. Dimensions in m. c) Pressure coefficient distribution along the floor of the test section: ++ Measurement, −− NACA 2412 airfoil (2.5 m, 18°), — Potential flow calculation.

Instantaneous velocity fields were acquired by PIV in streamwise/wall-normal planes at three streamwise positions covering the APG region between the suction peak and the detachment point. The *xy* dimensions of the measurement planes were chosen to be about $3\delta \times 1.3\delta$ at each position. To increase spatial resolution, measurements were made simultaneously with two overlapping large *xy* planes (two cameras). Table 2 presents the interrogation window width as well as the most relevant boundary layer parameters for three stations where the spatial resolution is identical in outer units (0.018δ). The grid resolution is however 0.009δ since the image interrogation was done with 50% overlap of the interrogation zones. The last station is very close to the position where $C_f = 0$, which is $x = 1615$ mm. A coherent structure identification technique similar to that of Adrian et al. [3] was used to detect the presence of hairpins and hairpin packets. Fifty instantaneous velocity fields at each streamwise position were inspected. Fifty fields of the ZPG TBL database of Adrian et al. [3] were analyzed in the same manner in order to ensure an equivalent basis of comparison.

2 Results and Discussion

Despite the presence of a very different pressure environment in this flow in comparison to the ZPG TBL, the gross features of the hairpin vortices and hairpin packets remain similar, even as separation is approached. In both flows, three hairpin packets are found on average in a single instantaneous velocity field covering about $3\delta \times 1.3\delta$. In each of these packets, the average number of hairpins is four (range of 3 to 12). Note that packets are not necessarily straigth and perfectly aligned in the streamwise direction [1]. As a result, the number of hairpins per packet can be underestimated when analyzing *xy*-planes since some hairpins may be missed. Moreover, a single packet may also sometimes be interpreted as two packets in a *xy*-slice.

Table 2 summarizes some average properties of hairpins and packets for $y > 0.2\delta$, as well as the associated random uncertainties due to the limited number of

Table 1 PIV interrogation window width ($\Delta y = \Delta x$) and boundary layer parameters for the reference streamwise positions. ZPG TBL data from ref. [3].

	x (mm)	$\Delta x/\delta$	Δx^+	δ^+	Re_θ	H	u_τ/U_e
ZPG TBL	—	0.014	28.4	2000	6845	1.38	0.0367
present exp.	1156	0.018	13.2	765.9	5329	2.03	0.0250
"	1392	0.018	15.2	836.7	8638	2.86	0.0159
"	1600	0.018	—	$\approx$0	12095	3.65	$\approx$0

Table 2 Average properties of hairpins and packets for $y > 0.2\delta$.

	x (mm)	$\langle\alpha\rangle$ (deg)	$\langle\gamma\rangle$ (deg)	$\langle x_{sep}\rangle/\delta$
ZPG TBL	—	61.8 ± 3.0	5 ± 2	0.204 ± 0.019
present exp.	1156	68.6 ± 3.0	12 ± 2	0.145 ± 0.013
"	1392	69.5 ± 2.5	11 ± 2	0.150 ± 0.014
"	1600	66.9 ± 2.4	11 ± 2	0.126 ± 0.010

realizations. The hairpin properties are schematically defined in the inset of figure 2. The neck of the hairpin vortices is found to be slightly more inclined with respect to the wall when the TBL is subjected to a strong APG. In the present flow, the neck angle α varies between 25° and 90° with an average value around 68°. In the ZPG TBL, the average α is around 62°. These differences are consistent with the fact that the rates of streamwise contraction ($\partial U/\partial x < 0$) and wall-normal extension ($\partial V/\partial y > 0$) are much more important in the present strongly decelerated TBL. Two hairpin parameters were not included in Table 2 because they are sensitive to both mesh and spatial resolutions. The effective diameter of hairpin heads D is mostly sensitive to mesh resolution while the swirl intensity of the hairpin heads $\bar{\lambda}$, defined as the average value of the swirling rate λ per vortex, is mostly sensitive to spatial resolution. The average of these parameters was computed on six streamwise intervals for each xy-plane to show these sensitivities. Representative results expressed in outer units are plotted in figs. 3a and 3b, where U_{ZS} is the Zagarola-Smits outer velocity scale. When accounting for mesh and spatial resolution effects, both the average diameter and the average swirl intensity of the hairpin heads are found to remain constant in the present flow, within experimental uncertainty, and comparable to their values in the ZPG TBL case.

Packets of hairpins are found throughout the boundary layer. In the present flow, the average of the growth angle γ of the hairpin packets in the streamwise direction is more important (11° vs 5°). Since the higher, most downstream hairpins in a hairpin packet are also normally the older ones, the increased titling and wall-normal stretching in the present flow implies that the growth angle should be larger, assuming identical time scales and hairpin generation mechanisms. The average of the streamwise separation between hairpins, x_{sep}/δ, decreases with increasing velocity defect. This result agrees with the increase in the streamwise contraction rates (when normalized by U_{ZS}/δ) from ZPG to APG conditions, as well as with

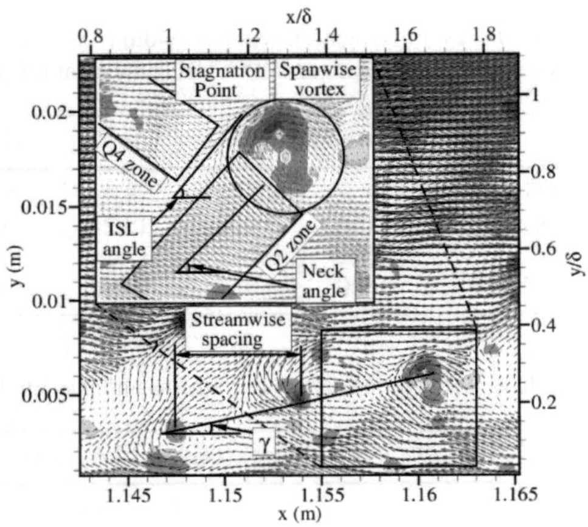

Fig. 2 Schematic definition of parameters and portion of an instantaneous velocity field in region $x = 1.128 - 1.185$ m with Galilean decomposition $(u - 0.5U_e, v + 0.5V_e)$. Contours of the swirling rate λ. One vector out of 4 in large plot for clarity. Inset: zoom on one hairpin signature.

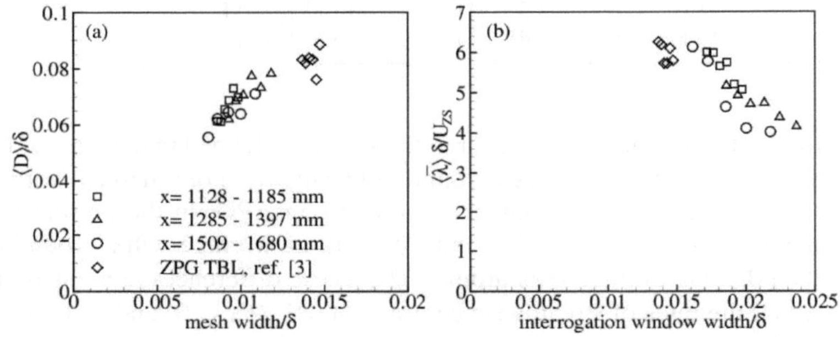

Fig. 3 Hairpin head parameters in outer units for $y > 0.2\delta$: a) Average effective diameter as a function of mesh width. b) average of $\bar{\lambda}$ as a function of interrogation window width.

downstream distance in the present flow. But the streamwise spacing also depends on the mutual repulsion of neighbouring hairpins and on the self-induction of the hairpins. The latter two depend in turn on the shape and strength of the hairpins, properties for which we only have partial information with xy-plane measurements.

Acknowledgement. Financial support from NSERC and CFI of Canada is gratefully acknowledged by the authors. MHSM wishes to thank the Ministry of Science, Research and Technology of the Islamic Republic of Iran for its financial support via its scholarship program.

References

1. Adrian, R.J.: Hairpin vortex organization in wall turbulence. Phys. Fluids (2007), doi:10.1063/1.2717527
2. Maciel, Y., Rossignol, K.S., Lemay, J.: A study of a turbulent boundary layer in stalled-airfoil-type flow conditions. Exp. Fluids 41, 573–590 (2006)
3. Adrian, R.J., Meinhart, C.D., Tomkins, C.D.: Vortex organization in the outer region of the turbulent boundary layer. J. Fluid Mech. 422, 1–54 (2000)

Signature of Varicose Wave Packets in the Viscous Sublayer

Ch. Brücker

Abstract. Experimental results are presented on the waviness of the wall-shear stress field of a two-dimensional transitional boundary layer in a zero-pressure gradient flow channel. Arrays of flexible micro-pillars are used as a dense grid of sensors to measure the surface distribution of tangential vorticity (longitudinal and transverse component) over time. The results show a quasi-periodic passage of varicose wave-like disturbances with strong anti-correlation of the transverse component. These are related to packets of 3-4 vortex loops representing the near-wall state of secondary varicose instability of the streaks.

Introduction

Quantitative visualization of the surface distribution of WSS and its fluctuation over time is possible with a new method using arrays of flexible micro-pillars sensing the WSS by their bending in the flow [1, 2]. Measurements with this technique are given here in for a two-dimensional transitional boundary layer at zero pressure gradient. The flow studies were carried out in a low-turbulence oil flow channel which was designed for high-resolution optical measurements of turbulent boundary layer flows, see Stoots et al. (2001) [3]. A thin flat plate (total length of 3 meter) with a sharp leading edge was placed in the middle of the test section. A tripping wire (diameter $1.5\,mm$) was fixed on the plate $0.15\,m$ downstream of the leading edge to trigger the evolution of Tollmien-Schlichting waves (TS). The sensor array of with an equidistant spacing of $500\,\mu m$ was placed at a location of $1.75\,m$ downstream of the leading edge with the base plane of the micro-pillars flush mounted with the surface of the plate. The Reynolds-number based on the free stream velocity $U_\infty = 2.5\,m/s$ and the streamwise position of the sensor was $Re = 4 \times 10^5$. The

Prof. Dr.-Ing. habil. Ch. Brücker
TU Bergakademie Freiberg, Institute of Mechanics and Fluid Dynamics
e-mail: Bruecker@imfd.tu-freiberg.de

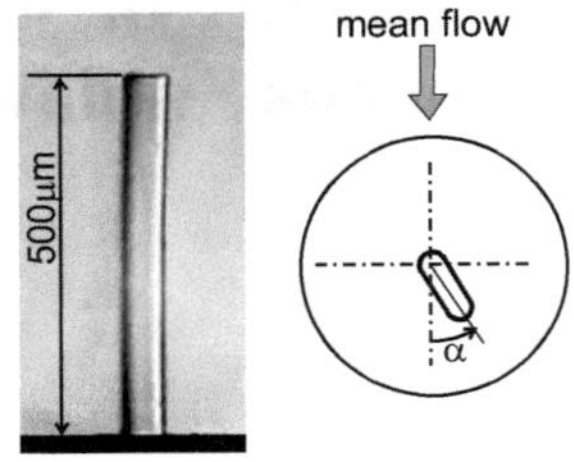

Fig. 1 Single micro-pillar in side-view and definition of the orientation of the WSS vector as indicated by the bending of the pillar.

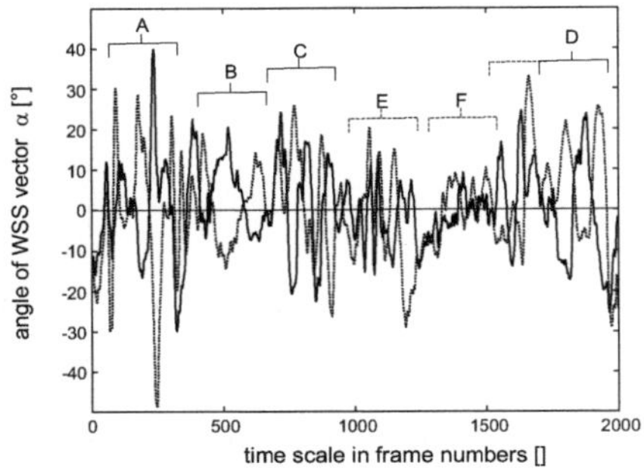

Fig. 2 Time-traces of the angle α of the WSS vector in a turbulent boundary layer flow at two measurement locations with a spanwise distance of $\Delta z^+ = 50$ (plotted on the ordinate is the WSS angle α in °, abscissa belongs to the frame number in the recorded sequence, recording frequency $3\,kHz$).

micro-pillars (diameter $70\,\mu m$, height $500\,\mu m$) have a uniform circular cross-section representing homogeneous circular beams with a planar base and tip, see Fig.1. They are made of PDMS (Poly-Dimethylsiloxane) which has a density of $\rho = 1050\,kg/m^3$ and a Young's modulus of $E = 1.6 \cdot 10^6 N/m^2$ after being cured. The high viscosity of the fluid ($\nu = 11 \times 10^{-6} m^2/s$ at room temperature) ensures that non-linear effects in the pillar response are damped out (Q-factor of 0.48, see equations 13 in Brücker et al 2007). A digital high-speed camera (Photron APX-RS, recording rate $3000\,fps$) equipped with a telecentric lens system (M=3) was used to record the pillar motion. Standard image processing routines were used to measure the pillar tip displacement with high resolution.

The following quantities shown with a + superscript are made dimensionless with the local mean *friction velocity* $u_\tau = \sqrt{\tau_W/\rho}$ as a scale representative of the turbulent fluctuations. The streamwise as well as spanwise spacing of the pillars is $\Delta x^+ = \Delta z^+ = 50$ wall units, which corresponds roughly to the typical width of the low speed streaks close to the wall (Iuso et al. measured a typical width of $\Delta z^+ = 50$ at a wall-normal distance of $\Delta z^+ = 20$) [4]. Fig.2 shows a typical time-trace of 2 neighboring sensors with a spanwise distance of $\Delta z^+ = 50$. The profiles display the angle α of the WSS vector in the plane of the wall (the streamwise direction corresponds to $\alpha = 0°$, spanwise direction is $\alpha = \pm 90°$) over time t in frame numbers (the recording frequency was $3000\,fps$). The total period of the recording sequence is $666\,ms$ or $t^+ = 732$ time units. Some characteristic events depicted as A-D are highlighted figure. The time-traces in Fig.2 let recognize several characteristic wave-like patterns with a clear anticorrelation of the angle of the WSS vector at the neighboring measurement locations, i.e. the WSS vector orientation of one of the neighboring pillars is approximately reflectionally symmetric to the other one.

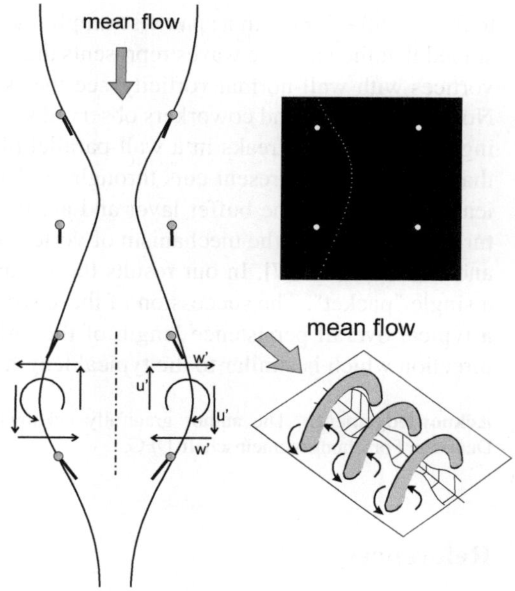

Fig. 3 Spatio-temporal reconstruction of pillar motion and the path of a single particle in a wall distance of approximately $y^+ \approx 5$ as shown in the visualization; the sketch on the right-hand side illustrates the corresponding disturbance pattern as rows of arc-like vortex loops in the very near wall region.

The wave-like patterns consist of more than 2-3 wave-cycles. Interestingly, sometimes the orientation of the WSS vector is highly tilted towards the spanwise direction with angles of about $|\alpha| \approx 40°$ (phase A) which indicates a strong spanwise shearing motion. Additionally the rapid change of orientation in one cycle hints on a strong wall-parallel "meandering" or "swirling" motion in the viscous sublayer. This is an indication of a significant component of the wall-normal vorticity therein. The wavy disturbances appear as a varicose modulation of the near-wall flow motion in spanwise direction, therefore we denote the typical antisymmetric pattern in the following as "varicose" waves according the definition given in Skote et al (2002)[5].

The number of events per unit time, deduced from the time-traces in Fig.2 yields a non-dimensional frequency $F = 2\pi f \nu / U_\infty^2$ in the range $220 \times 10^{-6} < F < 420 \times 10^{-6}$ which is roughly a factor or 2-4 of the fundamental frequency $F_I = 105 \times 10^{-6}$ of the Tollmien-Schlichting waves in accordance with linear stability theory. With increasing Reynolds-number, the range becomes smaller and the mean value approaches a value of $4 \times F_I$. Interestingly, Bake et al. (2002)[6] found in their DNS of Klebanoff type transition of the flow over a flat plate a typical frequency of 3-4 times the fundamental frequency in form of "spikes" in the fluctuating streamwise velocity near the wall. They related the spikes to the induction effect of ring-like vortices which evolve continuously in a number of 3-4 vortices at the tip of the Lambda-vortices before breakdown to randomization and turbulence.

Fig.3 demonstrates a spatio-temperal contruction of the pillar motion as part of the variocose wave overlayed onto a Lagrangian path of an individual tracer particle in the viscous sublayer which was present there by luck in the otherwise clean fluid (Fig.3 upper right). The streamwise wavelength is estimated from the wave-profile to $\lambda^+ \approx 150 - 200$ with a spanwise amplitude of the order of 10 wall units. It is assumed that the varicose waves represents the passage of symmetric counter-rotating vortices with wall-normal vorticity, see the sketch on the right-hand side in Fig.3. Note that Onorato and coworkers observed such counter-rotating vortex pairs flanking the low-speed streaks in a wall-parallel plane at $\Delta y^+ = 20$ [4]. They assumed that the vortices represent cuts through the legs of hairpin vortices which are typically observed in the buffer layer and are the main contribution to the non-linear turbulence cycle by the mechanism of vortex regeneration as described by Tomkins and Adrian (2003)[7]. In our results these waves appear in numbers of 3-4 within a single "packet". The succession of these structures along the same path results in a typical overall persistence length of the order of 1000 wall units in streamwise direction which is similar to the typical length scale of sublayer streaks.

Acknowledgement. The author gratefully acknowledges the support of this project by the Deutsche Forschungsgemeinschaft DFG.

References

1. Brücker, C., Spatz, J., Schröder, W.: Feasibility study of wall shear stress imaging using microstructured surfaces with flexible micro-pillars. Exp. Fluids 39, 464 (2005)
2. Brücker, C., Bauer, D., Chaves, H.: Dynamic response of micro-pillar sensors measuring fluctuating wall-shear stress. Exp. Fluids 42(5), 737 (2007)
3. Stoots, C., Becker, K., Durst, F., McEligot, D.: A large-scale matched index of refraction flow facility for LDA studies around complex gemetries. Exp. Fluids 30(4), 391 (2001)
4. Iuso, G., DiCicca, G.M., Onorato, M., Spazzini, P.G., Malvano, R.: Velocity streak structure modifications induced by flow manipulation. Phys. Fluids 15(9), 2602 (2003)
5. Skote, M., Haritonidis, J.H., Henningson, D.S.: Varicose instabilies in the turbulent boundary layer. Phys. Fluids 14(7), 2309 (2002)
6. Bake, S., Meyer, D.G.W., Rist, U.: Turbulence mechanism in Klebanoff transition: a quantitative comparison of experiment and direct numerical simulation. J. Fluid Mech. 459, 217 (2002)
7. Tomkins, C.D., Adrian, R.J.: Spanwise structure and scale growth in turbulent boundary layers. J. Fluid Mech. 490, 37 (2003)

Entrainment Reduction and Additional Dissipation in Dilute Polymer Solutions

Markus Holzner, Beat Lüthi, Alexander Liberzon, Michele Guala, and Wolfgang Kinzelbach

Abstract. We present a comparative experimental study of a turbulent flow developing in clear water and dilute polymer solutions (25 and 50 wppm polyethylene oxide). The flow is forced by a planar grid that oscillates vertically in a square container of initially still fluid. The two-component velocity fields are measured in a vertical plane passing through the center of the tank by using time resolved Particle Image Velocimetry (PIV). We obtain a lower entrainment rate for polymer solutions as compared to clear water. Extending arguments based on similarity and fractal theory to the case of dilute polymer solutions, we derive a relation between the entrainment rate and the fraction of input energy dissipated by the polymers.

1 Introduction

The reduction of drag in a turbulent pipe flow resulting from the addition of a tiny amount of polymer is known for about 60 years now and has been studied extensively. On the other hand, the effects of such additives in the bulk of flows, which are not bounded by walls is much less documented in literature, despite its fundamental importance. Recently such configurations have been studied experimentally (e.g., [7, 8, 10]), as well as numerically (e.g., [1, 3, 11]). In these studies, alterations by polymers were identified both, at the scale of energy containing eddies and at the smallest scales of motion, where the energy contained in the larger scales ε_T is

Markus Holzner · Beat Lüthi · Wolfgang Kinzelbach
Institute of Environmental Engineering, ETH Zürich, Wolfgang-Pauli-Str. 15, CH-8093 Zürich, Switzerland
e-mail: holzner, luthi, kinzelbach@ifu.baug.ethz.ch

Alexander Liberzon
School of Mechanical Engineering, Faculty of Engineering, Tel Aviv University, Israel
e-mail: alexlib@eng.tau.ac.il

Michele Guala
California Institute of Technology, Pasadena, CA, USA
e-mail: guala@caltech.edu

dissipated partly by the viscous eddies ε_v and partly by the polymers ε_p. However, the relation between macroscopic effects and the direct action of the polymer molecules on the small scales of turbulence remains unclear.

There is an interesting analogy between the effect of polymers in turbulent flows and the entrainment phenomenon observed in free shear flows (e.g., jets, wakes or boundary layers). The entrainment is known to be governed by large scale flow parameters, but the viscous scales are inferred to be responsible for the diffusion of vorticity [2]. More precisely, the interface separating the turbulence from the irrotational fluid, the so-called 'turbulent-nonturbulent interface' (TNTI), advances into the ambient with a propagation velocity, v_e, that is proportional to the velocity scale of the energy-containing turbulence in proximity of the interface, i.e., $v_e = \beta v_{\mathrm{rms}}$, where β is the entrainment constant. On the other hand, entrainment across the TNTI involves small scale processes characterized by the presence of a thin layer, called "viscous superlayer" by [2], through which vorticity "diffuses" from the turbulent to the irrotational side. [5, 6] showed quantitatively that the local entrainment velocity across the interface is proportional to the Kolmogorov velocity u_η. It is natural to ask, whether the nature of these viscous superlayers is altered by the presence of polymers. We present a comparative experimental study of a turbulent flow developing in clear water and dilute polymer solutions and show that polymer additives affect both, the overall propagation velocity and the local entrainment velocity.

2 Experimental Method

The flow is forced by a planar grid that oscillates vertically in a square container of initially still fluid. The oscillating grid apparatus and the forcing parameters are the same as the one used in the previous investigations conducted by the authors, see [6] and references therein. The two-component velocity fields are measured in a vertical plane passing through the center of the tank by using time resolved Particle Image Velocimetry (PIV), see [4] for details. In the experiments, after the forcing is initiated, a turbulent region develops and grows in time through entrainment of surrounding fluid until the fluid in the whole container is in turbulent motion. Six runs each were performed for clear water flow and for aqueous dilute poly(ethylene oxide) solutions of Polyox WSR 301, MW = 4×10^6 g/mol, at two concentrations of 25 and 50 wppm. We detect the TNTI by using the method based on the out-of-plane vorticity component described in [4].

3 Results and Discussion

We compare the propagation of the TNTI in water and dilute polymer solutions. The first result in Fig. 1a shows the evolution in time of the depth of the turbulent layer $H(t)$. Three curves are plotted representing water, 25 wppm and 50 wppm polymer

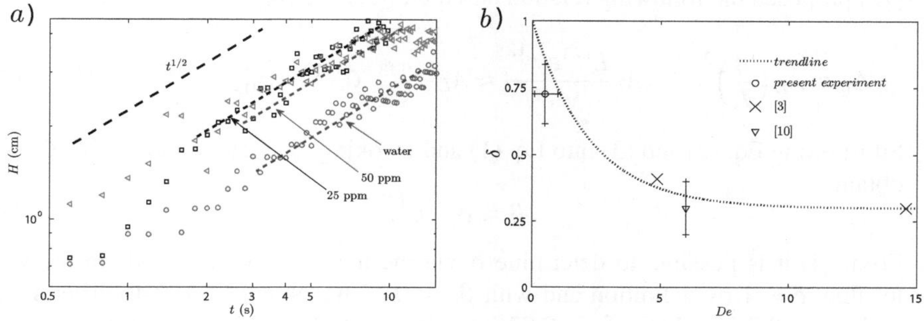

Fig. 1 Depth of the turbulent layer, $H(t)$, in water and dilute polymer solutions shown as symbols, while curves represent the best fit of the form $H = (Kt)^n$ (a). Dependence of $\delta = \varepsilon_v/\varepsilon_T$ on the Deborah number (b). 'x' marks the numerical results of [3], a triangle is used for [10], and the circle denotes the present estimate. The error bars display the experimental uncertainty in the case of the present experiment and the variation of concentration in case of the experiment of [10].

solution, respectively. Each curve is an average over 6 experimental runs, the scatter represents the variance. [9] analyzed the diffusion of turbulence driven by the planar energy source in time and predicted a power law behavior, $H = (Kt)^n$, where the parameter K is the so-called grid action. [4] recently verified this prediction for clear water and obtained $H = \sqrt{Kt}$. In the present study we extend the same analysis to dilute polymer solutions and obtain $n = 0.47$, 0.52 and 0.48 ± 0.05 for the three cases respectively, and henceforth use $n = 0.5$ for all cases. Next, we estimate the entrainment constant, $\beta = v_e/v_{\mathrm{rms}}$, as follows. The propagation velocity of the TNTI is taken from the PIV measurements as $v_e = 1/2K^{1/2}t^{-1/2}$. For the estimate of v_{rms}, data points were used at $y = H(t)$ in turbulent flow regions only, i.e. excluding the irrotational regions. For water, 25 ppm and 50 ppm data we obtain $\beta_w = 0.77$, $\beta_{25ppm} = 0.68$ and $\beta_{50ppm} = 0.70 \pm 0.07$, respectively.

Now we develop some considerations based on similarity principles and fractal theory to bring our results in context with other findings in literature. We can express the energy contained in the larger scales as $\varepsilon_T \propto v_{\mathrm{rms}}^3/L$. In the case of clear water this energy is equal to dissipation through viscosity by the smallest scales of turbulence, $\varepsilon_v = 2\nu\langle s_{ij}s_{ij}\rangle$ (s_{ij} is the fluctuating rate of strain tensor). When polymers are introduced, ε_T is partly dissipated through the deformation of flexible polymer molecules ε_p, while only a fraction δ is dissipated through the viscous eddies and we write $\varepsilon_v = \delta \cdot \varepsilon_T$ or $\varepsilon_p = (1-\delta)\cdot\varepsilon_T$. The entrainment flux can be expressed using a projected area A_L or the area A_η resolved up to the smallest length scales, e.g., [12]:

$$v_e \cdot A_L = u_\eta \cdot A_\eta. \tag{1}$$

With $\varepsilon_v \approx C \cdot v_{rms}^3/L$ the small scale quantities u_η and A_η can be expressed as

$$u_\eta = \nu^{1/4} \cdot \varepsilon_v^{1/4} \approx C^{1/4} \cdot v_{\mathrm{rms}}^{3/4} \cdot \nu^{1/4} \cdot L^{-1/4}. \tag{2}$$

[12] proposed the following relation based on fractal theory with $d = 7/3$:

$$A_\eta \approx A_L \left(\frac{\eta}{L}\right)^{2-d} \approx A_L \frac{L^{1/3}\varepsilon_\nu^{1/12}}{\nu^{1/4}} \approx A_L \cdot L^{1/3} \cdot C^{1/12} \cdot v_{\mathrm{rms}}^{1/4} \cdot L^{-1/12} \cdot \nu^{-1/4}. \quad (3)$$

Substituting Eq. (2) and (3) into Eq. (1) and invoking $v_e = \beta v_{rms}$ and $C = \delta \cdot C_*$ we obtain

$$\beta \approx \delta^{1/3} C_*^{1/3}. \quad (4)$$

From (4) it is possible to determine δ via the measured values of β. In the water flow $\delta = 1$ by definition and with $\beta_w \approx 0.77$ we obtain $C_* \approx 0.46$. Therefore, with $\beta_p = 0.7$ we obtain $\delta_p = 0.72$ implying that 72% of the energy input is dissipated by viscous eddies, while 28% is dissipated by the polymers. An important dimensionless number that characterizes the effects of polymers in turbulence is the Deborah number $De = \tau_R/\tau_\eta$, the ratio of the polymer characteristic time scale to the Kolmogorov time scale. Figure 1b shows the plot of δ versus De in which we add our experimental estimate to the data from numerical and experimental studies [3, 10]. At $De = 0$ the trendline has to cross unity (clear water), and at large Deborah numbers we assume the relation to level off asymptotically. We note that our measurements are consistent with the conjectured dependence of δ on De.

References

1. Berti, S., Bistagnino, A., Boffetta, G., Celani, A., Musacchio, S.: Small-scale statistics of viscoelastic turbulence. Europhys. Lett. 76(1), 63–69 (2006)
2. Corrsin, S., Kistler, A.L.: The free-stream boundaries of turbulent flows. NACA, TN-3133, TR-1244, 1033–1064 (1955, 1954)
3. De Angelis, E., Casciola, C., Benzi, R., Piva, R.: Homogeneous isotropic turbulence in dilute polymers. J. Fluid Mech. 531, 1–10 (2005)
4. Holzner, M., Liberzon, A., Guala, M., Tsinober, A., Kinzelbach, W.: Generalized detection of a turbulent front generated by an oscillating grid. Exp. Fluids 41, 711–719 (2006)
5. Holzner, M., Liberzon, A., Nikitin, N., Kinzelbach, W., Tsinober, A.: Small scale aspects of flows in proximity of the turbulent/non-turbulent interface. Phys. Fluids 19, 071702 (2007)
6. Holzner, M., Liberzon, A., Nikitin, N., Lüthi, B., Kinzelbach, W., Tsinober, A.: A Lagrangian investigation of the small scale features of turbulent entrainment through particle tracking and direct numerical simulation. J. Fluid Mech. 598, 465–475 (2008)
7. Liberzon, A., Guala, M., Lüthi, B., Kinzelbach, W., Tsinober, A.: Turbulence in dilute polymer solutions. Phys. Fluids 17(3), 031707 (2005)
8. Liberzon, A., Guala, M., Kinzelbach, W., Tsinober, A.: On turbulent kinetic energy production and dissipation in dilute polymer solutions. Phys. Fluids 18(12), 125101 (2006)
9. Oberlack, M., Günther, S.: Shear-free turbulent diffusion, classical and new scaling laws. Fluid Dynamics Research 33, 453–476 (2003)
10. Ouellette, N., Xu, H., Bodenschatz, E.: Modification of the turbulent energy cascade by polymer additives, in review [arXiv:0708.3945v2] (2008)
11. Peters, T., Schumacher, J.: Two-way coupling of finitely extensible nonlinear elastic dumbbells with a turbulent shear flow. Phys. Fluids 19(6), 065109 (2007)
12. Sreenivasan, K.R., Ramshankar, R., Meneveau, C.: Mixing, entrainment and fractal dimensions of surfaces in turbulent flows. Proc. R. Soc. Lond. A 421, 79–108 (1989)

Mixing at the External Boundary of a Submerged Turbulent Jet

A. Eidelman, T. Elperin, N. Kleeorin, G. Hazak, I. Rogachevskii,
O. Sadot, and I. Sapir-Katiraie

Abstract. We study experimentally and theoretically mixing at the external boundary of a submerged turbulent jet. In the experimental study we use Particle Image Velocimetry and an Image Processing Technique based on the analysis of the intensity of the Mie scattering to determine the spatial distribution of tracer particles. An air jet is seeded with the incense smoke particles which are characterized by large Schmidt number and small Stokes number. We determine the spatial distributions of the jet fluid characterized by a high concentration of the particles and of the ambient fluid characterized by a low concentration of the tracer particles. In the data analysis we use an approach that is based on analysis of the two-point second-order correlation function of the particle number density fluctuations generated by tangling of the gradient of the mean particle number density by the turbulent velocity field. This gradient is formed at the external boundary of a submerged turbulent jet. We demonstrate that the two-point second-order correlation function of the particle number density does not have universal scaling and cannot be described by a power-law function. The theoretical predictions made in this study are in a qualitative agreement with the obtained experimental results.

1 Experimental Set-Up

We investigate theoretically and experimentally mixing at the air jet interface using the incense smoke produced by incense particles sublimation. The experimental set-up includes the chamber with the transparent plexiglass walls, and the cylindrical tube with a jet nozzle mounted at the tip. A submerged air jet discharges into the chamber. The cross-section of the channel with transparent walls is 47×47 cm^2, a

---

A. Eidelman · T. Elperin · N. Kleeorin · I. Rogachevskii · I. Sapir-Katiraie
Department of Mechanical Engineering, Ben-Gurion University of the Negev, Beer-Sheva, Israel

G. Hazak · O. Sadot
Department of Physics, Nuclear Research Center, Beer-Sheva, Israel

diameter of the jet nozzle is $D = 10$ mm and the diameter of the jet in the probed flow region is about 35 mm. Flow measurements have been conducted with digital Particle Image Velocimetry (PIV) using a LaVision Flow Master III system to determine the jet velocity field. The light sheet is provided by a double PIV Nd-YAG laser, Continuum Surelite 2×170 mJ. A progressive-scan 12 Bit digital CCD camera with dual-frame-technique for cross correlation captures the images of 1280×1024 pixels with a size $6.7\,\mu$m $\times 6.7\,\mu$m. We also use an Image Processing Technique based on the analysis of the intensity of the Mie scattering to determine the spatial distribution of tracer particles. The submerged air jet, seeded with the incense smoke particles of sub-micron sizes with the mean diameter 0.7μm, flows into the still air inside the chamber.

We use images of a jet obtained in our experiments, whereby the interface between the jet and the ambient fluid is determined for a mean jet image averaged over 50 independent images that are acquired with a frequency of 2 Hz. The successive pairs of images are not correlated since the turbulence correlation time τ_0 is of the order of 10^{-3} s and the characteristic time of the mean flow in the image scale is of the order of 10^{-2} s. The estimate for the turbulence correlation time is obtained using the measured characteristic turbulent velocity $u_0 \sim (1-2) \times 10^3$ cm/s, the maximum scale of turbulent motions $l_0 \sim 1$ cm and $\tau_0 = l_0/u_0 \sim 10^{-3}$ s. We measure spatial particle distribution in a flow area of 18.4×18.4 cm^2 with a spatial resolution of 1024×1024 pixels that is 0.18 mm/pixel. Every recorded image is normalized by a light intensity measured just at the jet entrance into the chamber in order to eliminate the effects associated with a change of concentration of the incense smoke.

2 Theoretical Modelling of Turbulent Mixing and Comparison with Experiments

We use an approach that is based on the analysis of the two-point second-order correlation function of the particle number density fluctuations. These fluctuations are generated by tangling of the gradient of the mean particle number density, ∇N, by the turbulent velocity field (see, e.g., [1, 2, 3]). This gradient is formed at the external boundary of a submerged turbulent jet. The mechanism of mixing related to the tangling of the gradient of the mean particle number density is relatively robust. The properties of the tangling-induced fluctuations are not very sensitive to the exponent of the energy spectrum of the background turbulence. The requirements that turbulence should be an isotropic, homogeneous, and should have a very long inertial range, are not necessary for generating fluctuations by the tangling mechanism.

We use equation for the evolution of the two-point second-order correlation function of the particle number density. This equation includes the scale-dependent turbulent diffusion term and the source term of particle number density fluctuations at the jet-ambient fluid interface: $I = D_T\,(\nabla N)^2 \exp(-R/l_0)$, where $D_T \propto l_0\, u_0$ is the turbulent diffusion coefficient and $\mathbf{R}$ is the difference between the two point coordinates. This equation has been derived in [4] for a delta-correlated in time random velocity field and in [5] for a turbulent velocity field with a finite correlation time. Solution of this equation yields the two-point second-order correlation function of the particle number density.

In order to compare the theoretical predictions with the experimental results, the analysis of the experimental data for correlation function of the particle number density is performed with and without transformation of images into a binary form. In particular, the two-point second-order correlation function $\Phi(R)$ determined in our experiments is shown in Fig. 1. Figure 1 and our analysis demonstrate that the two-point second-order correlation function of the particle number density do not have universal scaling and cannot be described by a power-law function (see also [3]).

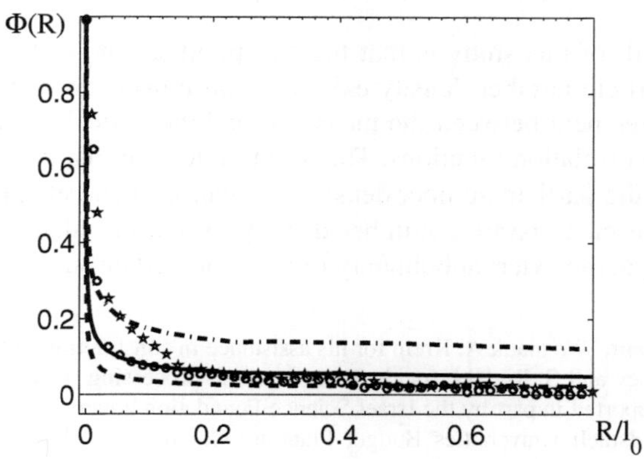

Fig. 1 Normalized two-point second-order correlation function $\Phi(R)$ for different values of parameter a_c: $a_c = 1.2$ (dashed-dotted line), $a_c = 2$ (solid line) and $a_c = 4$ (dashed line). Two-point second-order correlation function $\Phi(R)$ determined in our experiments with (unfilled circles) and without (stars) transformation of images into a binary form. The distance R between two points is measured along the mean boundary of the jet. Here a_c is the ratio of the characteristic scales of the compressible and incompressible parts of the turbulent diffusion tensor.

The main reason for the non-universal behavior of the correlation function of the particle number density is the compressibility of Lagrangian trajectories in a turbulent flow with a finite correlation time. For a delta-correlated in time random velocity field the power-law behavior of the correlation function of the particle number density is possible. The power-law behavior of the correlation function can be possible when the characteristic scale of the inhomogeneity of the mean particle number density is larger than the maximum scale of turbulent motions. However, in the vicinity of the jet boundary the characteristic scale of the inhomogeneity of the mean particle number density is much smaller than the maximum scale of turbulent motions.

We have found that there is a qualitative agreement between theoretical and experimental results obtained in this study. However, the theoretical results obtained in this study cannot be valid in the most general cases since we adopted a number of simplifying assumptions about the turbulence. We considered a homogeneous, isotropic, and incompressible background turbulence. The main contribution to the level of fluctuations of particle number density is due to the mode with the minimum damping rate. This mode is an isotropic solution of the equation for the two-point second-order correlation function of the particle number density. Consequently, it is plausible to neglect the anisotropic effects. This assumption is also supported by our measurements of the two-point second-order correlation function.

3 Conclusions

The main result of this study is that the two-point second-order correlation function of the particle number density exhibit a non-power-law behaviour. There is a qualitative agreement between the measured and theoretically predicted two-point second-order correlation functions. This is an indication that the presented theoretical model of the particle number density fluctuations generated by tangling of the gradient of the mean particle number density by the turbulent velocity field, can mimic mixing at the external boundary of a submerged turbulent jet.

Acknowledgement. We thank A. Krein for his assistance in construction of the experimental setup, and I. Golubev and S. Rudykh for their assistance in processing the experimental data. This research was supported in part by the Israel Science Foundation governed by the Israeli Academy of Science, the Israeli Universities Budget Planning Committee (VATAT) and Israeli Atomic Energy Commission.

References

1. Elperin, T., Kleeorin, N., Rogachevskii, I.: Phys. Rev. E 52, 2617–2634 (1995)
2. Elperin, T., Kleeorin, N., Rogachevskii, I.: Phys. Rev. Lett. 77, 5373–5376 (1996)
3. Eidelman, A., Elperin, T., Kleeorin, N., Hazak, G., Rogachevskii, I., Sadot, O., Sapir-Katiraie, I.: Phys. Rev. E 79, 026311 (2009)
4. Kraichnan, R.H.: Phys. Fluids 11, 945 (1968)
5. Elperin, T., Kleeorin, N., L'vov, V.S., Rogachevskii, I., Sokoloff, D.: Phys. Rev. E 66, 036302 (2002)

Turbulence in Electrically Conducting Fluids Driven by Rotating and Travelling Magnetic Fields

Jörg Stiller, Kristina Koal, and Hugh M. Blackburn

Abstract. The turbulent flow driven by rotating and travelling magnetic fields in a closed cylinder is investigated by means of direct numerical simulations (DNS) and large eddy simulations (LES). Our model is based on the low-induction, low-frequency approximation and employs a spectral-element/Fourier method for discretisation. The spectral vanishing viscosity (SVV) technique was adopted for the LES. The study provides first insights into the developed turbulent flow. In the RMF case, Taylor-Görtler vortices remain the dominant turbulence mechanism, as already in the transitional regime. In contrast to previous predictions we found no evidence that the vortices are confined closer to the wall for higher forcing. In the TMF more than 50 percent of the kinetic energy is bound to the turbulent fluctuations, which renders this field an interesting candidate for mixing applications.

1 Introduction

Alternating magnetic fields are widely used to control the flow of molten semiconductors and liquid metals. Although electromagnetic stirring has been a topic of research for decades, high resolution measurements and ab initio simulations of the turbulent flow appeared only recently [6, 3, 7]. Moreover, these studies were basically confined to the near-critical, transitional regime. In the present work we address the developed turbulent flow at higher, practically relevant forcing parameters. Our approach relies on the so-called low-induction, low-frequency and low-interaction approximation [6, 7]. In this limit the induced magnetic field becomes

Jörg Stiller
TU Dresden, 01062 Dresden, Germany
e-mail: joerg.stiller@tu-dresden.de

Kristina Koal
TU Dresden, 01062 Dresden, Germany
e-mail: kristina.koal@tu-dresden.de

Hugh M. Blackburn
Monash University, Vic 3800, Australia
e-mail: hugh.blackburn@eng.monash.edu.au

negligible such that the body force can be precomputed by solving the decoupled Maxwell equations. We consider the generic case of an incompressible isothermal flow that is driven either by a rotating magnetic field (RMF) or a travelling magnetic field (TMF) in a closed cylinder of aspect (height to diameter) ratio one. In both cases the body force scales linearly with the forcing parameter, which is given by the magnetic Taylor number

$$Ta = \frac{\sigma \omega B^2 R^4}{2\rho \nu^2} \sim f_\varphi \tag{1}$$

in the RMF case (azimuthal force) and

$$F = \frac{\sigma \omega B^2 k R^5}{4\rho \nu^2} \sim f_z \tag{2}$$

for the TMF (axial force). Here ρ, ν, σ represent the fluid's density, viscosity and electrical conductivity, R the cylinder radius, and ω, B, k the angular frequency, induction and wave number of the magnetic field. For further details of the mathematical model we refer to [7].

2 Numerical Method

The flow equations are discretised using a spectral element method in the meridional semi-plane coupled with a Fourier spectral method in the azimuthal direction (see [2] for details). For LES the spectral vanishing viscosity technique was adopted from [8, 5]. A preliminary validation for the turbulent channel flow showed that the SVV competes with the dynamic Smagorinsky model (Fig. 1).

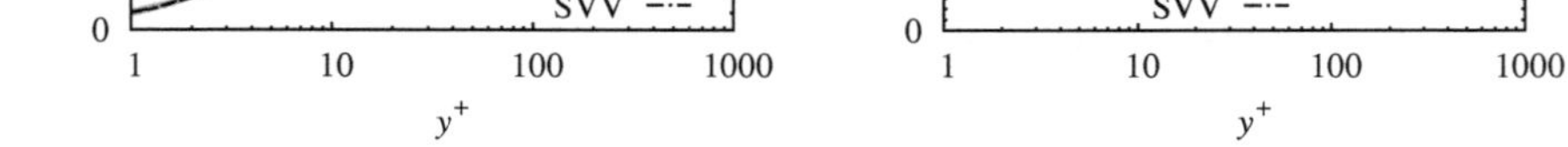

Fig. 1 Computed mean velocity (left) and rms value of the streamwise fluctuations (right) for a turbulent channel flow at $Re_\tau = 651$ in comparison with the DNS data of Moser et al. [4] and the LES of Blackburn and Schmidt [1] using the dynamic Smagorinsky model (DSM) with equivalent discretisation parameters.

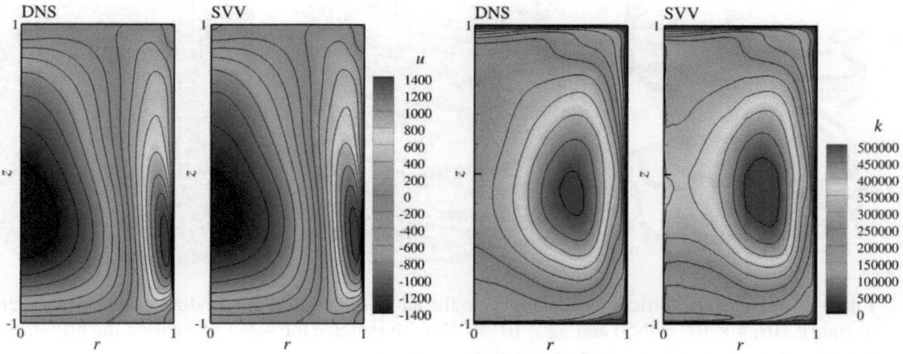

Fig. 2 Comparison of DNS and SVV for the turbulent flow driven by a travelling field at $F = 4 \times 10^6$: u represents the mean vertical velocity and k the turbulent kinetic energy.

Figure 2 provides a comparison between DNS and SVV results for the TMF with $F = 4 \times 10^6 \approx 40F_c$. The DNS was performed with a resolution of 50×25 quadrilateral elements of degree 11 and 160 Fourier modes, whereas only 10×5 elements and 32 modes were employed in the SVV run. This corresponds to a reduction of factor 125 in memory and up to 625 in computing time, while the results are still in very good agreement with the DNS. However, the increased turbulent energy in the centre indicates a weak instability of the SVV, which has to be addressed in future work.

3 Results and Outlook

A series of DNS and SVV-LES was performed up to $Ta \approx 320Ta_c$ for the RMF and $F \approx 620F_c$ for the TMF. In the RMF Taylor-Görtler vortices remain the dominant turbulence mechanism (Fig. 3). The vortices occupy the region $0.75 < r/R < 1$ independently from the forcing, instead of being squeezed to the wall with increasing Ta, which was previously assumed by several authors. While the fraction of turbulent kinetic energy remains close to one or two percent in the RMF, it exceeds more then 50 percent in the TMF. Moreover, the turbulence field tends to become almost isotropic with increased forcing (Fig. 4), thus rendering the TMF an ideal candidate for mixing applications.

Acknowledgement. Financial support from Deutsche Forschungsgemeinschaft in frame of the Collaborative Research Center SFB 609 is gratefully acknowledged. The computations were performed on an SGI Altix system based on a grant from ZIH at TU Dresden.

Fig. 3 Instantaneous vortices (λ_2 contours) in the RMF-driven flow for Taylor numbers (from left to right): 10^6, 4×10^6 (DNS) and 32×10^6 (SVV). To indicate the sense of rotation the brightness was chosen according to the sign of azimuthal vorticity.

Fig. 4 Instantaneous vortices in the TMF-driven flow for forcing parameters (from left to right): 4×10^6, 16×10^6 (DNS) and 64×10^6 (SVV)

References

1. Blackburn, H.M., Schmidt, S.: Spectral element filtering techniques for large eddy simulation with dynamic estimation. J. Comput. Phys. 186, 610–629 (2003)
2. Blackburn, H.M., Sherwin, S.J.: Formulation of a Galerkin spectral element – Fourier method for three-dimensional incompressible flows in cylindrical geometries. J. Comput. Phys. 197, 759–778 (2004)
3. Koal, K., Stiller, J., Grundmann, R.: Linear and non-linear instability in a cylindrical enclosure caused by a rotating magnetic field. Phys. Fluids 19, 088107 (2007)
4. Moser, R.D., Kim, J., Mansour, N.N.: Direct numerical simulation of turbulent channel flow up to $Re_\tau = 590$. Phys. Fluids 11(4), 943–945 (1999)
5. Pasquetti, R., Séverac, E., Serre, E., Bontoux, P., Schäfer, M.: From stratified wakes to rotor-stator flows by an SVV-LES method. Theor. Comput. Fluid Dyn. 22, 261–273 (2008)
6. Stiller, J., Fraňa, K., Cramer, A.: Transitional and weakly turbulent flow in a rotating magnetic field. Phys. Fluids 18, 074105 (2006)
7. Stiller, J., Koal, K.: Direct simulation of turbulence in the flow driven by rotating and traveling magnetic fields. In: 5th International Symposium on Turbulence and Shear Flow Phenomena, TU Munich, Garching, pp. 353–358 (2007)
8. Xu, C., Pasquetti, R.: Stabilized spectral element computations of high Reynolds number incompressible flows. J. Comput. Phys. 196, 680–704 (2004)

The Decay of Turbulence in Pipe Flow

Alberto de Lozar and Björn Hof

Abstract. It is well accepted that turbulence in pipe flow is transient for $Re \lesssim 2000$ and, after some time, always comes back to the laminar state. However, there is no current agreement about the behaviour of turbulence for higher *Re*: whereas some experiments indicate that turbulence becomes sustained after certain critical Reynolds number, Re_C, other studies show that the turbulent state keeps its transient behaviour, ruling out any critical point. Our experiments show that these different views are not caused by the different ways of generating turbulence or by some experimental noise, and therefore the turbulent state is always well defined. Our data also suggest that, in spite of the different interpretations, all experimental results presented up to date are compatible when some sistematic error is taken into account.

1 Introduction

The laminar profile in certain shear flows, such as Couette or pipe flows, is linearly stable for all Reynolds numbers. The transition to turbulence is only triggered when a perturbation of finite amplitude is introduced into the flow. For $Re < 2400$ turbulence appears in localised patches, called puffs. These puffs move approximately with the mean velocity, U, of the flow and at least at low Reynolds numbers they are of transient nature. Observations show that even after surviving for many thousand advective time units they can suddenly decay back to the laminar state. The decay

Alberto de Lozar
Max Planck Institute for Dynamics and Self Organization, Bunsenstr. 10, 37073 Göttingen, Germany
e-mail: adelozar@googlemail.com

Björn Hof
Max Planck Institute for Dynamics and Self Organization, Bunsenstr. 10, 37073 Göttingen, Germany
e-mail: bhof@gwdg.de

of the turbulent state is independent of its history so that the probability of a puff to decay after a time t follows an exponential law:

$$P(t, Re) = \exp(-(t - t_0)/\tau(Re)). \tag{1}$$

Here τ is the characteristic lifetime of the laminar state, which depends on the Reynolds number and t_0 is an initial relaxation period. This memoryless behaviour is typical of dynamical systems and it suggests that the turbulence corresponds to a chaotic saddle in state space[1]. The functional form of τ on Re sets the fate of the turbulent state. Many investigations have focused on the determination of this functional in the case of pipe flow. Whereas the first studies [2] found that τ follows $\tau \propto 1/(Re - Re_C)$ and thus lifetimes tend to infinity at Re_C, experiments with longer observational times [3] show an exponential growth of τ with Re ruling out a critical Reynolds number. As a consequence of the latter view the turbulent state always can be considered as a transient, at least formally.

However, the question of whether turbulence is sustained or not is still the subject of a current debate with experimental and numerical studies supporting both views[1]. The aim of this paper is to test whether different ways of triggering turbulence or weak experimental noise might generate the discrepancy between these studies. We show that the lifetimes are independent of initial conditions and noise, and we suggest that all experimental studies might be compatible when a shift in the Reynolds number is considered.

2 Lifetimes Results

We investigate the dependence of lifetimes on initial conditions. It is not a priori clear to what extend the dynamical behaviour encountered in the long time limit depends on the type of perturbation initially triggering turbulence. We have considered four kinds of perturbations to the flow: (a) injecting a small quantity of liquid into the flow through a hole in the pipe; (b) injecting and extracting the same quantity of liquid through two opposite holes; (c) creating the puff using the injection method at a higher Re (*e.g.* 2000) followed by a sudden decrease in flow rate to the Reynolds number we want to measure (*e.g.* 1850); and (d) generating intermitting puffs by placing an obstacle which blocks approximately one half of the pipe cross-section. All experiments where performed in a pipe with a diameter D = 4 mm and the observation times were t = 930 D/U and t = 269 D/U. The occurrence or not of the laminar state $P(t, Re)$ was monitored at the outlet of the pipe and statistics were taken for between 50 and 500 events. Details of these experiments are shown in [8]. The lifetimes are then calculated using equation 1. As shown in Fig. 1 (a) all decay rates collapse indicating that lifetimes are in fact independent of the initial perturbation. This confirms that all perturbations tested lead to the same chaotic state in

[1] For transient turbulence see [1] or [4]. For lifetimes of turbulence going to infinity, see [5], [6] or [7].

phase space. Furthermore the turbulent state rapidly loses all memory of how it was initially created which is typical of chaotic systems.

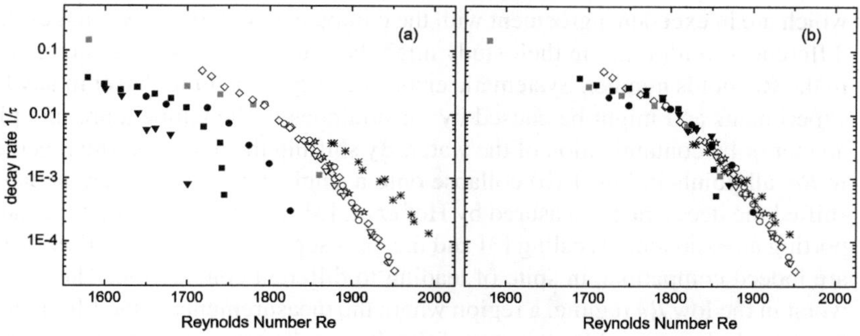

Fig. 1 (a) Decay rates measured in different experimental setups. Our results in a 4 mm pipe are represented by the open symbols and the symbol's shape indicates the form of perturbation used to trigger the turbulent puff: injection of fluid (◇), injecting and extracting liquid (○), sudden decrease of *Re* after generating the puff at a higher *Re* (▽) and puffs generated by an obstacle (⬠). ☆ were measured for *Re* altered periodically by ±5. The black solid symbols represent previous results obtained in a 20 mm pipe [5, 7]. Again the symbol's shape indicates the form of perturbation: injecting liquid through six small holes at an amplitude $A = 0.1$ (▼) and $A = 0.01$ (•) and by sudden decrease of *Re* as explained above (■). The amplitude, *A*, is defined as the ratio of the amount of fluid injected to the mean mass flux of the pipe. Results obtained in computer simulations [6] are represented by ■. Lifetimes measured in a very long pipe [3] are shown by ∗. This last study has lead to the view that turbulence is always of transient nature. (b) Same as in (a) but the literature data has been shifted in *Re*. The experiments in the long pipe (∗) were shifted by −2,5%. The experiments in the 20 mm pipe were shifted by 8.5% (▼), 3.75% (•) and 5.5% (■). The simulation results are unaltered.

The influence of weak noise in the lifetemes is also investigated in the same pipe. We investigated whether altering the pressure gradient periodically during the experiment would have any effect on the lifetimes. The Reynolds number is varied by ±5 with a frequency of 2 Hz. The turbulence was triggered by the injection method and no difference from the previous results was observed (see Fig. 1 (a)), supporting the robustness of the measurements.

Our results compare favorably with the high resolved computer simulations performed in a long pipe [6], represented in the graph by the gray squares[2]. However, our measured liftemines do neither follow an exponentinal scaling (as in [3]) nor a critical behaviour (as in [5] or [6]). The exact form for the lifetimes will be reported in another study [10].

[2] We have used the corrected result for the lowest *Re* as explained in [9].

For a comparison, the decay rates obtained in the previous experimental studies [5, 7, 3] are also presented in Fig. 1 (a). None of these results quantitatively agree neither with our measurements nor with the lifetimes obtained in the computer simulations. Contrary to our findings, one of the the conclusion of [5, 7] is that lifetimes depend on the initial conditions and indeed a different critical Reynolds number, Re_C, is found for each perturbation. However, motivated by our recent observations which are in excellent agreement with the computer simulations, we suggest that the differences in lifetimes in their study might be caused by possible systematic errors in the Reynolds number. Systematic errors are very difficult to detect in any kind of experiments and might be caused by uncertainties in the temperature, the pipe diameter or by contamination of the water. By shifting the data set a small percentage in *Re*, all points in Fig. 1 (b) collapse onto a single curve. Moreover, we have also shifted the decay rates measured by Hof *et al.* [3] which suggests that the data supporting an exponential scaling [3] and the ones supporting diverging lifetimes [5, 7] are indeed compatible, in spite of leading to different conclussion. The collapse is worst in the low *Re* regime, a region where the measurements of the lifetimes might be strongly influenced by inherent difficulties in determining the initial relaxation period t_0. A good determination of t_0 is always difficult because it might vary with the type of perturbation and *Re*.

References

1. Eckhardt, B., Schneider, T.M., Hof, B., Westerweel, J.: Turbulence transition in pipe flow. Annual Rev. Fluid Mech. 39, 447–468 (2007)
2. Faisst, H., Eckhardt, B.: Sensitive dependence on initial conditions in transition to turbulence in pipe flow. J. Fluid Mech. 504, 343 (2004)
3. Hof, B., Westerweel, J., Schneider, T.M., Eckhardt, B.: Finite lifetime of turbulence in shear flows. Nature 443, 55–62 (2006)
4. Schneider, T.M., Eckhardt, B.: Lifetime statistics in transitional pipe flow. Phys. Rev. E 78, 046310 (2008)
5. Peixinho, J., Mullin, T.: Decay of turbulence in pipe flow. Phys. Rev. Lett. 93, 094501 (2006)
6. Willis, A.P., Kerswell, R.R.: Critical behavior in the relaminarization of localized turbulence in pipe flow. Phys. Rev. Lett. 98, 014501 (2007)
7. Mullin, T., Peixinho, J.: Transition to turbulence in pipe flow. J. Low Temp. Phys. 145, 75 (2006)
8. de Lozar, A., Hof, B.: An experimental study of the decay of turbulent puffs in pipe flow. Phil. Trans. R. Soc. A 367, 589–599 (2009)
9. Hof, B., Westerweel, J., Schneider, T., Eckhardt, B.: Critical behaviour in the relaminarisation of localised turbulence in pipe flow (2007), `http://arxiv.org/abs/`
10. Hof, B., de Lozar, A., van Kuik, D.J., Westerweel, J.: Repeller or attractor? Selecting the dynamical model for the onset of turbulence in pipe flow. Phys. Rev. Lett. 101, 214501 (2008)

The Effect of Spanwise System Rotation on Turbulent Poiseuille Flow at Very-Low-Reynolds Number

Oaki Iida, K. Fukudome, T. Iwata, and Y. Nagano

Abstract. Direct numerical simulations (DNSs) with a spectral method are performed with large and small computational domains to study the effects of spanwise rotation on a turbulent Poiseuille flow at the very low-Reynolds numbers. In the case without system rotation, quasi-laminar and turbulent states appear side by side in the same computational domain, which is coined as laminar-turbulence pattern. However, in the case with system rotation, the pattern disappears and flow is dominated by quasi-laminar region including very long low-speed streaks coiled by chain-like vortical structures. Increasing the Reynolds number can not generate the laminar-turbulence pattern as long as system rotation is imposed.

1 Introduction

In the recent numerical studies on turbulent Poiseuille and Couette flows, the interesting turbulent structure is found to emerge at the critical Reynolds number where the stripes of turbulent and laminar regions appear side by side, which is coined as laminar-turbulence pattern [1]. The large-scale turbulent regions, where longitudinal vortical structures are concentrated, are tilted with respect to the streamwise direction. In the quasi-laminar region where the skin friction is significantly reduced, low-speed streaks become significantly longer in the streamwise direction without sinuous instability believed to trigger longitudinal vortical structure. Recently, Fukudome *et al.* [2] found from the budget of turbulent kinetic energy that the tilting of large-scale turbulent region is required for laminar-turbulence pattern to be maintained. The spatially asymmetric flow pattern strongly indicates the effects of mean spanwise vorticity.

Correspondence to Oaki Iida
Nagoya Institute of Technology, Gokiso-cho, Showa-ku, Nagoya 466-8555, Japan
e-mail: iida.oaki@nitech.ac.jp

Although mean shear is ultimately responsible for low-speed streaks as well as longitudinal vortical structures, the dynamical effects of mean shear can split into those of pure strain and spanwise vorticity. To clarify the effects of the spanwise vorticity on the laminar-turbulence pattern, turbulent Poiseuille flow under spanwise system rotation are numerically studied. Because flow becomes almost relaminarized in the suction side, our interest is in the pressure side where system rotation reduces the effects of mean spanwise vorticity.

2 Numerical Methods

In all cases, a spectral method is adopted for the incompressible Navier-Stokes equation with Fourier series in the x- and z-directions and a Chebyshev polynomial expansion in the y-direction, where coordinates system and flow configuration are shown in Fig. 1. For time integration, the second-order Adams-Bashforth and Crank-Nicolson schemes are adopted for the nonlinear and viscous terms, respectively. The second-order Adams-Bashforth is also adopted for the Coriolis term. $512 \times 65 \times 288$ grids points are used for the computational domain of $22\pi\delta \times 2\pi\delta \times 10\pi\delta$ in the x-, y- and z-directions, respectively (cases LB). We also performed the numerical simulations in much smaller computational domain of $4\pi\delta \times 2\delta \times 2\pi\delta$ (cases SB), which may elucidate the typical rotational effect on turbulent structures as in minimal channel flow. The Reynolds number Re defined by mean pressure gradient dP/dx, and channel half width δ, and kinematic viscosity ν is assumed as 80, 60, 50 and 40. In contrast, the rotation number Ro, defined as $2\Omega\delta/u_\tau^*$, is set to be 0.75 in all cases, where u_τ^* is $\sqrt{\delta(dP/dx)/\rho}$, and Ω is an imposed angular frequency.

3 Results and Discussion

Figures 2(a)-(d) show the instantaneous distribution of longitudinal vortical structures and low speed streaks. At $Re = 60$ without rotation (Fig. 2a), quasi-laminar and turbulent regions simultaneously appear side by side, i.e., laminar-turbulence pattern, or stripe of turbulent region. The turbulent regions are featured by the concentrated longitudinal vortices, while the quasi-laminar region is featured by the long and straight low speed streaks, which are punctuated at the turbulent regions. Imposing system rotation, however, drastically changes the flow structure at the same Reynolds number. It is noted in Fig. 2b that the large-scale intermittent structure completely disappears, and the turbulent region is replaced by the quasi-laminar region where very long low-speed streaks of more than 3000 viscous length appear. Interestingly, the long low-speed streaks are often coiled by chain-like vortical structures with a different sign of the streamwise vorticity, as observed in the wall turbulence at the transition [3]. The similar vortical structures are also experimentally confirmed by Alfredsson and Persson [4]. It is also noted that the close association

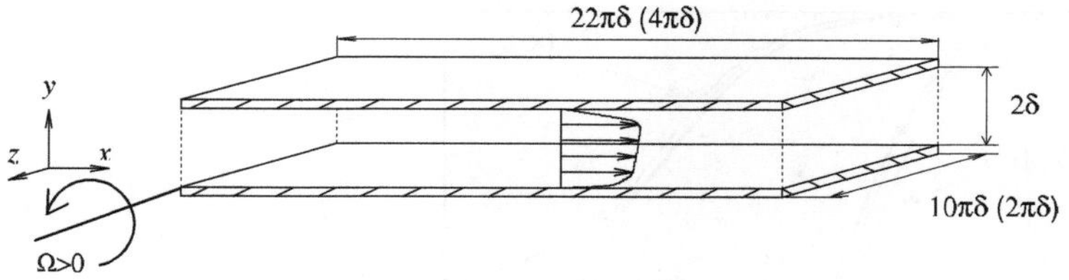

Fig. 1 Coordinate system and flow geometry of turbulent Poiseuille flow under system rotation.

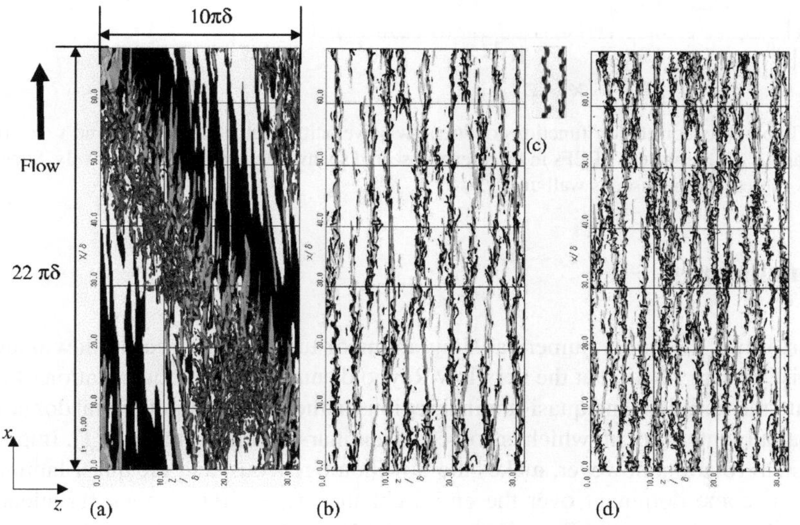

Fig. 2 Instantaneous distribution of streamwise vortical structure represented by positive value of second invariant of deformation tensor $II^* = -u^*_{i,j}u^*_{j,i}$(dark gray isosurfaces) and low speed streaks represented by $u^* = -3$(bright gray isosurfaces), where superscript * represents nondimensionalization by mean pressure gradient and kinetic viscosity. All figures represent top view from suction side. (a) $Re = 60$, $Ro = 0$, $II^* = 0.06$, Black regions represent low-speed streaks at $y^* = 10$, case LB [2], (b) $Re = 60$, $Ro = 0.75$, $II^* = 0.02$, case LB, (c) $Re = 60$, $Ro = 0.75$, $II^* = 0.02$, case SB, (d) $Re = 80$, $Ro = 0.75$, $II^* = 0.04$, case LB.

between the coiled vortical structure and long low-speed streak is more markedly observed in small computational domain (Fig. 2c). Even at the higher Reynolds number, the long low-speed streaks are still dominant in the entire region, and the laminar-turbulence pattern can not be observed, though vortical structures are sporadically concentrated (Fig. 2d). This clearly indicates that system rotation enhances the quasi-laminar region, and hence disturbs the generation of large-scale turbulent region.

It is also found in Fig. 3 that the two-points-correlation function of streamwise velocity fluctuations, i.e., R_{uu} does not change with increase in the Reynolds number, indicating the robustness of long low-speed streaks, which are lined with a spanwise period of $z^+ = 100$.

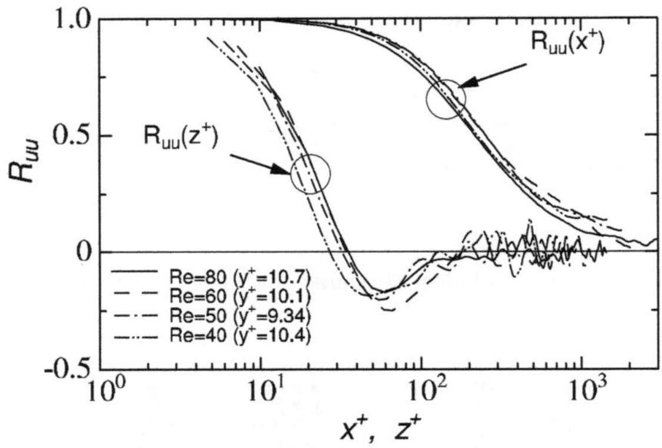

Fig. 3 Two-points-correlation functions of streamwise velocity fluctuations R_{uu} around $y^+ = 10$. $R_{uu}(x)$ and $R_{uu}(z)$ represent TPCFs in the streamwise and spanwise directions, respectively. Superscript + represents the viscous wall unit.

4 Conclusions

We performed the direct numerical simulations of turbulent Poiseuille flow under spanwise system rotation at the very low Reynolds numbers. Without rotation, flow split into the turbulent and quasi-laminar regions in the same computational domain, and spatially intermittent, which is coined as laminar-turbulence pattern [1]. Imposing system rotation, however, makes the flow homogeneous, and the quasi-laminar regions become dominant over the entire channel, where the laminar-turbulence pattern almost disappears. Even increasing the Reynolds number can not generate the laminar-turbulence pattern, indicating that system rotation enhances the quasi-laminar region, and hence disturbs the generation of large-scale turbulent region. The quasi-laminar region contains the extremely long low-speed streaks with a spanwise period of $z^+ = 100$ and chain-like vortical structures coiled around streaks, which are more uniquely identified in the smaller computational domain.

References

1. Barkley, D., Tuckerman, L.: Mean flow of turbulent-laminar patterns in plane couette flow. J. Fluid Mech. 576, 109–137 (2007)
2. Fukudome, K., Iida, O., Nagano, Y.: The Turbulent Structures of Poiseuille Flows at Low-Reynolds Numbers. In: Proceedings of the 7th JSME-KSME Thermal and Fluid Engineering Conference, Sapporo (2008)
3. Iida, O., Nagano, Y.: The relaminarization mechanisms of turbulent channel flow at low Reynolds numbers. Flow, Turbul. Combust. 60, 193–213 (1998)
4. Alfredsson, P.H., Persson, H.: Instabilities in channel flow with system rotation. J. Fluid Mech. 202, 543–557 (1989)

LES of the Flow over a High-Lift Airfoil Configuration

Daniel König, Wolfgang Schröder, and Matthias Meinke

Abstract. A large-eddy simulation of the flow over a high-lift airfoil configuration consisting of a slat and a main wing is performed at a freestream Mach number M=0.16 and an angle of attack of 13°. The Reynolds number, based on the clean chord length and the freestream velocity, is $Re=1.4\cdot 10^6$. The results show similarities between the turbulent structures of the slat cusp shear layer and a free shear layer and an impinging jet. The periodical occurrence of rollers and streamwise orientated rib vortices contributes essentially to the generated sound.

1 Introduction

The noise exposure near airports will become an increasingly important issue in the future due to the growing air traffic. Hence, low-noise design is a crucial factor in the development process of new airplanes. The aircraft noise is mainly caused by engines and airframe parts like landing gears and high-lift devices. The progress in reducing engine noise by the application of bypass fan engines shifted the focus towards airframe noise. The development of low-noise aircraft requires an investigation of the sound generating mechanisms, which are mainly determined by the underlying turbulent flow field. For this purpose, a large-eddy simulation (LES) of a high-lift airfoil configuration consisting of a slat and a main wing is conducted to analyze the flow field in the slat area.

Daniel König · Wolfgang Schröder · Matthias Meinke
Institute of Aerodynamics, RWTH Aachen University, Aachen, Germany
e-mail: d.koenig@aia.rwth-aachen.de, office@aia.rwth-aachen.de, m.meinke@aia.rwth-aachen.de

2 Numerical Method and Computational Setup

The Navier-Stokes equations for three-dimensional unsteady compressible flow are solved by a large-eddy simulation using an implicit subgrid model, i.e., the MILES approach is applied [1]. The inviscid terms are discretized by a modified AUSM method of second-order accuracy. For the viscous terms a central difference of second-order is used. The temporal integration is done by a second-order accurate explicit 5-stage Runge-Kutta method. For a detailed description of the flow solver the reader is referred to Meinke et al. [3].

The computational mesh consists of 55 million grid points. The extent in the spanwise direction is 2.1% of the clean chord length. The mesh resolution near the surface is $\Delta x^+ \approx 100$, $\Delta y^+ \approx 1$, and $\Delta z^+ \approx 22$.

On the far-field boundaries of the computational domain boundary conditions based on the theory of characteristics are applied. On the walls, an adiabatic no-slip boundary condition is imposed and a zero pressure gradient normal to the wall is prescribed. In the spanwise direction periodic boundary conditions are used.

The computation is performed at a freestream Mach number of M=0.16 and an angle of attack of 13°. The Reynolds number based on the clean chord length and the freestream velocity is Re=$1.4 \cdot 10^6$.

3 Results

First, the Mach number distribution and some selected streamlines of the time and spanwise averaged flow field are depicted in Fig. 1. It is evident that the slat cove region is an area of very low Mach number, which is characterized by a strong recirculation being illustrated by the streamlines. This recirculation area is separated from the flow passing the slat gap by a shear layer, which emanates at the slat cusp and reattaches near the slat trailing edge.

The distribution of the turbulent kinetic energy $k=\frac{1}{2}\left(u'^2+v'^2+w'^2\right)$ is depicted in Fig. 2. High k values occur in the shear layer, the recirculation area, and in the wake of the slat trailing edge. The major magnitudes of k are produced by the reattaching shear layer where the flow is decomposed. This is in good agreement with the results presented by Choudhari and Khorrami [2].

In the following, we will have a closer look at the unsteady turbulent structures in the slat region which are visualized by λ_2 contours.

Figure 3 reveals areas of turbulent flow located in the boundary layers of the slat and main airfoil, downstream of the slat trailing edge, and in the slat cove region. The grey scales mapped onto the λ_2 contours visualizes the Mach number distribution. The turbulent flow in the slat cove is bounded by the turbulent shear layer mentioned above. Also visible in Fig. 3 are the vortical structures in the shear layer and the slat cove. The structures in the slat cove rotate in a counter-clockwise direction around the center of the recirculation area.

The shear layer causes the formation of predominantly two-dimensional, spanwise vortex structures in the following referred to as rollers. They result from the velocity profile in the shear layer and the associated Kelvin-Helmholtz instability [4]. Figure 5 shows the λ_2 contours in the near slat cusp region, where the early rollers occur. The mixing and interaction of these shear layer structures with patterns from the recirculation area seem to lead to instabilities, which enhance the development of streamwise orientated vortical structures between two rollers. Similar structures, which are termed rib vortices, have been described e.g. by Rogers et al. [4] and Sakakibara et al. [5]. Figure 6 shows some rollers and rib vortices shortly before the shear layer reattaches. Note the sinusoidal appearance of the rollers. It seems that the rollers develop a slightly curved or wavy shape, respectively, due to their interaction with the rib vortices.

The vortical structures of the shear layer in the reattachment area are compared to those of a plane impinging jet. One great difference between the slat generated shear layer and the plane jet is the missing symmetry. However, for a first analysis of the vortical structures in the reattachment region the plane impinging jet seems to be appropriate. Figure 4 illustrates some more pronounced vortices being generated in the reattachment region. Their axes are aligned with the streamwise direction. Similar vortical structures have been observed by Sakakibara et al. [5], who called them wall ribs. In the case of the jet the wall ribs are formed by the impinging successive ribs and the cross ribs. In the present shear layer no cross ribs occur due to the missing symmetry, which is required for their development [5]. However, it can be seen that the successive ribs correspond to the streamwise ribs of the present solution. Unlike the jet rollers the shear layer rollers have a contribution to the wall ribs. This is due to the sinusoidal rollers in the spanwise direction and the acceleration of the flow passing through the slat gap. It is obvious that the parts of the rollers pointing in the direction of the slat gap undergo a stronger acceleration leading to a pronounced distortion of the rollers such that they finally collapse. The remaining structures are predominantly aligned with the streamwise direction. This explains the periodically changing strength and location of the wall ribs.

Figures 3 and 4 show also the turbulent wake of the slat trailing edge. This wake consists of the structures of the turbulent boundary layer on the suction side of the slat and the vortical structures, which are convected through the slat gap and are generated in the slat cove area.

References

1. Boris, J.P., et al.: New insights into large eddy simulation. Fluid Dynamics Research 10, 199228 (1992)
2. Choudhari, M.M., Khorrami, M.R.: Slat cove unsteadiness: Effect of 3d flow structures. In: 44th AIAA Aerospace Sciences Meeting and Exhibit. AIAA Paper 2006-0211 (2006)
3. Meinke, M., et al.: A comparison of second- and sixth-order methods for large-eddy simulations. Computers and Fluids 31, 695718 (2002)
4. Rogers, M.M., Moser, R.D.: The three-dimensional evolution of a plane mixing layer: The kelvin-helmholtz rollup. Journal of Fluid Mechanics 243, 183226 (1992)
5. Sakakibara, J., et al.: On the vortical structure in an plane impinging jet. Journal of Fluid Mechanics 434, 273300 (2001)

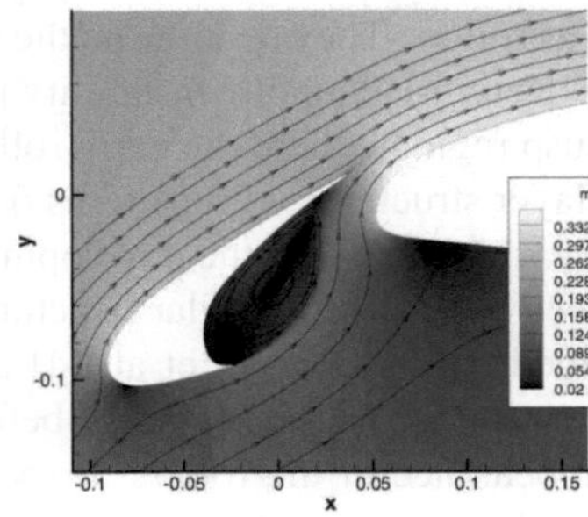

Fig. 1 Streamlines and Mach number contours of the time and spanwise averaged LES flow field data in the slat area.

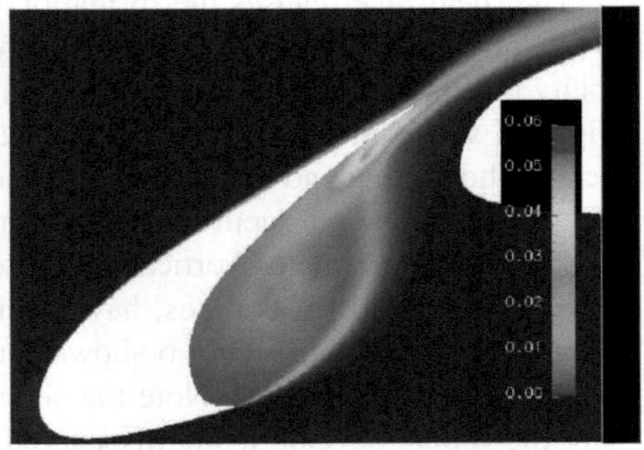

Fig. 2 Turbulent kinetic energy k nondimensionalized by u_∞^2 in the slat region.

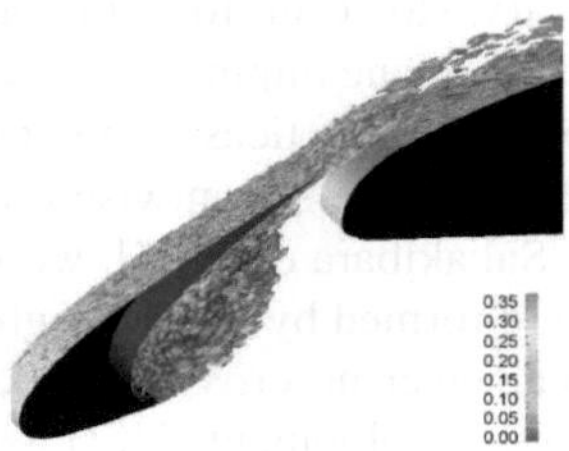

Fig. 3 Turbulent structures in the slat area visualized by λ_2 contours with mapped on Mach number distribution.

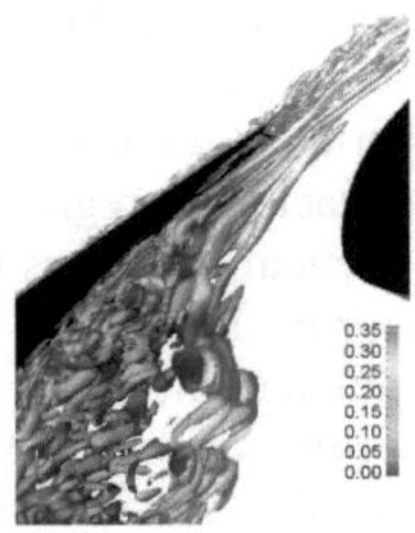

Fig. 4 λ_2 contours show vortical structures in the reattachment area of the slat cove shear layer.

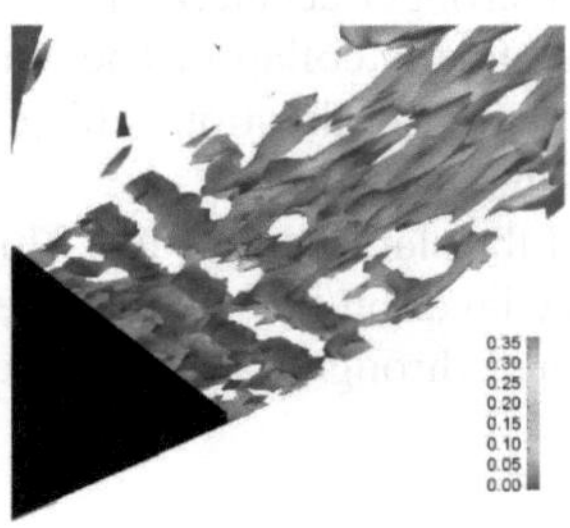

Fig. 5 Development of rollers downstream of the slat cusp and penetrating vortical structures from the recirculation area visualized by λ_2 contours.

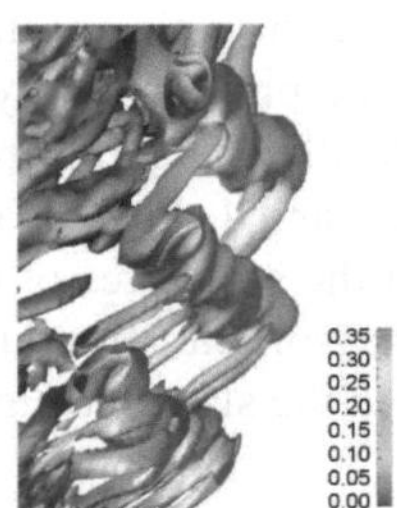

Fig. 6 λ_2 contours show rollers and streamwise rib vortices in the shear layer.

The Effect of Oblique Waves on Jet Turbulence

A. Segalini, R. Örlü, A. Talamelli, and P.H. Alfredsson

1 Introduction

For many years investigations have been conducted in order to understand the flow instabilities that lead to transition in jets. Among the earliest, the inviscid linear stability analysis of Batchelor and Gill [1] showed that immediately at the jet exit, where the velocity profile has a 'top-hat' behaviour, all the instability modes are able to be exponentially amplified while in the far field region only the helical mode seems to be unstable. The transition between these two different instability regions is still unclear and the analysis is complicated by the presence of several unstable modes embedded in the turbulence background. Therefore, a large amount of analytical theories, simulations and experiments have been done in order to highlight the role and the dynamics of a single or few modes in the evolution of the flow (cf. [3] and [7]).

Investigations in naturally and artificially excited jets have determined the importance of two instability lengthscales: one associated with the initial shear-layer thickness at the exit of the nozzle [7], and the other associated with the jet diameter which governs the shape of the mean velocity profile at the end of the potential core [4]. The instability modes in the first region develop through continuous and gradual frequency and phase adjustments to produce a smooth merging with the second region.

Axisymmetric excitation by means of acoustic forcing has been able to highlight several important aspects of the complex dynamics involved, like the role played by the so called shear layer [7] and jet column mode [4] acting in the near field of the

A.Segalini · A. Talamelli
Dept. of Mechanical and Aerospace Engineering (DIEM), University of Bologna, I-47100 Forlí, Italy
e-mail: antonio.segalini@unibo.it, alessandro.talamelli@unibo.it
R. Örlü · P.H. Alfredsson
Linné FLOW Centre, KTH Mechanics, Royal Institute of Technology, S–100 44 Stockholm, Sweden
e-mail: ramis@mech.kth.se, hal@mech.kth.se

jet at the nozzle exit and at the end of the potential core, respectively, as well as the presence of several nonlinear interactions between scales and their strong influence in the flow evolution. However, fewer works have been devoted to the investigation of higher azimuthal modes principally due to the higher complexity of the excitation facility (see e.g. [2] and [3]).

The motivation of this work was to investigate the possibility that non-linear combination of oblique waves generated by blowing and suction, in a manner similar to the one of Elofsson & Alfredsson [5], could generate streamwise streaks able to alter the dynamics of the flow field.

In this paper a preliminary analysis on the effect of this excitation is presented. A detailed investigation with different excitation amplitudes, frequencies and velocities has been performed in order to evaluate the effects on the flow dynamics. In the future the possibility of the complete description of this complex evolution will be further investigated.

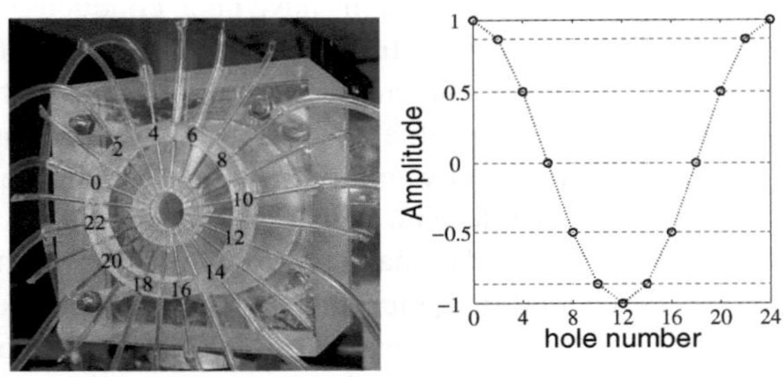

Fig. 1 Photo of the nozzle and excitation rig *(left)*. Excitation pattern *(right)*.

2 Experimental Set-Up

The experiments have been carried out in the Fluid Physics Laboratory at the Linné Flow Centre at KTH Mechanics in Stockholm. The air, driven by a centrifugal fan, passed through a pre-settling chamber placed 3 m downstream in order to reduce the incoming disturbances from the fan. Flow conditioning is performed by means of a honeycomb positioned in the settling chamber after which a plexiglass nozzle with 0.025 m exit diameter is mounted. A short straight plexiglass tube section to provide the acoustic excitation, equipped with 24 perpendicular aligned metal tubes, is fixed to the nozzle exit as shown in figure 1.

Velocity measurements have been performed by means of an in-house built X-wire probe using an *A.A. LAB AN-1003* anemometry system in CTA mode.

At the beginning of each set of experiments, the amplitude of each loudspeaker has been regulated in such a way that the streamwise velocity rms level around the excitation frequency, measured with a single hot-wire, followed the law depicted in figure 1.

3 Results, Discussion and Conclusions

An extensive parametric study in terms of frequencies and amplitudes for a selected range of Reynolds numbers was performed in order to focus on the most influencial parameter set. Since the aim of this work was to investigate the flow in the near field of the jet, a streamwise station of $x/D = 3$ has been chosen to investigate the effect of the excitation on the flow. Several combinations of Reynolds number and excitation frequency have been tested in order to evince the effect on the streamwise rms level. In figure 2 the relative rms (compared to the unexcited or natural case) against the dimensionless frequency $St_{\theta_0} = f\theta_0/U_J$, where θ_0 is the initial momentum thickness of the boundary layer at the nozzle lip and U_J the jet exit velocity, has been reported. Due to the non negligible scatter, a moving average has been performed to clarify the figure. It is interesting to note that the largest reduction in the turbulence intensity can be observed around $St_{\theta_0} \simeq 0.012$, which is close to the value found for an axisymmetrically excited jet [6].

In the following results for a Reynolds number of $Re_D = 25000$ will be presented where a frequency $f_{ex} = 1500$ Hz showed the strongest turbulence reduction (at least at $x = 3D$) and it will be referred to as the "excited case" in contrast to the natural case.

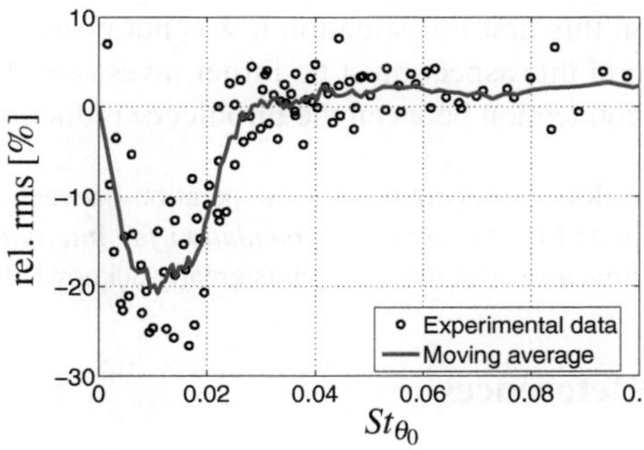

Fig. 2 Relative change in streamwise velocity rms at $x/D = 3$ between excited and unexcited case for different exit velocities.

Some important aspects can be deduced from the autocorrelation function or the velocity spectra, reported in figure 3. When the excitation frequency is close to the characteristic frequency of the shear layer, several nonlinear interactions appear within the first half diameter and then decay further downstream. At $x/D = 2$ the remaining peak is a subharmonic of the excitation frequency, which is somewhat similar to what was observed by Corke and Kusek [3]. It is worthwhile to note that the low frequency energy is lower than the unexcited case, meaning a strength reduction in the large scales due to the excitation at the jet exit.

The evolution of the spectra confirms that the excitation has a strong impact on the evolution of the whole flow field with the imposed periodicity given by the excitation itself. The scenario seems to be in agreement with the one depicted by Batchelor and Gill [1] with two unstable oblique modes that will combine to generate

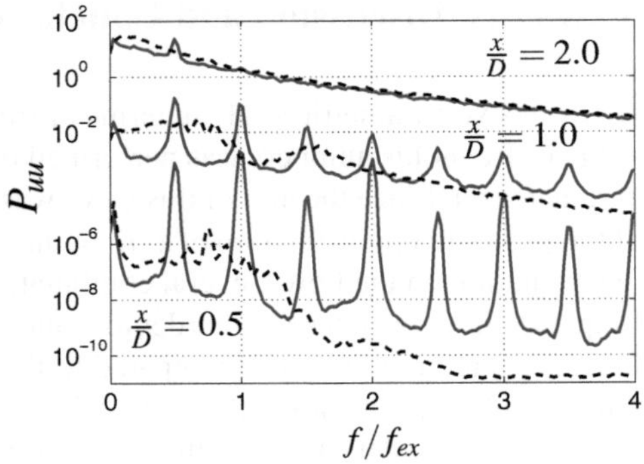

Fig. 3 Spectra in the shear layer region with highest turbulence intensity for $Re_D = 25000$ w/ excitation 1500 Hz (solid line) and w/o (dashed line).

several nonlinear interactions. This process saturates most of the energy of the shear layer at the various frequencies and starts to decay faster compared to the natural case, with only some modes that will decay slower than others (from the spectral analysis it seems to be the subharmonic of the fundamental).

This reduction of turbulence could be connected to the same phenomenon described by Elofsson and Alfredsson [5] about the effect of oblique waves in laminar boundary layers, where the authors showed that the interaction of two waves is able to generate streamwise streaks by means of nonlinear interaction. Unfortunately, in this first investigation it was not possible to detect the presence of such streaks, and this aspect must be furher investigated in the future in order to understand the connection between the turbulence reduction and the azimuthal forcing.

Acknowledgement. The cooperation between *KTH* and the *University of Bologna* is supported by *The Swedish Foundation for International Cooperation in Research and Higher Education* (STINT), which is greatly acknowledged.

References

1. Batchelor, G.K., Gill, A.E.: Analysis of the stability of axisymmetric jets. J. Fluid Mech. 14, 529–551 (1962)
2. Cohen, J., Wygnanski, J.: The evolution of instabilities in the axisymmetric jet. Part 2. The flow resulting from the interaction between two waves. J. Fluid Mech. 176, 221–235 (1987)
3. Corke, T.C., Kusek, S.M.: Resonance in axisymmetric jets with controlled helical-mode input. J. Fluid Mech. 249, 307–336 (1993)
4. Crow, S.C., Champagne, F.H.: Orderly structure in jet turbulence. J. Fluid Mech. 48, 547–591 (1971)
5. Elofsson, P.A., Alfredsson, P.H.: An experimental study of oblique transition in plane Poiseuille flow. J. Fluid Mech. 358, 177–202 (1998)
6. Husain, H.S., Hussain, A.K.M.F.: Experiments on subharmonic resonance in a shear layer. J. Fluid Mech. 304, 343–372 (1995)
7. Zaman, K.B.M.Q., Hussain, A.K.M.F.: Turbulence suppression in free shear flows by controlled excitation. J. Fluid Mech. 103, 133–159 (1981)

Turbulence Enhancement in Coaxial Jet Flows by Means of Vortex Shedding

R. Örlü, A. Segalini, P. H. Alfredsson, and A. Talamelli

1 Introduction

Over the past decades a variety of passive and active flow control mechanisms have been tested and applied in a variety of canonical as well as applied flow cases. An example for the latter is the coaxial jet flow, which has mainly been investigated regarding the receptivity to active flow control strategies (see e.g. [1]), probably due to the multitude of parameters characterising the complex flow field [2].

Physical and numerical experiments (see e.g. [3] and [5]) have established that the vortical motion in coaxial jet flows is dominated by the vortices emerging from the outer shear layer. The frequency of these vortices is related to the Kelvin-Helmholtz instability as predicted by linear stability analysis for single jets. The vortices in the inner shear layer, on the other hand, are trapped in the spaces left free between two consecutive outer shear layer vortices, and are therefore sharing the frequencies of the most amplified modes of the outer shear layer and do not relate to the values one would expect from linear stability analysis. This fact has become known as the "locking phenomenon", which describes the mutual interaction of both shear layers. Nevertheless it is believed that only the outer shear layer is able to significantly control the evolution of the inner shear layer [7], which may explain the focus of control strategies on the outer shear layer.

Besides the velocity ratio of the two streams and the Reynolds number, the thickness of the wall separating the inner and outer jet plays an important role in the evolution of transitional coaxial jets. Two trains of alternating vortices are found to

R. Örlü · P. H. Alfredsson
Linné FLOW Centre, KTH Mechanics, Royal Institute of Technology, S–100 44 Stockholm, Sweden
e-mail: `ramis@mech.kth.se`, `hal@mech.kth.se`

A. Segalini · A. Talamelli
Dept. of Mechanical and Aerospace Engineering (DIEM), Univeristy of Bologna, 47100 Forlí, Italy
e-mail: `antonio.segalini@unibo.it`, `alessandro.talamelli@unibo.it`

shed from both sides of the inner wall with a well-defined frequency, which scales with the wall thickness and the average velocity of the two streams [2].

In a recent study Talamelli and Gavarini [6] showed, by means of linear stability analysis, that the alternate vortex shedding behind the inner wall can be related to the presence of an absolute instability, which exists for a specific range of velocity ratios and for a finite thickness of the wall separating the two streams. The authors proposed that this absolute instability may provide a continuous forcing mechanism for the destabilisation of the whole flow field even if the instability is of limited spatial extent.

The proposed idea of Talamelli and Gavarini [6], namely to test if the absolute instability behind an inner wall of a coaxial jet nozzle with finite thickness can be utilised as a continuous forcing mechanism and hence as a passive flow control mechanism for the near-field of coaxial jet flows, has recently be confirmed experimentally by the present authors [4]. In the present paper the enhanced turbulence activity in both shear layers and hence the whole flow field will be documented in order to underline the opened possibilities.

2 Experimental Arrangement

The experiments were carried out in the *Coaxial Air Tunnel* (CAT) facility in the laboratory of the *Second Faculty of Engineering* at the *University of Bologna* in Forlí by means of hot-wire anemometry and flow visualisations (see [4] for details of the experimental set-up as well as the measurement technique).

In the present experiment, two different types of separation walls have been used. The first one has a thickness of $t = 5$ mm and ends in a rectangular geometry, whereas the second one ends with a sharp trailing edge making the wall thickness negligible ($t \approx 0$ mm) with respect to the sum of the side boundary layers thicknesses. These two separating walls will in the following be denoted as thick and sharp, respectively. The sharp and thick wall cases represent the flow cases in the absence and presence of the vortex shedding phenomenon, respectively, and enable therefore a selective investigation regarding the effect of the absolute instability as can be evinced from flow visualisation snapshots shown in figure 1.

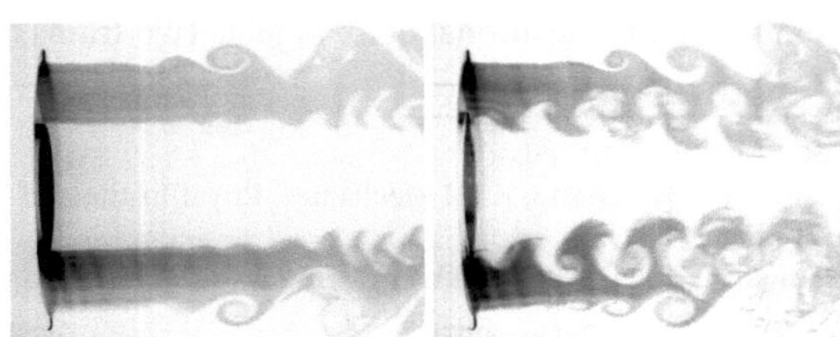

Fig. 1 Smoke flow visualisation at $U_o = 4$ m/s and $r_u = 1$ for the sharp *(left)* and thick *(right)* wall. Hereby U_o denotes the streamwise velocity of the outer stream and r_u the velocity ratio between the outer and inner streams.

3 Results

As mentioned in section 1, it is expected that the vortices in the inner shear layer are triggered by the evolution of the dominant vortices in the outer shear layer. This would establish itself, for instance, in a dominant frequency of the inner and outer shear layer which relates to the Kelvin-Helmholtz instability of the outer shear layer (particularly when the outer jet velocity exceeds the one of the inner).

However, as shown by linear stability analysis [6] and confirmed by hot-wire measurements [4], a different scenario is found in the presence of an absolute instability, i.e. the presence of vortex shedding behind the inner separating wall. Here, as clearly visible in figure 2(b), the dominant frequencies of both the inner and outer shear layers vortices scale with the thickness of the separating wall, t, and the average velocity of both streams, U_m, i.e. $St_{VS} = f_{max}t/U_m$.

It is important to note in this context that the observed "reverse locking phenomenon" opens doors to apply control strategies in the inner shear layer or on the inner separating wall to control whole near-field region.

The power-spectral density functions showed that the energy content of the flow, in the presence of the vortex shedding, is enhanced not only by means of the emergence of stronger organised structures, but also by a drastic increase in the incoherent background turbulence [4]. This trend can also be seen in terms of integral quantities like the rms of the radial velocity fluctuations along $r/D_i = 1$ for 4, 8 and 12 m/s as shown in figure 3 for the sharp (a) and thick (b) wall.

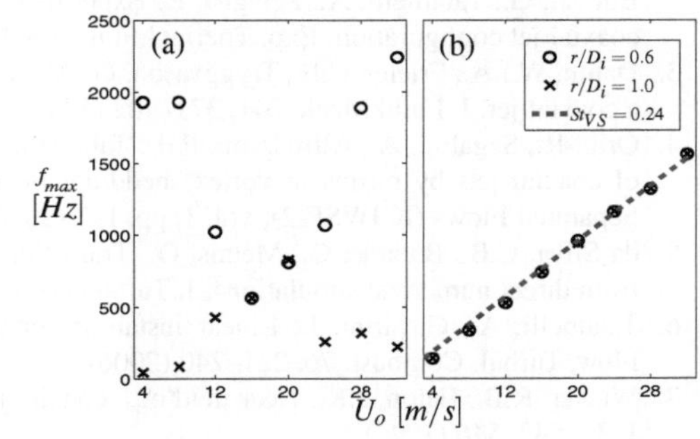

Fig. 2 Fundamental frequency as function of the outer jet velocity behind the inner and outer wall at $r_u = 1$ and $x/D_i = 0.26$ for the sharp (a) and thick (b) wall. Hereby D_i denotes the inner diameter of the inner nozzle.

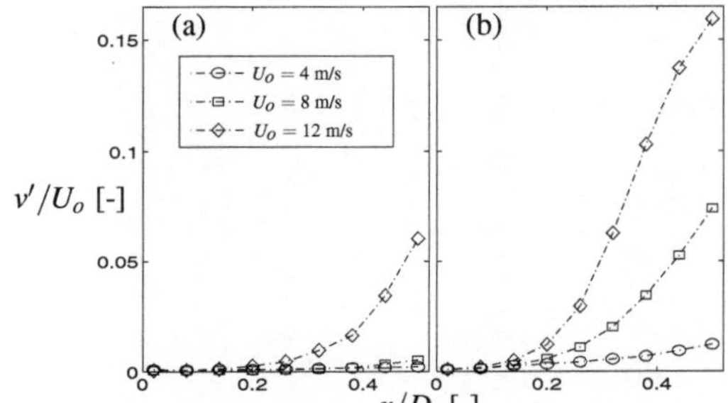

Fig. 3 Rms of radial velocity fluctuations along $r/D_i = 1$ at $U_o = 4$, 8 and 12 m/s for the sharp (a) and thick (b) wall. In the present paper r_u is kept constant at unity, i.e. $U_o = U_i = U_m$.

4 Conclusions

In summary we have demonstrated that it is possible to utilise the vortex shedding behind a finite thickness separating wall to control the evolution of the vortices in both shear layers and hence the whole near-field region, whenever the separating wall thickness and velocity ratio is within a specific range [6].

Furthermore the passive control mechanism was found to increase the turbulence intensity within the inner and outer shear layers and thereby the mixing between the two coaxial jet streams as well as the annular jet with the ambient fluid.

The present paper has shown that an anachronistic approach, i.e. a passive contrary active control utilising the inner shear layer to control the outer one contrary to the general trend to do the opposite, shows promising results in terms of flow control and turbulence enhancement.

Acknowledgement. The cooperation between *KTH* and the *University of Bologna* is supported by *The Swedish Foundation for International Cooperation in Research and Higher Education* (STINT), which is greatly acknowledged.

References

1. Angele, K.P., Kurimoto, N., Suzuki, Y., Kasagi, N.: Evolution of the streamwise vortices in a coaxial jet controlled with micro flap-actuators. J. Turbulence 6, 1–19 (2006)
2. Buresti, G., Talamelli, A., Petagna, P.: Experimental characterization of the velocity field of a coaxial jet configuration. Exp. Thermal Fluid Sci. 9, 135–146 (1994)
3. Dahm, W.J.A., Frieler, C.E., Tryggvason, G.: Vortex structure and dynamics in the near field of a coaxial jet. J. Fluid Mech. 241, 371–402 (1992)
4. Örlü, R., Segalini, A., Alfredsson, P.H., Talamelli, A.: On the passive control of the near-field of coaxial jets by means of vortex shedding. In: Proc. of the Int. Conf. on Jets, Wakes and Separated Flows (ICJWSF-2), vol. 1, pp. 1–7 (2008)
5. da Silva, C.B., Balarac, G., Métais, O.: Transition in high velocity ratio coaxial jets analysed from direct numerical simulations. J. Turbulence 4, 1–18 (2003)
6. Talamelli, A., Gavarini, I.: Linear instability characteristics of incompressible coaxial jets. Flow, Turbul. Combust. 76, 221–240 (2006)
7. Wicker, R.B., Eaton, J.K.: Near field of a coaxial jet with and without axial excitation. AIAA J. 32, 542–546 (1994)

Direct Numerical Simulation of Microbubble Dispersion in Vertical Turbulent Channel Flow

Dafne Molin, Andrea Giusti, and Alfredo Soldati

Abstract. In this work, direct numerical simulation of turbulence is coupled to lagrangian tracking to study the behavior of 220 μm bubbles in a vertical turbulent channel flow. Both one-way and two-way coupling approaches and both upward and downward flows are considered. For each simulation, the same external imposed pressure gradient is considered. In one-way simulations, this leads to a shear Reynolds number of Re = 150. In the coupled cases, the presence of bubbles increase/decrease the driving pressure gradient, respectively in upward/downward flow, thus yielding to an increase/decrease of the wall shear stress and of the shear Reynold number. For the considered bubble average volume fraction (α = 104), the corresponding shear Reynolds number are about $Re_{\tau,2U} = 174$ for the upflow case and $Re_{\tau,2D} = 121$ for the downflow case. Statistics of the fluid and of the bubble phase are presented. The interactions between bubble and the near-wall turbulence structures is also investigated and a preferential bubble segregation in high-speed/low-speed zones is observed for the upflow/downflow cases respectively. An attempt to describe the transfer rate between the gas and the liquid will be included with some preliminary results.

Dafne Molin
Centro Interdipartimentale di Fluidodinamica e Idraulica and Dipartimento di Energetica e Macchine, Universita' degli Studi di Udine, Via delle Scienze 208, 33100 Udine Italy
e-mail: dafne.molin@uniud.it

Andrea Giusti
Centro Interdipartimentale di Fluidodinamica e Idraulica and Dipartimento di Energetica e Macchine, Universita' degli Studi di Udine, Via delle Scienze 208, 33100 Udine Italy
e-mail: andrea.giusti@uniud.it

Alfredo Soldati
Centro Interdipartimentale di Fluidodinamica e Idraulica and Dipartimento di Energetica e Macchine, Universita' degli Studi di Udine, Via delle Scienze 208, 33100 Udine Italy
e-mail: alfredo.soldati@uniud.it

Introduction

Bubbly flows play an important role in a wide variety of areas, ranging from biomedical field, enviromental phenomena, industrial applications to chemical processes. In all these applications the presence of microbubbles which are non-uniformly distributed may significantly change transfer rate: the overall liquid-bubble interface controls gas-liquid tansfer, but complex bubble motion also have an influence on overall heat, momentum and mass transfer, playing a crucial role in many industrial and environmental processes. Due to their importance, bubbly flows have been extensively analyzed and several studies, mainly experimental, are available in the literature. However, the complete understanding of the dynamics of bubbly flows is still a challenging task, due to the large number of factors affecting the interactions between the bubbles and the surrounding fluid. The importance of the lift force effect on bubble behavior has been underlined in Giusti et al. [1], where numerical simulations of a swarm of microbubbles dispersed in a turbulent vertical close channel flow were considered. Mazzitelli et al. [2], simulated the behavior of microbubbles in isotropic turbulence. Regarding the two-way coupling effect, instead, Mazzitelli et al. [3] analyzed turbulence modification induced by microbubbles. Xu et al. [4] and Ferrante and Elghobashi [5] investigated microbubble power to reduce the skin friction drag, in horizontal turbulent boundary layers. The main object of the present work, instead, is to make a step foward into the understandings the physics behind bubble behaviour. Based on the previous paper of Giusti et al. [1] in which the importance of the lift force effect on bubble behaviour was analyzed in a numerical simulation of a swarm of microbubble dispersed in a turbulent vertical channel flow we apply the direct numerical simulation to study the two-way coupling effect on both the flow field as well the bubble behavior in the wall region of a vertical turbulent channel flow.

Details of the method have been published previously [6],[7], [1].

Results and Discussion

The main effect of bubbles in two-way coupling simulations, is to increase the liquid flow rate in the upflow and to decrease it in the downflow, as a consequence of the buoyancy force acting on bubbles. Significant flow rate variations were observed, even for the low void fraction value considered ($\alpha = 10^{-4}$), due to the low Reynolds number ($Re_\tau = 150$), i.e. to the low externally imposed pressure gradient, of the present simulations.

The two-way induced flow rate variations are related to variations of the wall shear. A very simple mono-dimensional model is presented to predict the wall shear modification as a function of the average void fraction . In the upflow case, the wall shear is expected to increase proportionally with the void fraction. Similarly, in the downflow case, a wall shear decrease, proportional to the void fraction, is expected.

The wall shear values resulting from present two-way coupled simulations are in agreement with the attended theoretical values.

Further informations can be achieved from the comparisons between the average streamwise fluid velocity profiles for the two-way coupled simulations (made non-dimensional with respect to the shear velocities referred to the "equivalent" shear Reynolds number) and the universal turbulent velocity profiles. In the downflow, the velocity profile obtained from the numerical simulation is in agreement with the theoretical velocity profile attended for a mono-phase turbulent flow at $Re_{\tau,DW} = 121$. This indicates that bubbles do not significantly modify the fluid turbulent structures. In the upflow, the velocity profile obtained from the numerical simulation shows lower values with respect to the theoretical velocity profile attended for a mono-phase turbulent flow at $Re_{\tau,UP} = 174$. Since we pointed out that bubbles, in two-way upflow case, reduce the fluid turbulent intensity, we should expect an increase of the average streamwise velocity profile. Then, to justify this unexpected result, we have to consider the non-uniformity of bubble distribution in the wall-normal direction, i.e. the strong bubble accumulation near the walls. This yields to the following consequences: a great part of the bubbly two-way push acting on the fluid is located in the near-wall region; this push is immediately counterbalanced by the wall shear; this push is not able to efficiently increase the fluid flow rate and the average streamwise velocity profile.

With regard to bubble behavior, we observe, both for the one-way and two-way simulations, bubble migration towards the wall in the upflow and away from the wall in downflow.

In a previous numerical study on bubbly flow [1], we pointed up that the lift force is the main factor determining bubble lateral migration in vertical ducts. It was also pointed up that, in the upflow, bubble migration towards the wall resulting from numerical simulations was overestimated with respect to the experimental observations. In the present simulations, we added a wall-induced lift force, using the model of Takemura and Magnaudet [8]; this resulted in a reduction of bubble migration velocity towards the wall, even if we still have higher values with respect to the experimental results. Furthermore, the obtained bubble concentration profiles (Fig. 1) show a continuous increase of the peak value at the wall, thus yielding to high local values of bubble volume fraction. Then, we had to stop our upflow simulations even if a steady state of bubble concentration profile was not reached yet, since our numerical code is based on the hypothesis of low void fraction.

The analysis of the distribution of the bubbles that are in the near-wall region showed the existence of bubble preferential segregation in low-speed or high-speed regions, respectively in the upflow or downflow case (both for one-way and two-way coupled simulations).

In Giusti et al. [1], a similar bubble behavior was observed for the downflow case, whereas, for the upflow, no bubble preferential segregation zones were found. thus suggesting that the bubble segregation in low-speed zones resulting from the present one-way and two-way upflow simulations, may be related to the wall induced lift (that was not considered in our previous paper). The interactions between bubbles and the near-wall turbulent structure as well the attempt to describe the transfer rate between the gas and the liquid will be briefly illustrated during the conference poster section.

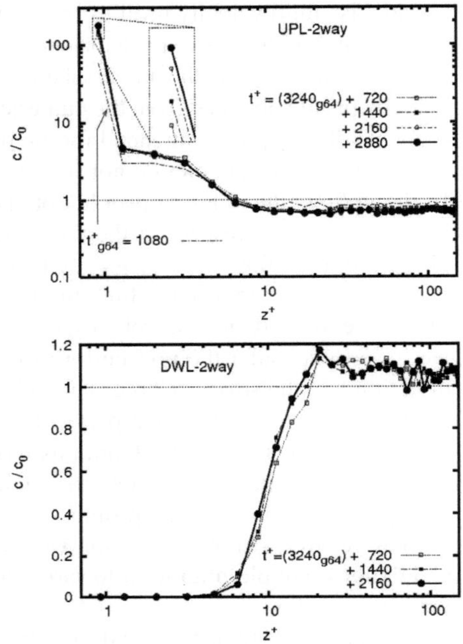

Fig. 1 Time evolution of microbubbles concentration profile for the upflow case (first figure) and the downflow case (second figure) for the two-way coupled simulation.

References

1. Giusti, A., Lucci, F., Soldati, A.: Influence of the lift force in direct numerical simulation of upward/downward turbulent channel flow laden with surfactant contaminated microbubbles. Chem. Eng. Science 60, 6176–6187 (2005)
2. Mazzitelli, I.M., Lohse, D., Toschi, F.: On the relevance of the lift force in bubbly turbulence. J. Fluid Mech. 488, 283–313 (2003a)
3. Mazzitelli, I.M., Lohse, D., Toschi, F.: The effect of microbubbles on developed turbulence. Phys. Fluids 15(1), L5–L8 (2003b)
4. Xu, J., Maxey, M.R., Karniadakis, G.E.: Numerical simulation of turbulent drag reduction using micro-bubbles. J. Fluid Mech. 468, 271–281 (2002)
5. Ferrante, A., Elghobashi, S.: On the physical mechanisms of two-way coupling in particle-laden isotropic turbulence. Phys. Fluids 15(2), 315–329 (2003)
6. Lam, K., Banerjee, S.: On the condition of streak formation in bounded flows. Phys. Fluids A 4, 306–320 (1992)
7. Soldati, A., Banerjee, S.: Turbulence modification by large scale organized electrohydrodynamic flows. Phys. Fluids 10, 1742–1756 (1998)
8. Takemura, F., Magnaudet, J.: The transverse force on clean and contaminated bubbles rising near a vertical wall at moderate Reynolds. J. Fluid Mech. 495, 235–253 (2003)

Experimental and Numerical Analysis of the Stability of the Vertical Water Jet with Rectangular Cross Section

Sergej Gordeev, R. Stieglitz, L. Stoppel, M. Daubner, T. Schulenberg, and F. Fellmoser

Abstract. This article describes experimental and numerical investigations of the stability of a water jet with a rectangular cross section aligned with the gravity field. This work focuses on the individual physical phenomena influencing the stability of the free surface jet flow like the contraction of the planar jet as well as the generation of capillary waves. In order to investigate the interaction between these two effects a series of experiments using water as model fluid are conducted in the Karlsruhe Liquid metal Laboratory KALLA at the research centre Karlsruhe. For the measurement of the cross-sectional shape of free water jet the Laser Doppler Anemometer (LDA) has been applied. Complementary to the experiments the numerical simulations have been performed using the commercially available code STAR-CD. The simulations show a reasonable agreement with the experimental measurements within the investigated parameter range.

1 Introduction

At the heavy-ion synchrotron Facility for Antiprotons and Ion Research (FAIR) of GSI [1], highly energized ion beams ($\triangle \Phi$=$2GeV$, e.g. uranium $U238$) with intensities of about 1012 particles/cycle are envisaged for basic physics experiments. A wide range of particle energies will be used for the production of secondary particles by projectile fragmentation at the fragment separator Super-FRS. For the highest power densities, a windowless vertical liquid-Li-jet-target with a rectangular cross section (16x70 mm) operated at high jet velocities of up to 10m/s is considered. In order to capture the entire U238beam dimensions within the lithium jet, the film dimensions in beam direction should not exceed more than 1% of its 70mm long flight path. Moreover, due to the extension of the energy deposition in gravity

Sergej Gordeev
Forschungszentrum Karlsruhe GmbH Institut für Kern- und Energietechnik (IKET), .O. Box 3640 D-76021 Karlsruhe, Germany
e-mail: Sergej.Gordeev@iket.fzk.de

vector direction the jet depth has to be maintained over a height of at least 50mm. But, fluid jets of this shape are influenced by contraction originating from the relatively high surface tension, which tends to form a cylindrically shaped contour and simultaneously accelerates the liquid. Additionally, capillary waves are generated at each of the rectangular shaped nozzle exit edges, which expand in horizontal direction. Dependent on the jet velocity, these capillary may merge to further modify the surface shape. Both phenomena occur simultaneously and interact. Due to the lack of experimentally validated turbulence models used in the CFD codes to describe free surface flows, this study is aimed to investigate the performance of CFD simulations in comparison with a complementary conducted water experiment at the KALLA laboratory of Research Centre Karlsruhe [2].

2 Experimental Set-Up

Figure 1 shows the schematic view of the test section corresponding to a 1 1 geometric scaled mock-up of the SUPER-FRS target. It consists of a flow conditioner, a nozzle and a vacuum glass container. Water is passing the flow conditioner, proceeds into vacuum glass container at the nozzle exit, where the jet is formed, and falls in the liquid-gas separation tank. Details may be taken from [2]. The cross-sectional shape of the free water jet is recorded by means of the Laser Doppler Anemometer (LDA). Acquired within this LDA mode is the amount of tracers passing the measurement volume at a given time interval. Hence, the result is probabilistic nature and the contour extracted in this mode represents a time averaged surface shape. The measurements are focussed to the small side of the

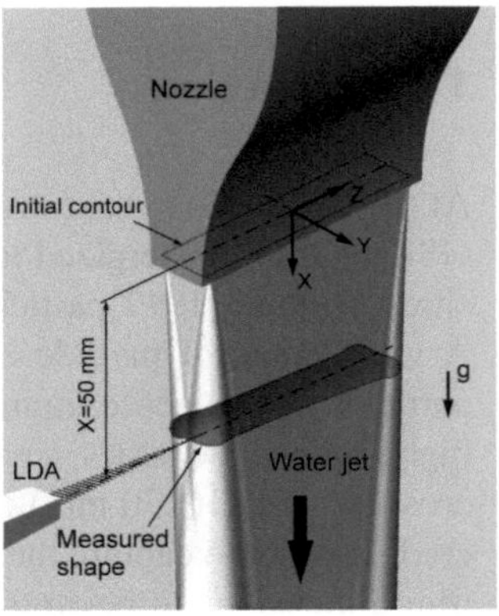

Fig. 1 Schematic view of the test section.

jet, where at first downstream the coalescence of the capillary can be expected. In this context data obtained in the $Y-Z$ plane $50mm$ downstream the nozzle exit are presented for mean flow velocities $1.5 \leq u_0(m/s) \leq 7$. This corresponds to Reynolds numbers Re $5.4x10^4 \leq Re \leq 2.55x10^5$ and Weber numbers Wb in the range $7x10^2 \leq Wb \leq 1.7x10^4$. Here, the Reynolds number and Weber number are defined by : $Re = \frac{u_0 L}{\nu}$, $W_b = \frac{u_0^2 L \rho}{\sigma}$. where L, ν, ρ and σ are the hydraulic diameter, the kinematic viscosity, the density and the surface tension, respectively. The interface data are recorded in form of 2D-matrix in a step width of 0.5 mm representing a physical domain of 12 x22 mm. Each element of the matrix represents a normalized quantity of the registered tracer particles. The extracted contour diagram, constructed in dimensional coordinates, shows the boundary of the water/air interface on the small side. On the diagram it is possible to distinguish the most probable free surface contours of the jets cross-section.

3 Measurement Results and Numerical Simulations

The figures 2 show the measured cross-section of the water air composition in the measurement plane 50 mm downstream the nozzle exit for different Reynolds numbers, in which the white domain describes the air domain while the dark grey regions denote the water phase. The intermediate grey scales represent time integrated fluctuations of the water/air interface. In case of $Re = 5.4x10^4$ the capillary waves merge upstream the measurement plane. The interaction of both capillary waves form a crest and thus the jet depth in the middle of the small side is larger compared to higher Re at the same position. An increase of the Reynolds number yields a reduction of the dispersion angle of the waves and hence the coalescence point is shifted downstream. Already at $Re = 9x10^4$ the coalescence point is located almost in the measurement cross-section. For $Re\,1.1x10^5$ two wave crests are observed. A further increase of Re leads to a growth of the distance between the wave crests and the surface shape in the centre of the small flattens.

In the numerical simulation, a quarter of the jet consisting of $2.5x10^6$ fluid elements taking advantage of symmetry conditions has been modelled. The computations have been conducted using STAR-CD CFD [3] code. For the simulation of the free surface flow, the Volume of Fluid (VOF) method and a SIMPLE-transient analysis option are used. Regarding the turbulent flow motion in the liquid domain, the v^2f - turbulence model [3] has been selected. The use of the v^2f - model for this nozzle flow type has been validated in [4] by a complementary experimental and numerical procedure. The figures 2 show a comparison of the experimental data and the calculated water-air interface (black lines) in the measurement plane. Concerning the global dimensions as well as the local topology of the particle distribution, the used turbulence v^2f model is able to reproduce the experimentally observed surface shape with a reasonable accuracy. The position of the capillary waves on the small side of the jet is predicted in all cases close to the measured data. As the figures 2 exhibit, the simulation slightly overestimates the influence of the surface tension on the interface deformation. This is expressed by a stronger rounding of the water/air interface at the outer boundaries.

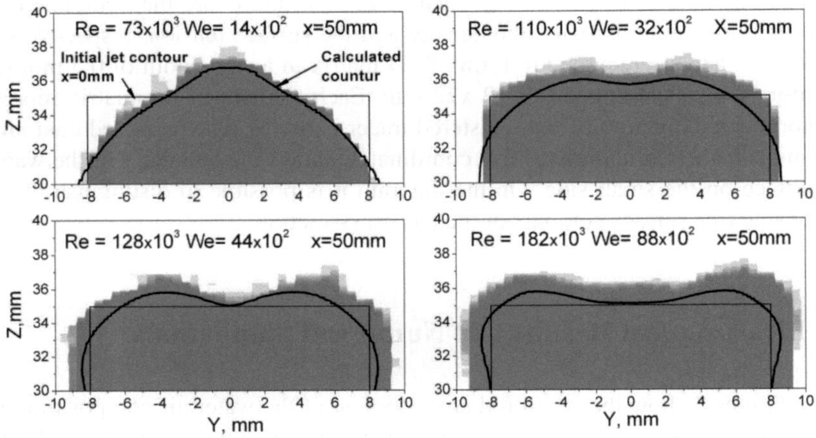

Fig. 2 Measured water/air interface shape in a plane X = 50mm downstream the nozzle exit for different Reynolds number. (dark grey = water, scaled gray = fluctuating water/air interface, white = air) compared with calculated water air interface (black line).

4 Conclusions

The presented work studies the fluid dynamic aspects of a free vertical liquid jet. This study involves the application of a LDA to record in a time integrated manner water/air interfaces and a validation of numerical CFD code necessary to design a liquid lithium target matching application conditions. The used LDA technique allows an exact measurement of the water air interface with an accuracy depending on the measurement volume size only. A comparison of the experimental data with the simulation results using a VOF method in combination with the $v^2 f$ - turbulence model exhibits a reasonable agreement.

Acknowledgments: This work has been performed in the context EU-program DIRAC secondary-Beams 515873.

References

1. An International Accelerator Facility for Antiproton and Ion Research, Conceptual Design Report, Gesellschaft für Schwerionenforschung (2001)
2. Stoppel, L., Gnieser, M., Daubner, S.M., Stieglitz, R.: Similarity Analysis for a Free Turbulent Jet. Jahrestagung Kerntechnik, 131–133 (2007)
3. Methodology, STAR-CD Version 3.20, CD Adapco Group (2004)
4. Gordeev, S.: Strömungsmechanische Analyse von Hochgeschwindigkeits-Flüssigmetall Targets mit freien Oberflächen für kerntechnische Anwendungen (in German), PhD Thesis University Karlsruhe-KIT (July 2008)

Control of Separated Flow Using an Oscillating Lorentz Force: Comparison of DNS, LES, and Experiments

Thomas Albrecht and Jörg Stiller

1 Introduction

The separated flow around inclined airfoils can be controlled by unsteady actuation near the leading edge (LE), increasing the maximum lift coefficient without the need for heavy and complex high lift devices such as flaps [3]. Zero net mass flow devices (ZNMF) are often used for this purpose. While certainly favourable for industrial application, actuation via ZNMF faces some problems. In particular, to independently control both amplitude and frequency of the excitation is considered a "great challenge" [4]. This is even more severe when the wave form of the actuation is non-sinusoidal, i. e., contains more than one frequency component.

Therefore, Cierpka et al. [2] applied a streamwise oscillating Lorentz force near the LE of an inclined flat plate. Driven by an electric current, arbitrary wave forms can be generated easily. Studying the effect of different wave forms, they found that the maximum lift gain is proportional to the *peak* momentum input rather than to its *mean* (RMS) value. However, the physical mechanism is not fully identified yet, hence, motivating the present numerical investigation.

While Large Eddy Simulations (LES) present an appropriate approach to compute separated flows, some uncertainty still remains about transitional flows, in particular, when combined with magnetohydrodynamic body forces. As accurately capturing the (forced) instability of the separated shear layer is essential for the investigated flow control method, we decided to conduct both LES and Direct Numerical Simulations (DNS). For the LES, the commercial flow solver FLUENT with the dynamic Smagorinsky-Lilly sub-grid scale model was used. DNS were carried out by means of the SEMTEX spectral element-Fourier code [1]. The first results of the ongoing research are presented in this paper.

Thomas Albrecht
Inst. for Aerospace Eng., Technische Universität Dresden
e-mail: thomas.albrecht@tu-dresden.de

Jörg Stiller
Inst. for Aerospace Eng., Technische Universität Dresden
e-mail: joerg.stiller@tu-dresden.de

2 Setup

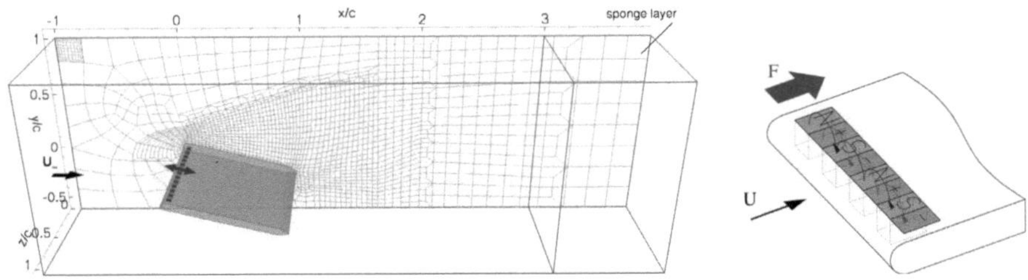

Fig. 1 Left: computational domain. Right: geometry of the Lorentz force actuator.

Modelling the experimental test section, Fig. 1 shows the computational domain with the flat plate at an angle of attack $\alpha = 16°$. While the flat plate is shown at its actual experimental dimensions (chord $c = 130$ mm, 140 mm span, 10 mm thickness), only a limited spanwise extend, but the full height, of the test section has been simulated. The inflow velocity was $U_\infty = 8$ cm/s, yielding a chord Reynolds number of $\mathrm{Re} = 10^4$. Details of the experimental setup are given in [2]. In the x,y-plane, we plotted the spectral element mesh used for the DNS, and for the upper left element, the underlying GLL mesh, as well. Both upper and lower boundaries – a free surface and a rigid wall in the experiments – are modeled by walls moving at the inflow velocity. A sponge layer protects the convective outflow boundary.

The actuator is located near the LE of the flat plate. It consists of streamwise aligned, alternating pairs of electrodes and magnets. The resulting body force decays approximately exponentially with increasing wall distance. Its amplitude, characterised by the (RMS) momentum coefficient $c'_\mu = 2.6\%$, is driven by the current fed into the electrodes. For the present study, the Lorentz force oscillates sinusoidal at a non-dimensional frequency $F^+ = fc/U_\infty = 1$.

3 Results

At the given Re and α, the laminar flow separates at the leading edge, becomes turbulent in the separated shear layer, intermittently re-attaches, and forms a turbulent wake (see Fig. 4). Hence, 2-D simulations are of little physical interest, however, they are feasible for grid resolution studies. To prepare the 3-D runs, we conducted a number of 2-D simulations with different polynomial orders (N) and number of spectral elements (NEL). Guided by low values of N often found in the literature, we started with a high element count mesh at N=5, which yields, however, large divergence errors. Gradually increasing N and, in turn, using less elements, reduced the divergence error to an acceptable level. The final mesh consists of 1757 elements at N=12, requiring only a slightly increased CPU time as compared to the initial mesh.

For the 3-D DNS, we also varied the spanwise extend L_z and the spanwise resolution. Results for $L_z/c = 0.1, 0.2$, and 0.4 are shown in Fig. 2. Computations using the smallest $L_z/c = 0.1$ predict large fluctuations of the lift and drag coefficients at a Strouhal number $\mathrm{St} = 0.2$, similar to those found in 2-D simulations (not shown). When doubling both the spanwise extend ($L_z/c = 0.2$) and number of spanwise Fourier modes m, the vortex shedding characteristics change suddenly at $t = 40$: the amplitudes of the predicted lift and drag fluctuations decrease, while St increases to ≈ 1. The same behaviour is observed when further increasing L_z and m, however, the transition to smaller amplitude/higher frequency fluctuations occurs earlier at $t \approx 30$. For the results reported in the following, $L_z/c = 0.2$ and $m = 32$ were used.

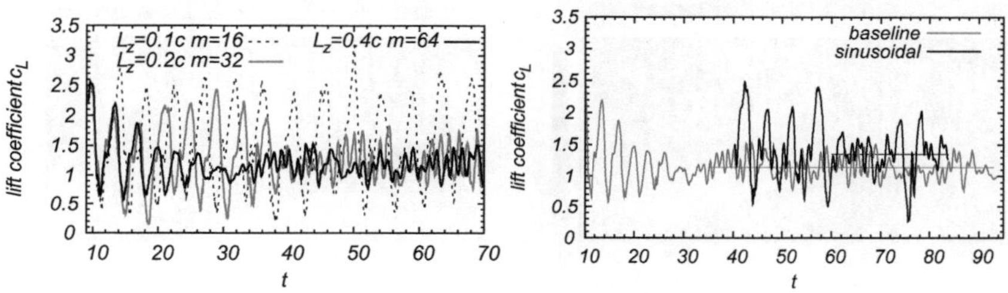

Fig. 2 History of lift coefficient. Left: variation of spanwise extend. Right: baseline vs. sinusoidal actuation.

Figure 3 compares the time-averaged streamwise velocity component obtained from DNS, LES, and experiments. For the baseline case shown on the left, the mean separation region is quite large and extends beyond the trailing edge. Both numerical predictions are in good agreement with the experimental results.

The experimental data of the controlled flow show a greatly reduced mean separation region, with its center shifted towards the leading edge. This is nicely captured by the LES. While the DNS also shows the shift towards the LE, the extend of the separated region is seriously overestimated. Probably, the wake is not sufficiently resolved, and/or, the streamwise extend is to small. Hence, the interaction between the wake and the separating vortices might not been captured correctly. Nevertheless, the mean lift (Fig. 2, right), but also the mean drag (not shown), computed by DNS, increase by 18% and 5%, respectively. As the actual reattached flow region is larger, the effect should even increase in refined simulations. Unfortunately, there are no lift/drag history data available from the experiments or the LES.

Finally, Fig. 4 compares vortex visualisations obtained from DNS for the baseline and controlled flow. The LE actuator obviously promotes transition to turbulence of the shear layer. The presence of two small vortices at the suction side has also been observed in the experiments at $F^+ = 1$, and is considered an optimum state for separation control. However, the DNS runs also showed instants at which both vertices merged prior to leaving the trailing edge, forming one big vortex and causing large lift fluctuations. Together with different excitation wave forms, this will be examined more closely in forthcoming, refined simulations.

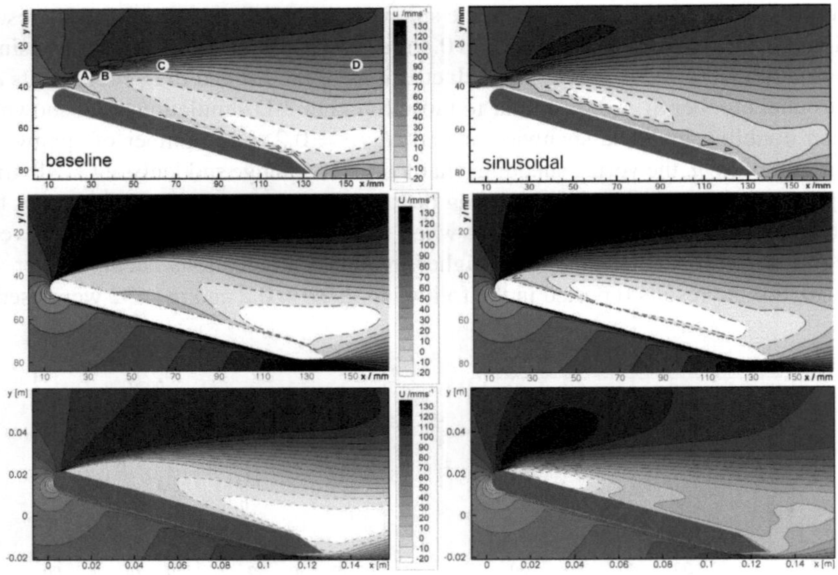

Fig. 3 Contours of the mean streamwise velocity component, obtained from experiments (upper row, [2]), DNS (middle row), and LES (lower row). Baseline (left) vs. sinusoidal (right) actuation.

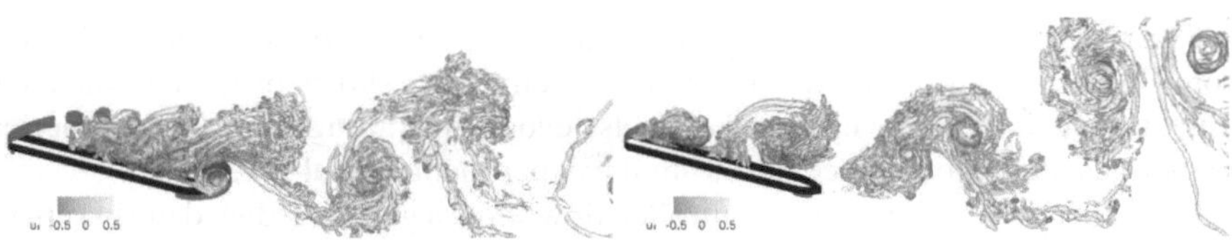

Fig. 4 Vortex visualisations (λ_2) obtained from DNS. Baseline (left) and controlled flow (right).

References

1. Blackburn, H.M., Sherwin, S.J.: Formulation of a galerkin spectral elementfourier method for three-dimensional incompressible flows in cylindrical geometries. J. Comput. Phys. 197, 759–778 (2004)
2. Cierpka, C., Weier, T., Gerbeth, G.: Electromagnetic control of separated flows using periodic excitation with different wave forms. In: King, R. (ed.) Active Flow Control. Notes on Numerical Fluid Mechanics and Multidisciplinary Design (NNFM), vol. 95, pp. 27–41. Springer, Berlin (2007)
3. Greenblatt, D., Wygnanski, I.J.: The control of flow separation by periodic excitation. Prog. Aero. Sci. 36(7), 487–545 (2000)
4. Rullan, J.M., Vlachos, P. P., Telionis, D.P., Zeiger, M.D.: Post-stall flow control of sharp-edged wings via unsteady blowing. AIAA Journal of Aircraft 43(6), 1738–1746 (2006), doi:10.2514/1.19495

Study of Effects of Wall-Normal Rotation on the Turbulent Channel Flow Using DNS

A. Mehdizadeh and M. Oberlack

Abstract. The effects of the wall-normal rotation on the turbulence channel flow have been studied. A series of direct numerical simulations have been performed with various rotation rates for Reynolds number 180 based on the friction velocity in the non-rotating case. All remarkable changes are discussed.

1 Introduction

Rotating channel flows are of great importance in many engineering applications. In these flows the structure of turbulence and mechanism of momentum transport is highly affected by additional body forces, namely, Coriolis and centrifugal forces. Turbulent channel flows with the streamwise and the spanwise rotations have been extensively studied by many authors [1],[2] and [4]. However, the turbulent channel flows with the wall-normal rotation have been rarely investigated. Since there is no possible experimental approach to the investigation of these flows, direct numerical simulation (DNS) is the only available method to examine them. In this work the pressure driven turbulent channel flow, rotating about wall-normal axis, is explored. A series of direct numerical simulations with various rotation numbers is carried out to establish the effects of the rotation on the flow. In the figure1 the flow geometry is shown. Note that all variables are non-dimensionalized by the friction velocity in the non-rotating case (U_{τ_0}) and the channel half height (h). For the entire analysis, the only wall-normal rotation i.e. Ω_2 or alternatively $Ro = \frac{2\Omega_2 h}{U_{\tau_0}}$ is to be varied.

2 DNS of Turbulent Wall-Normal Rotating Channel Flow

The numerical technique which was chosen is a standard spectral method with Fourier decomposition in the streamwise and the spanwise directions and Chebyshev decomposition in the wall-normal direction. The numerical code for channel flow was developed at KTH/Stockholm [5]. Additional features such as wall-normal rotation was has been added during the project.

A. Mehdizadeh · M. Oberlack
Department of mechanical engineering,Technische Universität Darmstadt, Darmstadt, Germany

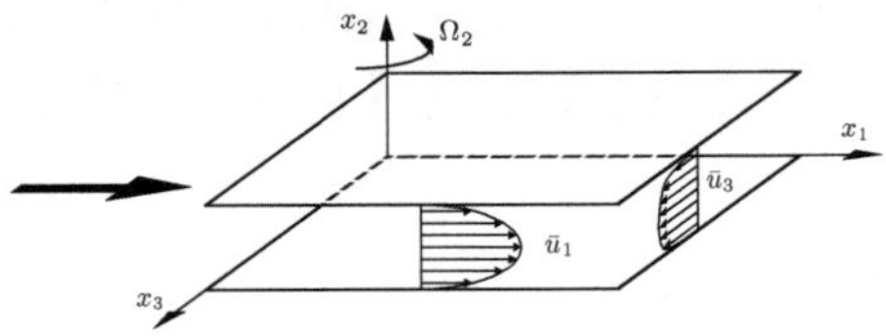

Fig. 1 Sketch of the flow geometry of turbulent channel flow with wall-normal rotation

Eight simulations have been presented to show the effects of the wall-normal rotation on the flow. The numerical results indicate that the flow is very sensitive to wall-normal rotation. A slight rotation can induce a strong secondary motion in the spanwise direction and reduce the streamwise mean velocity substantially. Due to increase in the rotation rate, the streamwise mean velocity decreases, but the spanwise mean velocity first increases, reaches its maximum at around $Ro = 0.054$ and then decreases by further increase in the rotation rate, because the mechanism of the generation of turbulence is suppressed apparently by Coriolis force effects, (Figure.2). Further increase of the rotation rate causes that the flow reaches a quasi-laminar region, where relaminarization effects can be observed. In this region flow exhibits different behavior in the spanwise and the streamwise directions compared to fully turbulent region, (Figure.3), with elongated laminar-like coherent structures (Figure.4), whose existence based upon the theory proposed by Brown [3], is due to the inflection point in the velocity profile. By further increasing the rotation rate ($Ro \geq 0.546$), turbulence disappears and the flow reaches a fully laminar state.

3 Conclusion

The general purpose of the present investigation is to study the effects of the wall-normal rotation on the turbulent channel flow. It has been established that the flow is very sensitive to this rotation and is highly affected even with a very small rotation rates and the effects can be easily observed in the mean velocities and the flow structures.

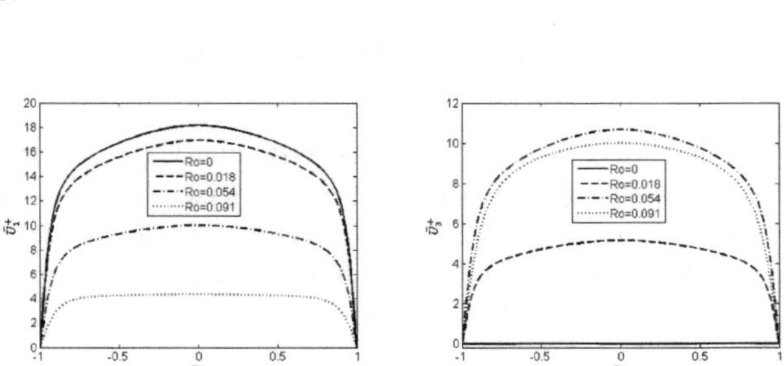

Fig. 2 Velocity profiles in the streamwise (left) and the spanwise (right) directions in the fully turbulent region.

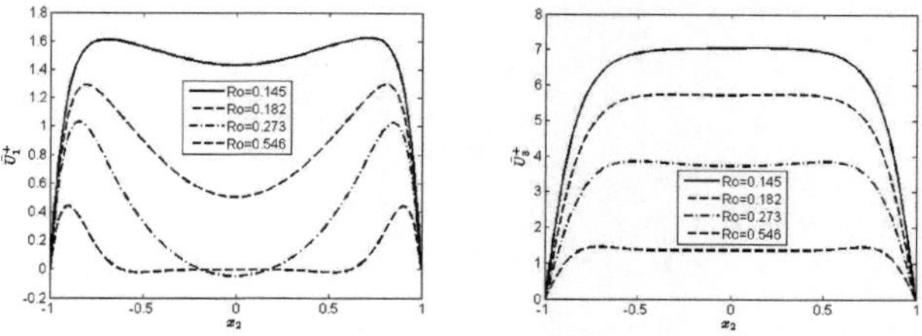

Fig. 3 Velocity profiles in the streamwise (left) and the spanwise (right) directions in the quasi-laminar region.

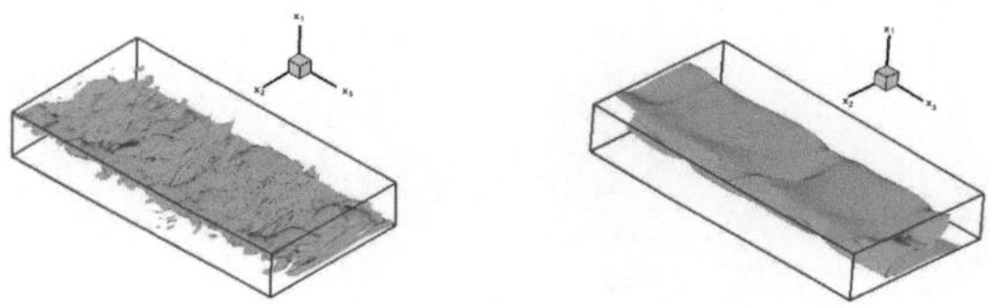

Fig. 4 Elongated coherent structures for $Ro = 0.145$ (left) and $Ro = 0.273$ (right).

References

1. Johnson, J.P., et al.: Effects of spanwise rotation on the structure of two-dimensional fully developed turbulent channel flow. J. Fluid Mech. 56, 533–557 (1972)
2. Grundestam, O., Wallin, S., Johansson, A.V.: Direct numerical simulation of rotating turbulent channel flow. J. Fluid Mech. 598, 177–199 (2008)
3. Brown, R.A.: Analytical methods in planetary boundary-layer modeling. Wiley and Sons, Chichester (1974)
4. Oberlack, M., et al.: Group analysis, DNS and modelling of a turbulent channel flow with streamwise rotation. J. Fluid Mech. 562, 383–403 (2006)
5. Lundbladh, A., Henningson, D., Johanson, A.: An efficient spectral integration method for the solution of the Navier-Stokes equations. Aeronautical Research Institute of Sweden, Bromma, FFA-TN 1992-28 (1992)

A Langevin Equation for the Turbulent Vorticity

Michael Wilczek and Rudolf Friedrich

Abstract. The vorticity field of fully developed turbulence displays a complex spatial structure consisting of a large number of entangled filamentary vortices (see illustration). As a consequence, the PDF of the vorticity shows a highly non-Gaussian shape with pronounced tails. In the present work a kinetic theory for the turbulent vorticity is presented. Under certain conditions the arising equation may be interpreted as a Fokker-Planck equation giving rise to a Langevin model.

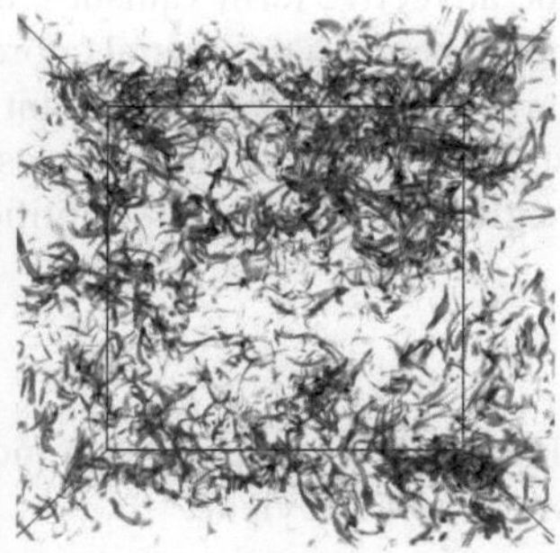

The appearing unknown conditional averages are estimated from direct numerical simulations. The Langevin model is shown to reproduce the single point vorticity PDF.

1 Kinetic Theory for the Turbulent Vorticity

The temporal evolution of the vorticity $\boldsymbol{\omega}(\boldsymbol{x},t)$ is governed by the vorticity equation,

$$\frac{\partial \boldsymbol{\omega}}{\partial t} + \boldsymbol{u} \cdot \nabla \boldsymbol{\omega} = \boldsymbol{S} \cdot \boldsymbol{\omega} + \nu \Delta \boldsymbol{\omega} + \boldsymbol{F}, \tag{1}$$

Michael Wilczek
Institute for Theoretical Physics, Wilhelm-Klemm-Str. 9, 48149 Münster, Germany
e-mail: mwilczek@uni-muenster.de

Rudolf Friedrich
Institute for Theoretical Physics, Wilhelm-Klemm-Str. 9, 48149 Münster, Germany
e-mail: fiddir@uni-muenster.de

where $\boldsymbol{u}(\boldsymbol{x},t)$ denotes the velocity field and $S_{ij} = \frac{1}{2}\left(\frac{\partial u_i}{\partial x_j}(\boldsymbol{x},t) + \frac{\partial u_j}{\partial x_i}(\boldsymbol{x},t)\right)$ denotes the rate-of-strain tensor. Focusing on incompressible fluids, the velocity field can be obtained from the vorticity field via Biot-Savart's law. ν denotes the kinematic viscosity and $\boldsymbol{F}(\boldsymbol{x},t)$ is an external forcing necessary to achieve a statistically stationary state. Taking the total temporal derivative of the fine-grained PDF $\hat{f}(\boldsymbol{\Omega};\boldsymbol{x},t) = \delta(\boldsymbol{\omega}(\boldsymbol{x},t) - \boldsymbol{\Omega})$ with subsequent averaging yields a kinetic equation for the turbulent vorticity PDF $f(\boldsymbol{\Omega};\boldsymbol{x},t) = \langle \hat{f}(\boldsymbol{\Omega};\boldsymbol{x},t)\rangle$ [1, 2, 3, 5],

$$\frac{\partial}{\partial t} f + \nabla \cdot \{\langle \boldsymbol{u}|\boldsymbol{\Omega}\rangle f\} = -\nabla_{\boldsymbol{\Omega}} \cdot \{\langle \boldsymbol{S}\cdot\boldsymbol{\omega} + \nu\Delta\boldsymbol{\omega} + \boldsymbol{F}|\boldsymbol{\Omega}\rangle f\}. \tag{2}$$

The right hand side of this kinetic equation for the vorticity PDF reveals the different dynamical influences: the average vortex stretching term, vorticity diffusion and the forcing conditioned on the sample space vorticity. Taking into account homogeneity, the advective term vanishes, as both the conditional average as well as the PDF do not depend on the spatial coordinate. Given a sufficiently high Reynolds number, the large-scale forcing should not affect the smallest scales of the flow. As the coherent structures live on these scales, the conditional average of the forcing term may be neglected. With these simplifications the kinetic equation reads

$$\frac{\partial}{\partial t} f = -\nabla_{\boldsymbol{\Omega}} \cdot \{\langle \boldsymbol{S}\cdot\boldsymbol{\omega} + \nu\Delta\boldsymbol{\omega}|\boldsymbol{\Omega}\rangle f\}. \tag{3}$$

Taking into account the homogeneity of the flow, calculating the Laplacian of the vorticity PDF yields

$$\frac{\partial^2}{\partial x_i^2} f = 0 = -\frac{\partial}{\partial \Omega_j}\langle \frac{\partial^2 \omega_j}{\partial x_i^2}|\boldsymbol{\Omega}\rangle f + \frac{\partial^2}{\partial\Omega_j \partial\Omega_k}\langle \frac{\partial \omega_j}{\partial x_i}\frac{\partial \omega_k}{\partial x_i}|\boldsymbol{\Omega}\rangle f. \tag{4}$$

In the following we will focus on the stationary PDF. Combining the kinetic equation (3) with the homogeneity relation (4) and by a change of variables, $t^* = -t$, the temporal evolution of the vorticity PDF is in case of homogeneous flows described by

$$\frac{\partial}{\partial t^*} f = -\frac{\partial}{\partial \Omega_i}\left(D_i^{(1)} f\right) + \frac{\partial^2}{\partial\Omega_i\partial\Omega_j}\left(D_{ij}^{(2)} f\right) \tag{5}$$

with $D_i^{(1)} = -\langle S_{ij}\omega_j|\boldsymbol{\Omega}\rangle$ and $D_{ij}^{(2)} = \langle \nu\left(\frac{\partial\omega_i}{\partial x_k}\frac{\partial\omega_j}{\partial x_k}\right)|\boldsymbol{\Omega}\rangle$. That means, we have a Fokker-Planck type equation for the backward-time evolution of the turbulent vorticity, directly derived from the equation of motion. The corresponding Langevin equation reads

$$\frac{\mathrm{d}}{\mathrm{d}t^*}\Omega_i(t^*) = D_i^{(1)}(\boldsymbol{\Omega},t^*) + \sqrt{2D_{ij}^{(2)}(\boldsymbol{\Omega},t^*)}\Gamma_j(t^*) \tag{6}$$

with $\langle\Gamma_i(t)\rangle = 0$ and $\langle\Gamma_i(t)\Gamma_j(t')\rangle = \delta_{ij}\delta(t-t')$.

For comparison with the numerical data we will consider a single component of the vorticity. Equation (5) then simplifies to $\frac{\partial}{\partial t^*} f = -\frac{\partial}{\partial\Omega_x} D^{(1)} f + \frac{\partial^2}{\partial\Omega_x^2} D^{(2)} f$, with

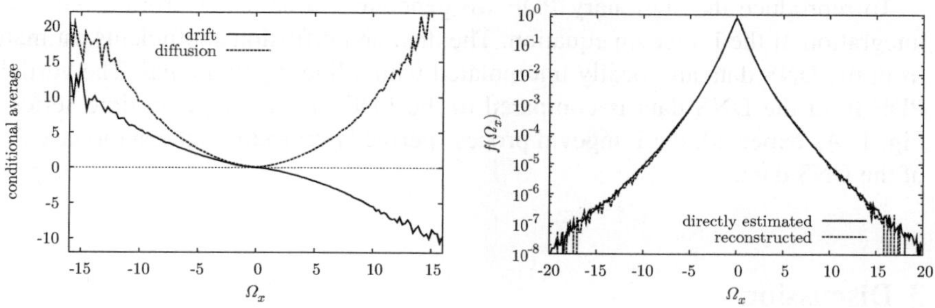

Fig. 1 Left: drift and diffusion coefficient obtained from the simulation data. While the drift coefficient is in leading order linear, the diffusion coefficient turns out to be approximately quadratic. Right: Vorticity PDF estimated directly from the simulation data and the PDF reconstructed from the Langevin time series.

$D^{(1)}(\Omega_x, t^*) = -\langle (\boldsymbol{S}\cdot\boldsymbol{\omega})_x | \Omega_x, t^*\rangle$ and $D^{(2)}(\Omega_x, t^*) = \langle \nu(\nabla\omega_x)^2 | \Omega_x, t^*\rangle$, which yields the stationary solution

$$f(\Omega_x) = \frac{\mathcal{N}}{\langle \nu(\nabla\omega_x)^2|\Omega_x\rangle} \exp\left(- \int_{-\infty}^{\Omega_x} \mathrm{d}\Omega'_x \frac{\langle (\boldsymbol{S}\cdot\boldsymbol{\omega})_x|\Omega'_x\rangle}{\langle \nu(\nabla\omega_x)^2|\Omega'_x\rangle} \right) \tag{7}$$

with a normalization constant $\mathcal{N}$. This shows, that in a stationary homogeneous flow the vorticity PDF is determined by the dynamical effect of vortex stretching and the vorticity gradient. The corresponding one-dimensional Langevin equation we will consider in the following reads $\frac{\mathrm{d}}{\mathrm{d}t^*}\Omega_x(t^*) = D^{(1)}(\Omega_x, t^*) + \sqrt{2D^{(2)}(\Omega_x, t^*)}\Gamma_j(t^*)$.

2 Numerical Results

The numerical results are generated with a standard Fourier-pseudospectral code [4] simulating statistically stationary three-dimensional fully developed turbulence on a triply periodic domain. The presented simulation has a grid resolution of 512^3 collocation points and achieves a Taylor-based Reynolds number of $\mathrm{Re}_\lambda = 164$.

We estimate the vorticity PDF $f(\Omega_x)$ and the conditional averages $\langle(\boldsymbol{S}\cdot\boldsymbol{\omega})_x|\Omega_x\rangle$ and $\langle \nu(\nabla\omega_x)^2|\Omega_x\rangle$ from the simulation data, the results are shown in Fig. 1. The conditionally averaged vortex stretching term, which (up to a sign) corresponds to the drift coefficient, shows a strong positive correlation with the vorticity. Physically speaking this means that a strong vortex is subject to a stronger vortex stretching than a weak one. The conditionally averaged squared vorticity gradient, which corresponds to the diffusion coefficient, exhibits a strong Ω_x-dependence. In terms of the Langevin model this means that the highly non-Gaussian form of the vorticity PDF comes due to a multiplicative noise.

To reproduce the stationary PDF we generate a time series from a numerical integration of the Langevin equation. The drift and diffusion coefficients estimated from the DNS data are locally interpolated with a linear polynomial. The vorticity PDF from the DNS data is compared to the PDF of the Langevin time series in Fig. 1. As expected, the Langevin process perfectly reproduces the stationary PDF of the DNS data.

3 Discussion

We have presented a general procedure to obtain a Fokker-Planck like equation from a kinetic theory of the turbulent vorticity. In order to find this correspondence a change of variables, $t* = -t$, is necessary, which means that the Fokker-Planck equation has to be interpreted backwards in time. It has to be assumed that this correspondence is of formal nature. The backward time series of the Langevin equation are not supposed to correspond to real vorticity time series. This becomes clear, when comparing higher order statistics like for example the joint PDF for two different times. The Langevin model, which only contains information at a single point in time is not capable of reproducing these higher order statistics. However, we would like to emphasize two important points, which can be learned from this approach. First, the presented procedure delivers a straight-forward derivation of Langevin models which by construction yield the correct statistics. The approach can easily be extended to other turbulent variables like velocity increments or the velocity gradient tensor. An extension to higher order statistics is also possible. Second, this procedure explicitly shows, that when modeling turbulent time series with stochastic processes, non-Gaussian PDF's are naturally obtained by the use of multiplicative noise.

Acknowledgement. We thank Oliver Kamps and Michel Voßkuhle for useful discussion. Computational resources are allocated by the LRZ Munich under project h0963 and on the BOB cluster at the RZG.

References

1. Lundgren, T.S.: Distribution functions in the statistical theory of turbulence. Physics of Fluids 10(5), 969–975 (1967)
2. Novikov, E.A.: A new approach to the problem of turbulence, based on the conditionally averaged Navier-Stokes equations. Fluid Dynamics Research 12(2), 969–975 (1993)
3. Pope, S.B.: Turbulent Flows. Cambridge University Press, Cambridge (2000)
4. Canuto, C., Hussaini, M.Y., Quarteroni, A., Zang, T.A.: Spectral Methods in Fluid Dynamics. Springer, Berlin (1987)
5. Wilczek, M., Friedrich, R.: Dynamical Origins for Non-Gaussian Vorticity Distributions in Turbulent Flows. arXiv.org:0812.2109

Application of Helical Characteristics of the Velocity Field to Evaluate the Intensity of Tropical Cyclones

G. Levina, E. Glebova, A. Naumov, and I. Trosnikov

Abstract. The paper presents results of numerical analysis for helical features of velocity field to investigate the process of tropical cyclone formation, namely, the downward helicity flux through the upper boundary of the viscous atmospheric turbulent boundary layer has been calculated. The simulation was carried out by use of the regional atmospheric ETA model and NCEP reanalysis global data. Calculations were performed for two tropical cyclones – Wilma (Atlantic basin, 2005) and Man-Yi (North-West Pacific, 2007). It has been found, that the chosen helical characteristic reveals an adequate response to basic trends in variation of such important meteorological fields as pressure and wind velocity during the hurricane vortex evolution. The analysis carried out in the paper shows that the helicity flux can be used as an illustrative characteristic to describe the intensity and destructive power of tropical cyclones.

1 Brief History of the Problem

In 2006 two papers appeared which presented research results in different fields of investigation. The authors of [4] proposed a new scenario of tropical cyclogenesis and showed by near-cloud-resolving simulations (2-3 km horizontal grid spacing) how a mesoscale tropical depression vortex could develop from cumulonimbus convection as a result of upscale vorticity cascade. In these simulations the growth of flow scales happened by multiple mergers of small-scale convective elements. They

G. Levina
Institute of Continuous Media Mechanics, Ural Branch/Russian Academy of Sciences, Perm, Russia
Space Reseach Institute/Russian Academy of Sciences, Moscow, Russia
e-mail: levina@icmm.ru

E. Glebova · A. Naumov · I. Trosnikov
Hydrometeorological Center of the Russian Federation/Federal Agency of Hydrometeorology and Environmental Monitoring, Moscow, Russia
e-mail: Ek.Glebova@gmail.com

first formed a number of vortical hot towers, each of 10-30 km horizontal scale, which eventually resulted in an intense mesoscale helical vortex. Paper [2] demonstrated a very similar way of formation of large-scale spiral vortex structures from small-scale convective cells in the Rayleigh-Bénard convection affected a special helical forcing. The obtained results [2, 4] were brought in together for common discussion in seminars of Montgomery Research Group at Colorado State University (Fort Collins). This paved the way to introducing the analysis of helical characteristics of the velocity field in investigations of tropical cyclones by use of mesoscale atmospheric modeling systems.

2 Helicity Flux as a Measure of Tropical Cyclone Intensity: Theoretical Ground

In the atmosphere non-uniform heating results in the formation of convective circulations of different scale which under the Earth's rotation become helical. Helicity is a fundamental hydrodynamic characteristic and, the general helicity balance equations are well known in the literature, see e.g. [1] and references therein.

In paper [1] by use of the concept of velocity-field helicity the derivation of the helicity balance equations was briefly reproduced to obtain on their basis a so-called helicity index, which is the downward helicity flux through the upper boundary of the viscous atmospheric turbulent boundary layer. The helicity flux was represented in a mathematical form convenient for the study of stationary vortices. It is suggested [1] that the helicity index can be treated as a measure of the intensity and destructive power of intense atmospheric vortices, including tropical cyclones, tornadoes, and dust devils.

To obtain a formula for the helicity index, in paper [1] the case was considered where the kinetic energy (intensity) of toroidal circulation was many times greater than the intensity of meridional circulation. The beta effect was disregarded as well, assuming that the Coriolis parameter f was constant. Finally, the author [1] took into account that, for the central part of developed tropical cyclones, except for their remote periphery, the effect of the Earth's rotation was negligibly small compared to the contribution of the relative motion of air. All above mentioned gives a following quantitative criterion for an axisymmetric vortex in cylindrical coordinates with the center at the vortex's axis

$$SI = (8\pi/3) \int_0^\infty V^3 \, dr \tag{1}$$

where $V(r)$ is the azimuthal wind velocity dependent on the radius r. Thus, the helicity index SI is determined by the product of the cubed maximum wind speed and the width of the strip swept by maximum winds during a vortex movement over the ground surface. For tropical cyclones an important similarity was noted in paper [1], that is between the helicity index (1) and the classical formula used to evaluate the destructiveness of hurricanes. This gives grounds to consider formula (1) as a worth characteristic to calculate in numerical modeling of tropical hurricanes.

3 Calculations of Helicity Index for Tropical Cyclones

The helicity index has been calculated for the two tropical cyclones (TCs), Wilma in The Caribbean Sea and The Gulf of Mexico (2005) and Man-Yi in the North-West Pacific (2007), during their whole lifetimes. The tracks of the hurricanes are shown in figure 1.

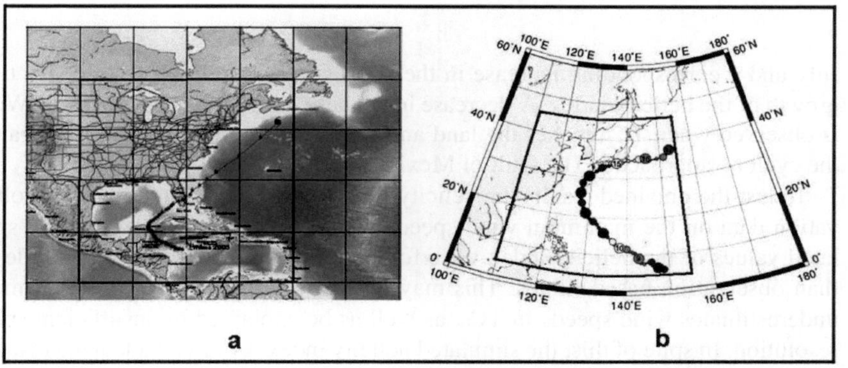

Fig. 1 Tracks of hurricanes (NOAA Archive): a – Wilma, 15-25 October 2005; b – Man-Yi, 7-20 July 2007

The mesoscale atmospheric ETA model, developed by team of F. Mesinger [3], was applied in our numerical simulation. The used modification of the ETA model includes 45 vertical levels, the upper of which corresponds to 25 hPa. In all calculations the horizontal grid increment was chosen equal to 0.2°, which is about 22 km. Velocity field in an area of 4500×4500 km was examined. NCEP reanalysis data with resolution 1° was used to form initial and boundary conditions.

As a result of the postprocessing of the ETA model output the tangential velocity field has been obtained. The helicity index has been calculated from formula (1) at height $h = 10$ m over the ground surface where the maximum wind speed of hurricane vortex was expected to find. The results for both tropical cyclones are shown in figure 2.

At the stage of tropical depression, from which simulations for both TCs usually starts, there already exist well pronounced local maxima of vorticity and wind velocity. So that, the helicity index is found to be higher than the background one in the tropical atmosphere. As the TC develops, the helicity index continues increasing. In the typhoon Man-Yi a rapid decrease in pressure by 22 hPa per 18 hours during 9-10

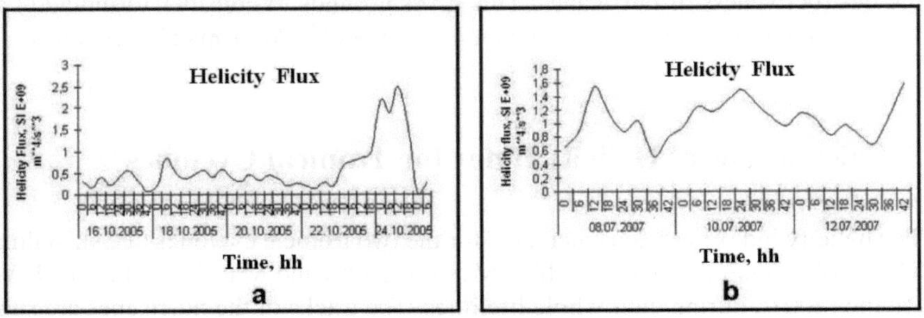

Fig. 2 Helicity flux during hurricane evolution: a – Wilma, 15-25 October 2005; b – Man-Yi, 7-20 July 2007

July and a corresponding increase in the wind speed is followed by the 7-8 times growth of the helicity index. A decrease in the helicity index of the hurricane Wilma is observed when TC reaches the land and weakens, whereas, it increases again as the cyclone approaches The Gulf of Mexico.

To test the obtained results the helicity flux was also calculated by use of observation data on the maximum wind speed. The comparison showed that the simulated values of the helicity index were found to be by 2-3 order of magnitude less than observation-based values. This may result from the fact that the ETA model underestimates wind speeds in TCs, as well as be explained by insufficient spatial resolution. In spite of this, the simulated helicity index reveals an adequate response to basic trends in variation of such important meteorological fields as wind velocity and pressure during the hurricane vortex evolution. The evolution of helicity flux in both cases under consideration truly demonstrates all changes in the intensity of developed hurricane vortices.

Acknowledgement. The authors are thankful to M.V. Kurgansky for useful discussions. This work is supported by the Russian Foundation for Basic Research, Project 07-05-00060.

References

1. Kurgansky, M.V.: Vertical helicity flux in atmospheric vortices as a measure of their intensity. Izv. Atmos. Ocean. Phys. 44, 67–74 (2008)
2. Levina, G.V., Burylov, I.A.: Numerical simulation of helical-vortex effects in Rayleigh-Bénard convection. Nonlin. Processes Geophys. 13, 205–222 (2006)
3. Mesinger, F.: Numerical Methods: The Arakawa Approach, Horizontal Grid, Global, and Limited-Area Modeling. In: Randall, D.A. (ed.) General Circulation Model Development: Past, Present and Future. International Geophysics Series, vol. 70, pp. 373–419. Academic Press, Cambridge (2000)
4. Montgomery, M.T., Nicholls, M.E., Cram, T.A., Saunders, A.B.: A vortical hot tower route to tropical cyclogenesis. J. Atmos. Sci. 63, 355–386 (2006)

An Experimental Study of Turbulent Vortex Rings

L. Gan and T.B. Nickles

Abstract. The phenomenon of vortex rings can be observed in some aquatic animals producing propulsion, in oil drilling and gaseous pollution releases. Vortex rings in this laboratory study are generated by pushing a piston through a tube with an orifice opening in water. In this paper, turbulent vortex rings are studied by means of Stereoscopic Particle Image Velocimetry (PIV). Typical turbulence quantities e.g. the turbulent stress distribution can be visualized after some mathematical processes.

1 Introduction

In this research activity, turbulent vortex rings are studied as a contribution to the subject of coherent structures in turbulent flows. Turbulent vortex rings here are simply at high Reynolds numbers when they are produced. To the author's knowledge, there is only one systematic experimental study of turbulent vortex rings by Glezer & Coles [1] in which they studied the turbulence quantity distributions and presented them in their well established self-similarity theory. They used LDA to record point information in the flow field and adopted a profound statistical algorithm to filter out the 'faulty' information at the point. (Turbulent vortex rings, due to their highly exited nature, are strongly dispersed at the position of their testing points.) This algorithm is a disadvantage of LDA and this can be overcome by PIV technique.

L.Gan
Cambridge University, Engineering Department, Trumpington Street, Cambridge CB2 1PZ, UK
e-mail: lg331@cam.ac.uk

T.B.Nickels
Cambridge University, Engineering Department, Trumpington Street, Cambridge CB2 1PZ, UK
e-mail: tbn22@cam.ac.uk

2 The Experiment

Vortex rings of this study are produced in a water tank of $750mm \times 750mm \times 1500mm$. A perspex nozzle of $144mm$ and $1000mm$ long is mounted vertically above the tank. A perspex plate of $500mm \times 500mm$ is attached to the exit of the tube to make it orifice-like. A PVC piston is driven by a stepper motor through the nozzle, and when the piston is given a impulse to move downwards, a vortex ring is formed.

A criterion to separate laminar and turbulent vortex rings is given in Glezer [2] (their Fig.6) by means of flow visualizations. Our experimental conditions are set based on that figure such that turbulent vortex rings are produced at $Re = 40,000$, and $L/D = 2$ which are far on the right away from the dashed line of that figure and hence can give quite turbulent rings.

Determined by the nature of turbulent vortex rings (not a steady turbulence), ensemble averaging rather than time averaging needs to be adopted and 55 rings are produced. In the first set of experiments, PIV field of view is aligned with the center plane of the ring trajectory. The PIV calibration plate is positioned at three stations along the streamwise direction and is in the position of the left core of the ring. The purpose of focusing on only one core is to make the spatial resolution as high as possible while still covering the main mean structure. (By ensemble averages, turbulence quantities of the core area should be roughly axis-symmetric, and our earlier experiment where the entire ring's cross section was recorded proves this hypothesis.) PIV recording rate is set at $450Hz$. In the second set of experiments, the field of view is such that the normal direction of the view plane is aligned with the streamwise direction, thus, by Taylor's hypothesis the three dimensional ring can be constructed.

3 Results

Only the results of the second station (0.85 to $2.24L/D$ downstream from the tube exit) of the first set of experiments are presented and a rough idea of how turbulence quantities are distributed in and around core area can be obtained. For the three dimensional ring construction, because of some limitations of experimental conditions, the results are only approximate.

3.1 Turbulence in and Around Core Area

A simple convergence test which plots the quantity $\frac{1}{N}\sum_{i=1}^{N} u_i'^2$ against N, where N is from 1 to 55, is firstly operated. It tests at two different points, one is at core center, the other is at core edge. The test shows that 55 realizations should be enough for ensemble averaging purpose.

Turbulent vortex rings experience dispersion as early as our field of view position. This dispersion is partly due to the turbulent nature of the ring and partly due to their imperfect formation and need to be discarded. A method to discard suspected wrong formed rings (a filtering process) is then developed but not to be discussed here. The turbulence quantities are then calculated using the ensemble averaged data after this filtering process.

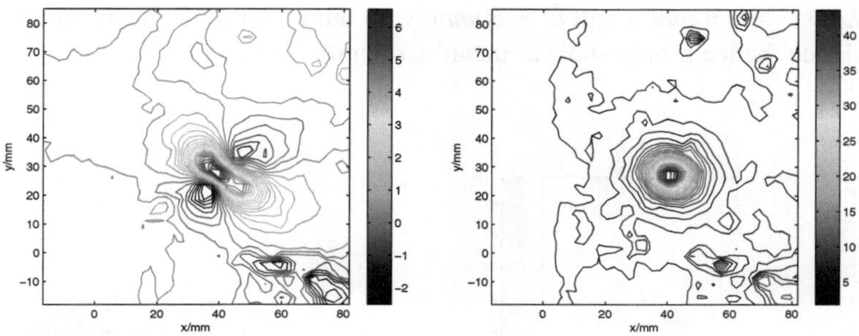

(a) (b)

Fig. 1. (a) Reynolds shear stress field; (b) Turbulence kinetic energy field at 1.84*sec* after ring formation.

One of the three Reynolds shear stresses $-\rho\overline{u'_i u'_j}$ and turbulence kinetic energy $\rho\overline{u'_i u'_i}$ are presented in Fig. 1. Although it is not shown here, the azimuthal component $\rho\overline{w'w'}$ only contributes about 15% of the total turbulence kinetic energy and it's intensity distributes quite evenly in the entire field rather than concentrating in the core area. The other two shear stresses related to w' are also not shown here. Not surprisingly, stresses and energy are all concentrated in the core area where strongest shear locates. The spatial resolution of this set of experiment is 2.46*mm*. If we refer to the relation for homogenous turbulence $\lambda/\delta \sim Re_\delta^{-1/2}$ and $\eta/\delta \sim Re_\delta^{-3/4}$, where δ, λ, η are integral scale, Taylor microscale and Komogorov scale respectively, it is clear that, at a length scale of roughly λ or 13η, the flow field is far away from local isotropic. Turbulence production is written as $-\rho\overline{u'_i u'_j}e_{ij}$ where e_{ij} is the strain-rate tensor for mean velocities. This term consists of 9 symmetric terms among which 7 can be determined with our experimental arrangement except for 2 quantities involving $\partial u'/\partial z$ and $\partial v'/\partial z$. The result contour for 7 terms is shown in Fig. 2. Dissipation, however can only be calculated accurately when Kolmogorov scale is resolved, hence is not computed here.

3.2 Three Dimensional Structure Construction

In this set of experiments, PIV calibration plate is fixed horizontally at about $1L/D$ downstream from the tube exit. Data are recorded at $650Hz$, but only one of every 13 images are used to construct the 3D image by Taylor's hypothesis, hence the virtual frequency is $54.1Hz$. Taylor's hypothesis requires that $u'/U \ll 1$ in order to 'freeze' the turbulent flow, where u' is the turbulent quantity and U is the speed of the moving station which in our case, is just the navigating speed of the ring. At $1L/D$ $U \approx 150mm/s$ and $u' \approx 50mm/s$, so our condition is far from the 'frozen turbulence' hence it only show a 'mean' structure as in Fig. 3.

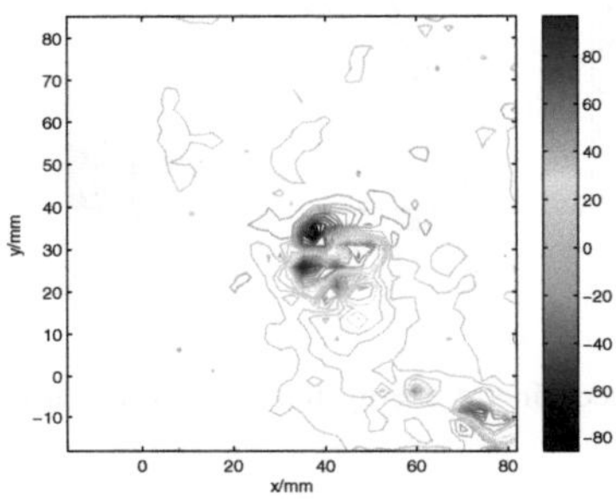

Fig. 2. Turbulence production field at $1.84sec$ after ring formation.

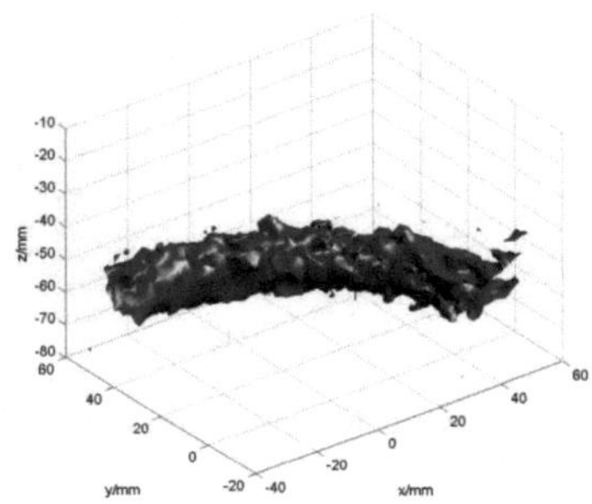

Fig. 3. A quarter of the ring recorded at effectively $54.1Hz$ at a station about $1L/D$

4 Conclusion

At $Re = 40,000$ and a spatial resolution of 1 Taylor microscale, the turbulent vortex ring field is highly inhomogenous and is also far from local isotropic. Turbulence kinetic energy and production are intensely concentrated in the mean core area where strongest shear locates. Future work will increase the spatial resolution by another factor of 2 to see how the turbulence quantity distributed.

References

1. Glezer, A., Coles, D.: An experimental study of a turbulent vortex ring. J. Fluid Mech. 211, 243–283 (1990)
2. Glezer, A.: The formation of vortex rings. Phys. Fluids 31, 3532–3542 (1988)

Author Index